Women, Gender, and Technosciences, 1900–2020

This innovative volume analyzes the historical entanglement of gender, technosciences, and government/governance.

Situated at the crossroad of women and gender studies, science and technology studies, and political sociology, this volume shows the ever-accumulating gendered mechanisms that have determined the careers of scientific women and their access to power positions. It underlines on different scales—from the lab to international organizations or states—how the masculine culture of technoscientific practices has assigned women to subaltern institutional positions, while social practices of legitimization and recognition ended up granting some women access to leadership positions outside of institutions. With a broad geographic, political, and disciplinary scope, the contributors draw on a variety of new sources including interviews, private collections, and archives to examine the institutions, structures, and policies that shaped the technosciences, as well as the individuals who developed practices and environments that gained agency for themselves and their contemporaries.

This book will be of interest to students and scholars alike interested in women and gender studies, political studies, STS, history, and sociology of science and technology.

Grégory Dufaud is Professor of contemporary history at Université Polytechnique Hauts-de-France, France. He is a historian of twentieth-century Russia who specializes in history of science. His latest book, *Une histoire de la psychiatrie soviétique* (2021), was awarded the Prix Jean Garrabé from the Société de l'Évolution psychiatrique.

Isabelle Lémonon-Waxin, a physicist and historian of science, is an associate researcher of Cermes3 and Centre François Viète (France). Her Ph.D. dissertation *La Savante des Lumières françaises* (EHESS, France) has been awarded a DHST Dissertation Prize in 2021. She is an officer of the Commission on Women and Gender in History of Science, Technology and Medicine of the DHST-IUHPST.

Routledge Studies in the History of Science, Technology and Medicine

Transforming American Science
Universities, the Government, and the Cold War
Jonathan Engel

Human Extinction
A History of the Science and Ethics of Annihilation
Émile P. Torres

Tore Godal and the Evolution of Global Health
Conrad Keating

William Blake, the Single Vision, and Newton's Sleep
A History of Science, Poetry, and Progress
Keith G. Davies

Pestilence, Insanity, and Trees
How Stephen Smith Changed New York
John M. Harris Jr.

Communication Maintenance in *Longue Durée*
Edited by Gabriele Balbi and Roberto Leggero

Women, Gender, and Technosciences, 1900–2020
A Beard to Govern
Edited by Grégory Dufaud and Isabelle Lémonon-Waxin

For more information about this series, please visit: https://www.routledge.com/Routledge-Studies-in-the-History-of-Science-Technology-and-Medicine/book-series/HISTSCI

Women, Gender, and Technosciences, 1900–2020

A Beard to Govern

Edited by
Grégory Dufaud and
Isabelle Lémonon-Waxin

NEW YORK AND LONDON

First published 2025
by Routledge
605 Third Avenue, New York, NY 10158

and by Routledge
4 Park Square, Milton Park, Abingdon, Oxon, OX14 4RN

Routledge is an imprint of the Taylor & Francis Group, an informa business

ISBN: 978-1-032-87982-6 (hbk)
ISBN: 978-1-032-91323-0 (pbk)
ISBN: 978-1-003-56259-7 (ebk)

DOI: 10.4324/9781003562597

Typeset in Sabon
by codeMantra

Contents

Figures

Contributors

Pnina Geraldine Abir-Am has published widely on the history of women in science and the history of molecular biology. Her article "The Women Who Discovered RNA Splicing", *American Scientist*, 2020, received a SIIA-Excel-2021 Award. She is a resident scholar at WSRC, Brandeis University (Greater Boston, US).

Anna Cabanel, a Postdoctoral Researcher at KU Leuven (Belgium), focuses on gender history and the social history of science, exploring issues of diversity and inclusion in the scientific community. She authored *Femmes de sciences. Le mouvement des university women dans l'entre-deux-guerres* (2025).

Katja Doose is Assistant Professor of Environmental History at the University of Lyon 2 (France). She is the author of the monograph *Tectonics of the Perestroika. The Political New-Ordering of Armenia after the 1988 Earthquake*. Her research focuses on environmental and science history in Russia and Central Asia since the nineteenth century and in particular on the history of the Earth sciences.

Grégory Dufaud, is Professor of contemporary history at Université Polytechnique des Hauts-de-France (France). He is a historian of twentieth-century Russia who specializes in history of science. His latest book, *Une histoire de la psychiatrie soviétique* (2021), was awarded the Prix Jean Garrabé from the Société de l'Évolution psychiatrique.

Angeline Durand Vallot is Associate Professor of American studies and gender studies at University Lyon 1 (TRIANGLE). Her research specializes in American women's history, focusing on feminist activism, political strategies, reproductive rights, the history of science, and women's education.

Loukas Freris is a Ph.D. candidate within the ERC project "Living with Radiation", at the Friedrich-Alexander University (Erlangen-Nuremberg, Germany). He specializes in the history of international organizations (in particular the IAEA and UNESCO) as crucial actors in post-war science. Freris focuses on Greece, examining the political and diplomatic aspects of science in that country.

Irvy Gledhill is Honorary Adjunct Professor in flow physics at the University of the Witwatersrand (South Africa). Recent publications include Ndebele and Gledhill (2023). She is committed to improving the work environment for all scientists. She is Vice-President of the Network of African Science Academies.

Rachel Ivie is Director of Higher Education Programs and Grants at the American Association of Physics Teachers (USA) and has a Ph.D. in sociology from UNC-Chapel Hill. Rachel is Previous Director of the AIP Statistical Research Center and Fellow of the American Astronomical Society. Publications are located at www.linkedin.com/in/rachelivie.

Aleksandra Kasatkina is Associate Professor in the Department of History, HSE University (St Petersburg, Russia). Her research interests are linguistic anthropology and ethnography of speaking. She is a co-author of the article "Thrown into Collaboration: An Ethnography of Transcript Authorization" (2018, with Zinaida Vasilyeva and Roman Khandozhko).

Sally Gregory Kohlstedt, Professor Emerita at the University of Minnesota (USA), currently studies science-public interactions, many facilitated by women in science. See "Qualified Mentorship: Josephine Tilden and the Minnesota Seaside Station" (2022) and *Teaching Children Science: Hands-On Nature Study in North America* (2010).

Alexei Kojevnikov is Professor in the Department of History, University of British Columbia (Vancouver, Canada). He works on the history of modern physical sciences, from Einstein to Dr. Strangelove, and on the cultural and social histories of Soviet science. His current project is "Space-Time, Death-Resurrection, and the Russian Revolution".

Isabelle Lémonon-Waxin, a historian of science and physicist, is an Associate Researcher at Cermes3 and Centre François Viète (France). Her Ph.D. dissertation *La Savante des Lumières françaises* (EHESS, France) was awarded the DHST Dissertation Prize in 2021. She is an officer of the Commission on Women and Gender in History of Science, Technology and Medicine of the DHST-IUHPST.

Ygor Martins is a historian and currently a doctoral student at the PPGHCS (FIOCRUZ, Brazil). His central investigative interest focuses on the history of Brazilian mental medicine, with an emphasis on the trajectory of women in this discipline along the evolutions of institutions. His research focuses on the republican period.

Valentine Mercier is a Ph.D. student in history at the University of Lyon 2 (France) and at the University of São Paulo (Brazil). She specializes in the history of contemporary Brazil. After a master's thesis on the political commitment of women between 1945 and 1961, she is now working on the first women psychiatrists there.

Donald L. Opitz is Associate Professor at DePaul University (Chicago, USA). His research explores the interrelationship of women, gender, sexuality, and science in the context of the British Empire. He is an editor of two published anthologies and is currently writing a monograph about the international movement to advance women in agriculture and horticulture in the period 1870–1920.

Galina Orlova is Associate Professor at the School of History, HSE University (Moscow, Russia). Her work on the techno-social history of the Soviet atomic age started with the participation in the Obninsk project. She is a co-author of "The Soviet Energy Imaginary: Electricity, Atom, Oil" (2022, with Ilya Kalinin and Natalia Nikiforova).

Silvina Ponce Dawson, Departamento de Física, FCEN-UBA and IFIBA, and UBA-CONICET has a Ph.D. in physics, UBA. She is Full Professor at Universidad de Buenos Aires higher researcher at CONICET, and President of the International Union of Pure and Applied Physics (2025–2027). Her research interests include biological physics and nonlinear dynamics. She has done extensive work to help reduce the gender gap in STEM.

Judith Rainhorn is a Professor of modern history at University Paris 1 Panthéon-Sorbonne (France). Her research interests include urban history, the history of medicine and health, and environmental and labor history in France and the United States since the nineteenth century, on which she has published extensively. Her last book, *Blanc de plomb. Histoire d'un poison legal* (2019), was awarded three academic prizes.

Maria Rentetzi is Professor and Chair of Science, Technology, and Gender Studies at FAU Erlangen-Nürnberg (Germany) and an ERC grantee. Her research focuses on nuclear history, gender science studies, and science diplomacy. Her latest monograph is *Seduced by Radium: How Industry Transformed Science in the American Marketplace* (2022). She is the editor of a book series on science diplomacy, and the journal *Almagest* at Brepols.

Kirill Rossiianov is Senior Research Associate at the S.I. Vavilov Institute for the History of Science and Technology of the Russian Academy of Sciences (Moscow, Russian Federation). His research interests include the social history of Russian science, the history of primatology, genetics, and animal experimentation in the twentieth century. He now works on the biography of Nadezhda Ladygina-Kohts, the Russian zoopsychologist and primate scientist.

Ekaterina Rybkina is a Researcher and Lecturer at the Chair of Modern and Contemporary History with a focus on the history of Eastern Europe at Friedrich-Alexander-Universität (Erlangen-Nürnberg, Germany). Her research focuses on the history of technology and communications in the Russian Empire and the Soviet Union.

Gwendoline Torterat is an Assistant Professor in Social Anthropology at the University of Picardie Jules Verne (UR 4287, Habiter le Monde, France). She examines the production of knowledge about archaeological heritage through citizen participation and its forms of hidden economy.

Olga Valkova, historian, is currently Chief Research Fellow at the S.I. Vavilov's Institute for the History of Science and Technology (Moscow, Russia). Her main research focuses on the history of women scientists, and she is an author of the monograph *Storming the Citadel of Science: Women Scientists of the Russian Empire* (2019).

Brigitte Van Tiggelen is Director for International Affairs and resident scholar at the Science History Institute (Philadelphia-Paris, US-France). Her research topics include chemical sciences, women, couples, and gender in the sciences as well as scientific heritage. She edited *Women in Their Element* (2019) (with A. Lykknes).

Acknowledgments

This book originates from exchanges organized by the editors and Valérie-Burgos-Blondelle at the *Governing Science and Technology, Governing through Science and Technology: What Was at Stake for Women? (From the Late 19th to the Early 21st Century)* conference held online in 2021 due to the pandemic.[1] This conference was organized in memory of the historian Larissa Zakharova (1977–2019) to whom this essay collection is dedicated.

Our thanks go to the Centre Alexandre Koyré, the Centre d'études des mondes russe, caucasien & centre européen, the Centre de recherche médecine, sciences, santé, santé mentale, société (Cermes3), the Deutsches Historisches Institut, the Écoles des hautes études en sciences sociales, and the Commission on Women and Gender Studies in the History of Science, Technology and Medicine (IUHPST) for sponsoring the event. This event and the resulting editorial work were financially supported by Cermes3 and the Centre d'études franco-russes, to whom we express our sincere thanks for their unwavering support.

We would like to express our gratitude to the members of the scientific committee who participated in bringing together the contributors of this book: Alain Blum, Patrice Bret, Valérie Burgos-Blondelle, Françoise Daucé, and Liliane Hilaire-Pérez. Our gratitude also goes to the contributors who trusted us throughout the editorial process and to the many researchers who guided our reflection and discussed the paths taken: Pnina G. Abir Am, Arthur Clech, Jean-Christophe Coffin, Laurent Coumel, Hélène Gispert, Donald L. Opitz, Bibia Pavard, Natalia Pushkareva, Maria Rentetzi, Nathalie Richard, Marsha Richmond, Lucile Ruault, Anna Stogova, Brigitte Van Tiggelen, and Annette Vogt.

The scientific reflections that guided the construction of this book owe a great deal to Valérie Burgos-Blondelle, to whom the editors offer their sincere thanks. We also benefited from her advice and encouragement, along with Donald L. Opitz and Emmanuelle Retaillaud, when writing our introduction. We are grateful to them.

Our special thanks go to Sally Gregory Kohlstedt and Donald L. Opitz who, with kindness, generosity, and all the erudition for which they are renowned, gave full meaning to the word "empowerment" during the editing of this book.

This endeavor would not have been possible without the invaluable assistance of Robert S. Fyke and Alexander Bainbridge for their painstaking translation work. We are also indebted to Robert S. Fyke's skillful and careful proofreading of the entire manuscript, who never threw in the towel on those "very long French sentences with multiple relative clauses".

Note

1 The genesis of this symposium dates back to a series of conferences held in Paris and in Moscow between 2016 and 2019, as part of a collective research project organized by Larissa Zakharova and Grégory Dufaud entitled *Governing Science, Governing through Science in the Soviet Union 1945–1991*. The three conferences of this project gave rise to two publications: Grégory Dufaud and Larissa Zakharova (eds.). 2019. "Science, Fiction and Power in the USSR". *Kritika: Explorations in Russian and Eurasian History*, 20(4), and Grégory Dufaud and Ksenia Tatarchenko (eds.). 2022. "The lives of late Soviet science, 1945-1991". *Cahiers du Monde Russe* 63(1). https://doi.org/10.4000/monderusse.13073. Questions about the role of gender, or about the representations of women and men that have emerged as a part of these exchanges, all created an opportunity to organize the symposium in 2021.

Foreword

Sally Gregory Kohlstedt

This ambitious volume with its international dimensions reminded me of how much engaged scholars gain in personal and professional insights through our research on the history of women in science even as we interpret their individual challenges and document their collective resourcefulness. In 1981, when enthusiasm for international cooperation ran strong beneath the overcast politics of the Cold War, an unlikely partnership of women from the United States and Hungary proposed a Commission on Women within the International Union for the History and Philosophy of Science.[1] The IUHPS (founded initially in 1947) had been organized on quite narrowly defined disciplinary fields and methods of science, and its individual Commissions largely controlled access to participation and the program. Women were sparsely represented. The proposal for a Commission on Women proved controversial, but a peculiar coalition of North American, Soviet, and multiple smaller delegations from around the world organized to pass it over strong resistance from British and Western European representatives who derided the idea and queried whether there was really anything to study. Founding members had garnered strategies from the research they had done and used their positioning and power to invite women from around the world to submit proposals. Writing acceptance letters enabled isolated scholars to attain funding from their home delegations and then showcase historical research on women in science; the participants also built new collaborative networks. This volume, which also grew out of conferences on women's history, is a manifestation of the synergy between historical narratives and contemporary intellectual and personal life.

Feminist scholars address a puzzling historical record: the number of women participating in technoscience over the past century has increased worldwide, but women's achievements have not produced comparable visibility and leadership. Women's representation in academia, government agencies, and industry has rarely translated into significant funding and well-positioned careers. The scholars in this volume explore that conundrum through demographic data and detailed case studies, and they focus not only on detecting but also on documenting the persistent and debilitating limits to women's recognition and advancement. Even when stated policies are promising and contemporaries express good will, historians find that the exercise of male authority, formal and informal, undermines the expectation of fair and equal treatment. Taken together, the articles in this volume argue that governance, an important but elusive idea and sometimes subtle practice, is at the heart of the problem. The powerful effects of masculine authority, as deliberately exercised in policies and institutional practices, are not a new finding, but these accounts contribute another important dimension to the discussion. Case by case, the authors probe deeper to describe and document the sometimes-subtle influences exercised by the behavior of men with power that can also, sadly, influence women who adapt to their circumstances, sometimes even adopting self-effacing demeanors that inhibit them from persisting in their work or from getting their due recognition.

A significant counterweight to the negative pressure of masculine governance, however, has been women's collaboration as they found power in articulating and addressing gendered assumptions that constrained them. The authors here uncover the social, emotional, as well as intellectual practices that pushed back against arbitrary constraints on women, offering sustaining encouragement in the face of self-doubt and outright discrimination. Small networks were crucial to offset daily challenges, as when a cohort of Brazilian psychologists consolidated their common concerns to establish some degree of social and political autonomy. Sometimes segregated into subfields by presumed feminine skills like fine manual dexterity or by differential pay scales, women's proximity could unite them as they adapted to their marginality and formulated different styles of work practices and even distinctive patterns of collaboration. Larger organizations crossed disciplines to energize affiliates and establish bases for collective action like the International Federation of University Women. Work in male settings taught women strategies for personal or collective advantage to challenge segregation and discrimination. In these circumstances, women could adopt "anti-heroic" styles of leadership stressing engagement and collaboration in which relationships emphasized personal as well as professional support. Aggregate numbers could contribute to an activist outlook but might also simply provide parallel paths for women's engagement and leadership.

Investigating governance in multiple settings, this collection of historical essays offers an implicit invitation to make comparisons over time and across geographical boundaries. Five contributors to this volume examine, directly or indirectly, the effect of official Soviet policy (post-1918) that proclaimed the equality of women. In impressive ways, the access for women working in medicine, technology, and the sciences expanded. The authors point out that women's particular experiences and interests led them to explore issues of human reproduction and, in other sectors, use their limited scientific capital to advocate for ecology and preservation. The interwar years provided opportunities that were unprecedented, leaving pockets of opportunities in particular research areas, and, as in the later atomic project, sometimes women were able to establish "gender-marked governance". But, despite official policy, the ideal of gender equality or neutrality proved elusive, and Soviet women endured segregation, exclusion, and satire, often feeling themselves objects of curiosity rather than colleagues in work settings. The Soviet example contrasts with policies in other countries, but even with a stated aim of equality, the persistence of male governance, whether formal or informal, has proven remarkably pervasive and persistent over time and geographies.

In other chapters, too, examples of access and progress for women could be remarkable where the philanthropy of a woman donor or federal agencies under wartime duress encouraged women's initiatives and provided funding for special training. Such circumstances provide inspiring evidence of times and places where women broke through hierarchies and traditions to realize personal ambitions and gain the recognition and leadership their efforts had earned. Women could be indispensable in major laboratories, sometimes gaining modest acknowledgment. Nonetheless, in official records of noteworthy results, their contributions failed to be fully recognized in the well-established male-dominated institutions that allocated awards and direct financial sponsorship. Too often, too, early opportunities eroded over time and acknowledgment was short-lived. The mainstream historical record lost sight of many individual or collective achievements, leaving a public perception that there were few women in science. Why? Scientific masculinity has overshadowed the work of women as it governed institutional and political records. Nonetheless, as authors in this volume demonstrate, women's lives and contributions are recoverable using demographic data, oral histories, closely examined institutional records, personal archives, memoirs, and even fiction.

Historically, men in science established the process of scientific inquiry as a high-status professional category and built formal and institutional frameworks that codified masculine assumptions about who should determine access to and credit for such work. Such authority, however, was never absolute. Gender proved not to be a defining characteristic for doing science. As history demonstrates, despite credentialing and delineation of lines of authority, male governance remained elusive because science and scientists are not easily governed in the inherently unstable universe of ideas and ever-changing technologies. These essays document not only the highly effective exercise of masculine authority but also resistance to it as women participated in technoscience, not always welcome but often insistently. In the recent past, positioning women in governance structures has become an important mechanism for advancing their visibility and agency, power that these "exceptions" can use to advance themselves, their underrepresented colleagues, and the ecological environments in which technoscience is practiced. A current example is geophysicist Marcia McNutt, now president of the National Academies of Science in the United States, who has made it a priority to investigate the sometimes very personal issues that restrain women and minorities. She empowers contemporary scientists to identify and address gender issues that limit equity through conferences, publications, and policy recommendations.

We become cautiously optimistic as we see that such women, working within the system and learning its mechanisms, have enabled the opportunities for women not only to gain access but also to advance a quite different agenda.

Governance exercises real power but is not invincible. Contributors to this volume go beyond data on inequality and stated policies, while acknowledging their pervasive influence, to reveal how cultural norms, informal authority, and gendered assumptions have operated in the long twentieth century. Through collective strategies, women gained access to higher education in most places of European influence early in the twentieth century and persisted in pursuing scientific research. While national differences are evident, shaped by economic and cultural norms as well as politics, they encountered strikingly similar resistance from masculine governance in sciences and engineering. The operation of male authority could and did constrain the engagement of women in science—but it was never absolute. The accounts here demonstrate that within that world of masculine governance, women scientists could be pragmatic, instrumental, and independent, creating spaces in which they stretched and reshaped technoscience models and practices. I found the vitality, insightfulness, and enterprise of the women who populate this volume inspiring as they faced pervasive and stubbornly persistent gender biases. History is instructive. Research and analysis reveal the depth and subtleties of masculine governance and, simultaneously, the multiple ways women around the world have managed to subvert, even invert, arbitrary authority to bring about genuine change as women continue to assert their rights in the world of science.

Note

1 Donald L. Opitz positions the foundation of the Commission on Women in the national and international movement challenging the hierarchical configuration of women in science and, indeed, those who study their history; see Opitz, "Gender History of Science," in Lukas M Verburgt, ed. *Debating Contemporary Approaches to the History of Science* (London: Bloomsbury Publishing Plc., 2024), 71. For a contemporary reflection see Margaret W. Rossiter, "Opening Remark," in *International Conference on the Role of Women in the History of Science, Technology, and Medicine in the 19th and 20th C. Proceedings, Veszprem, August 15–19, 1983*. 2 vols (Budapest MTESZ, 1984), 2: ix–xi.

Introduction

The Gendered Government of the Technosciences

Grégory Dufaud and Isabelle Lémonon-Waxin

Translated by Robert Fyke

According to a recent report in *Times Higher Education*, 43 of the top 200 world universities are directed by women.[1] Among the 24 highest-ranked universities, six women are highlighted, including Louise Richardson, the first woman ever named head of Oxford University, founded in 1096. In 2019, *Forbes* placed five women among the world's 20 Tech Leaders in their Power Ranking.[2] The same year, 21% of the positions in the "C-Suite" in the same field were held by women (Huang et al. 2019: 3). If the technosciences have been the crucible of an ideology of progress and emancipation, in the twentieth century, women have long held a subordinate position, remaining excluded from its direction. More recently, women participate widely in science, technology, engineering, and mathematics (STEM), but they still only exceptionally occupy the most prestigious positions and, as Donald Opitz and Brigitte Van Tiggelen underline in their chapter in this book, women still encounter inequities and "harmful gender stereotypes".[3] Bridging the gender gap in STEM will not only require us to introduce many kinds of female role models from the past and present, but also above all, we will need to be aware of the technoscientific practices and culture's gender biases inherited from the past. The intangible and material aspects of technoscientific culture were historically shaped by men as many historians have already demonstrated.[4] Since at least the seventeenth century in Western societies, having masculine attributes such as a beard or a mustache has been an advantage for access to knowledge (Schiebinger 2003). In order to expose the gendered inequities and prejudices that still exist in science today, a group of women scientists used these masculine attributes to create the Bearded Lady Project initiated in 2014. In a social game where technoscientific practices and ways of gaining legitimacy were constructed by men for men, women are positioned as outsiders in the race for leadership. Lisa Brush has underlined that "masculinity is the unexamined foundation of norms of leadership, government, and politics", highlighting the role of male domination in all acts of government (Brush 2003: 16). In other words, only through the lens of gender can the normative model of masculinity be highlighted in the government of the technosciences. As Joan Scott has noted, "gender is a primary way of

DOI: 10.4324/9781003562597-1

signifying relationships of power" (1986: 1067). A major goal of this book is thus to analyze the historical entanglement of technosciences, gender, and government. To fulfill this original approach to the technosciences, this essay collection will shed light on the ever-accumulating gendered mechanisms that have blocked, stopped, or sometimes favored the careers of scientific women (and other gendered bodies) and their access to positions of power.

How should we understand the "government of the technosciences"? If we view sciences as aiming to produce knowledges and to use technologies, at the same time technologies manufacture and transform activities using knowledge. This tangle of sciences and technologies, and the fact that "the knowledge produced is at the same time theoretical but technical" is precisely what the term "technosciences" is intended to highlight (Bensaude-Vincent et al. 2011: 367). The government of the technosciences encompasses both the government *of* science, and government *by* science (Leggett and Sleigh 2016: 2). It involves practices that structure, regulate, publish, fund, promote, or validate. It could be defined as "the ability to mobilize tools and ways of doing in order to act and to organize others" (Pestre 2014: 14). If we adopt this more open perspective, the idea of government not only refers to the institutions, structures, and policies which shaped the technosciences but also refers to the individuals who developed practices and environments that gain agency for themselves and groups over others.[5] It includes the notion of governance—a more recent and normative term that incorporates techniques of social management (Pitseys 2010).[6] From the point of view of the government of science, how should we analyze its entanglement with gender? Women are both *actors* and *targets* of this government, thus they have embarked on a long, slow process that has enabled them to win over leadership positions and gain greater influence. This process began at the end of the nineteenth century when women gained access to scientific careers made possible by the increased enrollment of young women in higher education in Europe and North America. For women, a primary way of combating male domination was to achieve legitimacy potentially enabling them to rise to positions of authority. Consequently, this book raises questions about gender roles in technoscientific decision-making. Through the legitimization process, even women scientists without positions of authority could shape their disciplines by driving new fields of research or shaping technoscientific policies. Several case studies in this book value individual paths, demonstrating that the government of science can also take place on a kitchen counter.[7] Once women scientists had won positions of authority, they developed policies that targeted women, as illustrated by the case of Ellen Gleditsch (1879–1968). A chemist, starting from a position as pharmacist's assistant obtained from a high school diploma, she became an associate professor at the University of Oslo in 1916, before founding the Norwegian Association of University Women in 1919. While presiding over the International Federation of University Women from 1926 to 1929, Gleditsch pushed for international mobility funds for women students (Lykknes, Kvittingen, and Børresen 2005).

Thus while, at the end of the nineteenth century, it was largely individuals and collectives that advocated for the advancement of women in the technosciences; more recent national interventions have standardized the financing of scientific projects, integrating gender and diversity analyses within the initial conception of research projects (Hunt and Schiebinger 2021).

Questioning the gendered dimension of the government of the technosciences therefore highlights the tension between two contradictory processes. On the one hand, cultural scientific practices assigned women to subaltern institutional positions that render their contribution invisible. On the other hand, social practices of recognition and legitimization within and beyond scientific institutions ended up granting women access to leadership positions. This tension underpins one purpose of this book by challenging gender dichotomies such as private/public, seemingly obvious associations such as progress as a source of emancipation, or education as the origin of social mobility, and revisiting classical analytical notions such as "gender performance", "margins" and "periphery". Bringing together contributions about several countries, this volume queries national, cultural, and political rifts, including universally accepted chronological divisions which, for example, Joan Kelly has challenged in "Did Women Have a Renaissance?" (Kelly 1977). Chapters in this book provide analysis on different scales—from the lab to the state to international organizations—highlighting the many ways gender policies carried out by multiple twentieth and twenty-first-century actors have promoted or hindered women's careers. Through this multi-scalar analysis, essays address the dynamics of the government of the technosciences through the lens of gender, revisiting persistent themes in the historiography, while extending our understanding of the relationship between the government of the technosciences, the status of gender, and the several paths women have taken within that relationship.

This introduction will start with a deeper analysis of the relevant historiographical themes, before presenting a historical framework for understanding the entangled relationship between women, gender, and the government of the technosciences. Within this framework, we propose to explore three persistent issues from women studies and gender studies in the sciences that run through the chapters of this book: the assignment of women to subordinate positions, the dynamics of women's advancement in the government of the technosciences, and the performance of gender. Finally, we will outline the structure of the volume and highlight the contributions of each chapter.

Gender in the Government of the Technosciences

Grasping women's status and gender relationships in the government of the technosciences implies a double displacement of our point of view. In the first place, a historiography inspired by the work of Michel Foucault has questioned the way state power has mobilized scientific knowledge and skills to govern populations and to control territories, making clear how situated in time and space this knowledge mobilization was (1975). Recently in the West, such research has helped to define and theorize the "regulatory regime" of scientific activities (Jasanoff 1990; Pestre 2014; Lamy 2015), as well as exploring the consequences of the government of the technosciences for the social functions of scientific disciplines, and the content of scientific knowledge. For a long time, these analyses have sidelined questions about gender despite the research provided by feminist philosophers, such as Sandra Harding, Judith Butler or Donna Haraway, who have questioned epistemological standards and the notion of objectivity (Haraway 1988; Butler 1990; Harding 1991). In the aftermath of these feminist criticisms, Delphine Gardey has pointed out that established forms of government such as "institutions (School, University, Military), produce more than their titles, roles, and functions, they also produce gendered and social characters and create varied forms of masculinity when applied to masculine institutions" (Gardey 2000: 33).

Second, this displacement of analytical attention concerns the historiography of women and gender in the technosciences. Such historiography is plentiful, but whether it asks about the subjectivity of scientific knowledge, the exclusion of women from Western science, the question of scientific authority, or feminist critiques of science and technology, in all cases it has not yet explicitly addressed the government of the technosciences (Fox Keller 1982; Harding 1986; Haraway 1988; Harding and O'Barr 1987; Laslett et al. 1996, Åsberg and Lykke 2010). By drawing on an enduring legacy of feminist studies, historians have spotlighted the presence of women in the technosciences. Their presence has increased since the end of the nineteenth century, but this was far from being the result of a continuous and inevitable process: women scientists' gains have been difficult and reversible, and various forms of discrimination have continued through to the present (Rossiter 1982, 1998, 2012; Jordanova 1993a; Kohlstedt 1995, 1997; Fox Keller 1985; Schiebinger 1999; Guillopé and Roy 2020).[8] The feminization of certain scientific disciplines has also involved their partitioning, often resulting in the confinement of women scientists to occupations considered less prestigious (Schiebinger 1989; Jordanova 1993b).

Based on this double displacement, this essay collection analyzes how gender functions across several societies in the access to, and exercise of the government of the technosciences. If during the twentieth century, gender relations have favored a masculine grip on positions of power, such relations have evolved during the second half of that century as women have conquered positions that had long been forbidden to them, thanks in particular to corrective policies toward women. Uncovering and highlighting the dynamics of integration and exclusion requires us to read into the hollowness of silences in written and oral sources that often hide as much as they reveal. As a result, contributors to this volume have adopted a wide variety of contexts, frameworks, and methods of analysis: biographical, prosopographical, sociological, and quantitative approaches. This diversity of approaches is a strength of this book, which analyzes its subjects in various countries (the

United States, France, Brazil, Greece, and the Soviet Union), and different political regimes (from Communist regimes to liberal democracies). In Communist countries, despite the egalitarian discourse promoted by authorities, women scientists were still relegated to inferior positions. The presence of articles dealing with the Soviet Union helps us to overcome familiarity within Western countries by questioning the specificities of the two types of political regime. Chapters present various situations, but a majority of them deal with the Global North and more than half of them originate from countries where the gender studies tradition is not very strong (Greece, France) compared to the United States, one where gender studies is under attack (Russia), and one from the Global South (Brazil). From this point of view, the book is a contribution to the historiography of women and gender studies in science in these countries. The scope of the book is also very wide in order to encompass the general features of technosciences as much as their individual specificity. The gendered enactment of government is at work from archaeology, genetics, engineering, medicine, ecology, psychiatry, atomic energy, to climate, and documentary sciences. Several contributions deal more specifically with medicine and biology in relation to reproductive sciences, historiographically emphasizing how crucial it has been for women to (re)gain control of their own bodies.

The Assignment of Women to Subordinate Positions

Given that the sciences were knowledge systems of their contemporary societies, the government of the technosciences did not exert itself from a single jurisdiction. Such government has been exerted at many levels, in different places and times, and implicating a wide range of actors: ministry offices (when they existed), laboratories, journals, international agencies and institutions, scientific societies, and NGOs. At the start of the nineteenth century, the technosciences were not gender-neutral: they were a man's world, practiced in laboratories attached to university chairs, to industry, and to national and international foundations. As countries gradually implemented science policies according to their own time frames, specific locations on the margins of learned institutions, such as scholarly or industrial societies, authorized scientific interactions between amateurs and professionals. These interactions drew the frontiers of institutionalized technosciences while they maintained the permeability of scientific knowledge and practices (Fages 2018; Guillemain and Richard 2016; Chapman 1998). As a minority, women scientists were often excluded from most institutions, or opted for the status of amateurs "by choice" (Pomata 2013). In the case of Soviet Russia, government authorities officially proclaimed gender equality and promoted women such as Natalia Freichko, who had started out as an amateur radio engineer. But women seldom occupied positions of authority (Rybkina, Chapter 12). Scientific positions that were either socially poorly recognized or perceived as feminine, and as a result were often ignored by men, were thus available to women scientists. Several chapters show how women could fill scientific positions neglected by men, or out of ideological convictions. In Alice Hamilton's case in the United States, she became an expert for occupational health after starting her medical career in obstetrics and women's health (Rainhorn, Chapter 10). Or, in medicine, Nise da Silveira's case in Brazilian psychiatry was driven by a communist *sensibilité* (Mercier and Martins, Chapter 5). In their respective fields, each scientist gained power and prestige to participate in the elaboration of scientific policies.

During the Cold War, the United States imposed itself as the technoscientific model for the Western world. Scientific investment grew and diversified the technical and material means of doing research, changes which gave birth to an organizational structure that more or less associated universities, industry, and the military (Pestre and Jacq 1996). Public agencies specializing in particular scientific fields judged to be strategically important were created to stimulate the economy. The rivalry between the two superpowers stimulated the development of strategic new scientific technologies, such as in aerospace and nuclear science, where women remained "hidden figures" (Howes and Herzenberg 1993; Light 1999; Orlova and Kasatkina, Chapter 7). In the United States, private enterprise insinuated itself as an important source of research, such as the pharmaceutical industry, which produced "The Pill"—the contraceptive Enovid—starting in 1960. In this case, however, the industry produced a substance it had not developed: the invention was researched from the end of the 1940s through the 1950s by a team organized by Margaret Sanger, campaigning for contraceptive access as a part of concerns about overpopulation (Durand-Vallot, Chapter 11). In the Soviet Union where the private sector was inexistent, the scientific system that emerged was commanded by an advanced, centralized and vertical logic. It was organized around what Loren Graham and Irina Dezhina have called the "three pyramids": a privileged military-industrial complex, the Soviet Academy of Science as source for fundamental research, and the universities that were mainly involved in teaching (Graham and Dezhina 2008: 1). Globally, major organizations such as UNESCO, or the NGOs, have formulated scientific research programs based on their recommendations for improving social development. For example, in 1961, the UN Food and Agricultural Organization (FAO) and UNESCO started the production of a *Soil Map of the World* in order to gain global environmental knowledge so as to manage natural resources rationally (Selcer 2015).

On a local scale, laboratories have brought together teams of tenured, post-doctoral, and doctoral researchers, supported by administrations, techniques, and equipment, all placed under the responsibility of directors accountable to their superiors. Women were increasingly present, particularly in emerging scientific fields such as nuclear, computer, or aerospace sciences. Nonetheless, they still represented a minority in this scientific system, and most often women worked in positions subordinate to men and within less prestigious disciplines.[9] For instance, in the United States, Margaret Rossiter has noted the sharp increase in women's enrollment in certain sciences between 1942 and 1946: a 131% increase in physics, 113% in mathematics, and 227% in chemistry. However, in 1948, only 6% of US scientists and engineers were women; and, in 1962, they still only represented 6.78% of scientists and engineers (Rossiter 1998: 25, 30, 98). Also, in laboratories, some directors profited from their positions to award themselves primary roles in new discoveries, while diminishing the roles of their female colleagues. In this way, the two principal female researchers and co-authors of joint articles on the discovery of gene splicing methods and DNA introns were sidelined for the 1993 Nobel Prize in Physiology in favor of the two male directors of their laboratory (Abir-Am, Chapter 9).

Over the past thirty years, the government of research has undergone systemic changes where women have had greater responsibilities, partly counteracting previous masculine distributions of power. These changes have expanded the number of actors, redistributed innovation around the world, broadened the definition of intellectual property, and reduced university autonomy. Moreover, the government of research has evolved alongside of scientific practices and disciplines particularly with the development of biotechnology, with recourse to mathematical and computational tools, with the return of scientific observation, and with the growth of earth system sciences (Pestre 2013). The administration of science and knowledge production has also seen a growth in citizen participation, attested by the emerging category of "citizen science" (Silvertown 2009). These evolutions have been accompanied by a slow increase of women's presence in STEM. In 2019, women represented 28% of STEM workers in the United States, compared with 25% in 1993.[10] UNESCO's data for 107 countries confirms this on a global scale revealing that 33% of researchers in STEM were women between 2015 and 2018.[11] Despite this slow increase, concerns about parity in the government of the technosciences remain. In 2021, women hold only 14% of executive positions in STEM industries and only 8% of STEM CEOs are women in the United States.[12] The recent report of the Gender Gap in Science Project initiated by the International Council for Science (Guillopé and Roy 2020) also pointed out this alarmingly modest progress. Having women in leadership positions should be crucial to promote gender equality and inclusion for as long as women's entry into the technosciences is still an issue. To promote this claim, scientific unions have long worked to tailor precise international statistical surveys that would help bridge the gap by reorienting technoscientific policies around gender (Ivie, Gledhill and Ponce Dawson, Chapter 2). Furthermore, several technoscientific journals have adopted policies respecting greater diversity throughout the editorial process, based on the International Committee of Medical Journal Editors' recommendations. Such standards collide with researchers being pressured to publish more, where one current estimate suggests 30,000 academic journals publishing two million articles annually.[13] However, these initiatives are still struggling to establish real gender parity in the technosciences.

The Fraught Dynamics of Women's Advancement in the Government of Technosciences

From the moment women gained access to technoscientific education at the end of the nineteenth century, women multiplied the strategies they used to participate in knowledge production: bypassing, diverting, or appropriating institutional and social frameworks. During the twentieth century, women gained access to technosciences, where they passed from the "status of scientific minors to that of scholarly *alter egos*" (Fauvel, Coffin, and Trochu 2019), but positions of power still remained closed to them. Some important aspects of the contributions to this volume shed light on those interstices women scientists slipped through, the practices they implemented to access leadership positions, and the limitations that opposed them. For example, in the United States, when the first laws concerning working conditions were adopted in 1910, Alice Hamilton opted to work in occupational health. She was attacked many times as a woman by men who questioned her analyses, their misogyny being all the more virulent as financial interests were at stake. When Hamilton was named Assistant Professor at Harvard in 1919, this certainly qualified as recognition for her research work, but she was confronted with a new type of discrimination: she was excluded from her peers' male-oriented social events and venues and was never advanced to a full professorship (Rainhorn, Chapter 10).

From the beginning of the 20th century, women scientists tried to gain allies, if not outside of the technoscientific world, at least outside of their own professional backgrounds. They have thus sought the support of men, as Ygor Martins and Valentine Mercier have described in their collective portrait of women psychiatrists in Brazil who began their careers in the 1920s, and wound up in hospital and clinic directorships and university chairs starting in the 1940s (Chapter 5). Their careers rested on the continued development of ever-widening networks of scientists, administrators and politicians from diverse backgrounds. Once in established positions, this cohort of women psychiatrists advanced other women's careers through professional associations. Their career paths also showed a political dimension, marked by participation in projects like temperance campaigns. This relationship to politics worked both ways: women psychiatrists used the support of male authority to access positions of scientific government and also used their status as psychiatrists to defend their political convictions. To reach leading positions, women also created institutions on an international scale where women could come together to accompany and support other women in their careers. Thus, the International Federation of University Women (IFUW), established in 1919, was an organization that raised major questions about gender inequality using scientific methods such as statistics, to reorient general scientific policies (Cabanel, Chapter 3).

Once women scientists had reached positions of authority, they were better able to develop their research, provided the political, social, and judicial circumstances permitted them to do so. For example, in the field of reproductive sciences, the techniques women scientists developed opened up access for other women to forms of autonomy and self-governance. More specifically in the Soviet Union, the 1917 October Revolution inaugurated a period of new scientific experimentation that closed up again in the 1930s. By promoting liberal politics toward abortion, family life, marriage, and homosexuality, the Bolsheviks created a social and intellectual environment favorable to biomedical research—with opportunities for women—that would not have been possible only a few years earlier. For example, Vera Danchakova, biologist and the first female professor in Russia in 1908, founded at the Timiryazev Institute for Scientific Research a morphogenetics laboratory where she studied the plasticity of embryonic tissue and the determination of sexual characteristics. Or Antonina Shorokhova, gynecologist in Tashkent, organized a gynecological service in that city, and researched artificial insemination, both based on her concepts that women had a right to have children, and a need to optimize their choice of sperm donor. To this end, she developed a sperm agility test (1929) which became widely used in the Soviet Union (Kojevnikov and Rossiianov, Chapter 4).

After the Second World War, new opportunities opened to women scientists did not erase gender asymmetries, they renewed them. In the Soviet Union, women provided scientific tasks—such as radio-chemical analysis or computational work within atomic research projects— which were judged to be auxiliary. Despite the importance of their work, and following the classic pattern of scientific knowledge production, women scientists were relegated to the role of "invisible technicians" (Shapin 1989); while physicists, or other scientists meant to be showcased, were almost without exception men (Orlova and Kasatkina, Chapter 7). Thus, the feminization of certain professions neither implied that male scientists had yielded governance to women, nor did it imply that sexism would not reignite whenever male scientists felt their authority wavering.

Not all women who led scientific lives wished to make a career out of it, and some of them chose to remain on the periphery of scientific institutions: they were involved in science, but they did not occupy an institutional position. In peripheral situations, women were free to conduct their own research and this freedom enabled them to shape their field of research. While it is difficult to assert whether that choice referred back to misogyny or other obstacles women scientists had faced, peripheral positions also appear to be related to the image of heteronormative couples and to the gendered division of work that women embodied in their personal experiences. For instance, the French archaeologist Arlette Leroi-Gouhran, wife of the famous archaeologist André Leroi-Gouhran, produced major epistemological and methodological innovations in the study of ancient plant pollen at home. She made significant advances in the emerging science of palaeopalynology inside of a gendered division of space and with the help of a male assistant from the

Centre National de la Recherche Scientifique (CNRS). A first paradox related to Leroi-Gourhan's trajectory was that she had a collaborator even though she did not have a position or a title. A second paradox was her renown in a new field of archaeology that did not translate into a new academic position (Torterat, Chapter 14). Another peripheral path women took to scientific government was to undertake projects outside of established knowledge institutions. In the United States, Margaret Sanger's proposal to develop a hormone-based contraceptive was supported by Katharine McCormick, whose inheritance from her husband's death in 1947 financed the research. The two women were joined by well-known specialists in the field, Gregory Pincus and John Rock, who perfected a method of hormonal inhibition to ovulation. The first oral contraceptive was approved by the US Food and Drug Administration in 1959 (Durand-Vallot, Chapter 11).

Conversely, on an international scale, women's collectives elaborated strategies to reach global gender equity in science and technology. For example, during the Cold War, in Greece, UNESCO used public libraries, primarily staffed by women librarians, as a means of generalizing and standardizing knowledge. Loukas Freris and Maria Rentetzi (Chapter 8) demonstrate the paradox of a politics that considered public libraries to be useful propaganda tools against communism, and entrusted them to women volunteers given little recognition and often asked to work for free. Likewise, professionalizing international women's collectives perfected statistical tools ensuring better integration of women in the technosciences. Following in the footsteps of the IFUW's concern for gender equality, the International Union of Pure and Applied Physics (IUPAP) developed specific surveys in the 2000s. In order to better understand the origins of the gender gap in the technosciences, measure gender inequality, and fight against it, the IUPAP, joined by other scientific unions, put together and conducted surveys measuring inequalities and formulating recommendations. At the heart of the IUPAP, government evolved toward greater parity based on micro and macro "change agents" working toward a "post-heroic leadership style based on collective action and engagement" (Ivie, Gledhill and Ponce Dawson, Chapter 2).

Performing Gender

The obstacles encountered by women scientists for admission to positions of power were more or less significant, depending on the profession and the country. In their careers, the veiled or explicit opposition was more or less powerful. Such obstacles crystallized into, for example, the famous metaphor of the "glass ceiling", whose durability lies in constantly changing its height above the floor in line with social and political circumstances.[14] As in other fields, the gendered division of the technosciences has had no justification; it expresses itself through practices and discourse that naturalize and reproduce it. Women scientists in any career choice and any level of achievement, often remained scrutinized through the prism of gender stereotypes relating appearance, intelligence, and behavior. Assumptions about presumed physical characteristics for the practice of different types of scientific work have often determined which fields are more suitable to one sex or the other. In the twentieth century, these stereotypes were still at work: Alice Hamilton, at the end of her occupational health career was still associated with supposed female qualities of care, while women calculators on atomic projects in the Soviet Union were associated with dexterity. Thus, the contributions in this volume demonstrate that there is no real difference between political regimes. The Soviet Union posed a useful historical nuance: official discourse about having achieved gender equality ironically made it more difficult to actually advance it. Women were certainly better integrated in the technosciences there than they were in Western Europe, but official doctrine blinded the state and most of society toward gender relations. As a result, in meteorology and climatology, Russian women scientists struggled to reach positions of responsibility, their sense of inferiority persisted, and the discrimination remained unspoken (Doose, Chapter 6).

Despite discrimination and stereotypes, women in the technosciences often fostered ambivalence toward feminism. Not all women who climbed the ladder up to positions of authority, overcoming self-identified gender roadblocks, or working toward reappropriating women's own bodies and sexualities, necessarily identified with feminism. Neither Hamilton, nor Danchakova, nor Shorokhova, were self-proclaimed feminists. On the other hand, Sanger represented herself both as a feminist and as a eugenicist, where these ideologies motivated her development of contraceptives. The scientific work of women was indeed part of the political and social movements of their time. Echoes of this relationship are found in the biologization and naturalization of social hierarchies, in the politicization of biological matters, and in the ideologies of progressive, technical, and scientistic movements. In the reproductive sciences, women often led efforts intended to both select individuals possessing desirable characteristics and improve the quality of populations. Brazilian psychiatry also fostered ambitions to scientifically engineer society and the nation, and this was animated by female supporters of the mental hygiene movement.

Regardless of the political and social contexts, women constructed their scientific careers by confronting male colleagues, employers, and laboratory directors, as well as state and private institutions. Each woman had to cope with the barriers encountered along the way. Some women respected those barriers as Arlette Leroi-Gouhran did; others transgressed them like Alice Hamilton did; still others played off of them as, in France, the physicist and chemist Marie Curie did (Des Jardins, 2010). Women scientists who playfully navigated those barriers, or broke free from them—even when they were supported by a network of male colleagues—still had to cope with those rules of behavior established by gendered social roles. Opitz and Van Tiggelen suggest analyzing scientific production as a "performance" emphasizing that "adopting the lens of gender in studies of governance, enables us to observe the fuller scope of knowledge production-as-performance, with particular emphasis on how communication both genders and governs technosciences" (Chapter 1, p. 28). Within this double perspective, gender representations conveyed by language, attitudes, and clothing shaped the personae women scientists portrayed in certain situations (Daston & Sibum 2003). So too did narratives concerning women scientists reflect the gender norms and stereotypes at work in society, as well as the emancipation movements women were navigating. For instance, women scientists were described as heroines of Soviet radio, carrying on their frail "girlish shoulders" the weight of many men's technical work (Rybkina, Chapter 12, p. 220); or they were depicted as "living computers", "girls" and machines, and "little suns" by the physicists working on the Soviet nuclear project (Orlova and Kasatkina, Chapter 7, p. 130–137). In each case, women coped, fought, or played with gender representations.

Gendered identities were double-edged, and they could thus be wielded to spur action. Networks of household or philanthropic socialization in which women were frequently confined by their supposed "benevolent nature" nevertheless allowed them to catalyze scientific projects, as McCormick did with her financial support for contraceptive research. Similarly, the nature of "exceptional women" conceived from a career retrospective, was eventually used as a tool of scientific and political engagement, as exemplified by the case of Vera Varsanofieva in the Soviet Union. Professor of geology and mineralogy at Moscow State University in 1935, Varsanofieva became affiliated with several scientific societies, was a party delegate, and was a member of the State administration in Moscow. Embarking on salvaging natural reserves that the Soviet government had decided to liquidate in 1951, Varsanofieva mobilized the scientific and political resources she had accumulated. Supported by influential people and institutions, she obtained the creation of a Principal Directorship for Hunting and Natural Reserves, under the direct authority of the Council of Ministers, and the return of Natural Reserve lands previously removed (Valkova, Chapter 13).

Cracking the glass ceiling

In the twentieth century, given the handful of women scientists who were successful, it may seem futile to raise historical questions about gender relations in technoscientific decision-making, and about women's participation in government and scientific policies. As in other domains, the technosciences have only delegated the most important positions to women in trivial numbers. If such subjects appear elusive, they nevertheless push us to reflect upon the obstacles that women scientists have resisted, and the ways they have mobilized resources that cracked open the "glass ceiling". New reflection about the domination of masculinity in scientific culture and about corrective policies that bridge the gender gap in science, both originated from their access to the government of the technosciences. As a focus for gender relations, such government brings out the asymmetry of power relations and opens the way to new perspectives.

As the contributors to this volume show, key historiographical concepts and methods are brought into play to effectively analyze the ways in which actors navigated and negotiated gendered structures in the government of technosciences. Thus, the first part of this volume begins with two analyses that extend the introduction and suggest some possible analytical paths based on studies of government and gender in the technosciences. These two chapters offer perspectives on the persistence of historical problems and methodological considerations that align with current realities about understanding and reversing gender inequalities in the technosciences. Taken together, these chapters offer a compelling basis for renewed scrutiny of the dynamics of government and gender, scrutiny which later contributions address more fully through specific case studies and in a variety of contexts. Opitz and Van Tiggelen bring in an historiographical perspective highlighting how historical studies could then help renew the ways historians categorize their actors as exceptional/victim, professional/amateur, etc., and provide tools to bridge the gender gap (Chapter 1). In turn, Rachel Ivie, Irvy M. A. Gledhill, and Silvina Ponce Dawson adopt a sociological approach to analyze how designing surveys to document gender inequalities and the gender gap in physics—and more widely in science—shaped new forms of IUPAP's governance through women's actions, reaching greater equity (Chapter 2).

The second part of the book interrogates the technoscientific policies that target women and gender relationships. To this end, the authors focus on the role played by different kinds of institutions and organizations—unions, federations, States, international organizations, political, and professional networks—on gender power relations in the technosciences. Following in previous researchers' footsteps, these authors demonstrate how institutions and organizations were places of power where roles and hierarchies masked gender relations, while at the same time implementing gender stereotypes and inequalities. From a fresh perspective, the authors in this volume also show how institutions and organizations may have been crucibles of greater epistemic justice in reducing gender inequalities. Anna Cabanel shows how the IFUW, shaped in the 1920s and 1930s its own international policy to scientifically demonstrate the gender gap and to integrate women into the management of academia (Chapter 3). Cabanel characterizes the necessary tools the IFUW implemented to reach greater equity: a new gender-neutral persona for the scientist, diplomas that defined new academic norms, funding that developed women's agency, networks and mentoring to empower them, and new publication spaces to give women a room of their own in the male-dominated scientific world.[15] Alexei Kojevnikov and Kirill Rossiianov (Chapter 4) situate in revolutionary Russia of the 1920s how new reproductive technoscientific practices implemented by women scientists and the advancement of women's own government of their body were perceived by the revolutionary state as advantages in Russian society. For Brazil, Valentine Mercier and

Ygor Martins (Chapter 5) analyze the roles of professional, political, and associative networks to empower women psychiatrists, from 1941 to 1970. In these networks, women psychiatrists surrounded themselves with scientific female colleagues, hiring associates that enabled them to establish their own innovative projects. Katja Doose argues that, from 1919 to 1991, Soviet gender policies claiming "equality led to blindness toward gender" that affected Russian society (Chapter 6, p. 107). State policies made it difficult for women to think about gender discrimination in the technosciences, giving rise to inferior self-perception as female scientists. After World War II, Soviet nuclear research rested in women's hands, but, as Galina Orlova and Aleksandra Kasatkina show in Chapter 7 "governing the Atomic Project was not women's business" (p. 124). Despite their crucial role in the atomic project, they could never reach official leadership positions. The structure of research labs was gendered, based on naturalization discourses that assigned women to long, tedious, repetitive tasks. Yet some women, among them, the largely female mathematical computing teams, were able to develop their own technoscientific culture in a male-dominated area. Loukas Freris and Maria Rentetzi (Chapter 8) expose how, during the Cold War, the Greek government and UNESCO used public libraries, perceived as feminine realms, to fight communism, and to create a national identity. The education and status of women librarians were mainly disregarded by the Greek state, and they had to settle their own training. But thanks to UNESCO support, one of them was able to reach leadership of the Greek program and to empower women librarians. Pnina Abir-Am (Chapter 9) enlightens how in the United States, through their institutional authority over women, lab directors were encouraged to erase women's contributions in major discoveries to secure their own reputation and career. Abir Am's chapter questions how the legitimacy of scientists could sometimes be partly built or denied in terms of gendered relations, and not in terms of scientific proofs. It also points out the ethical responsibility of the Nobel prize committee in awarding the prize.

The last section of the book explores the idiosyncratic paths some women have taken to access leadership. On the one hand, some chapters show the manifestations of the "glass ceiling", adding to an already substantial historiography on the topic. On the other hand, chapters reveal how women scientists found "room" to modify their research field and reshape the ways the technosciences were governed. Scrutinizing individual paths brings to light women scientist's agency, expressed as their self-awareness and their ability to act, conform, resist, and transform. Thus, in addition to designing the government of the technosciences, some women scientists challenged the gender norms of the scientific community. Judith Rainhorn (Chapter 10) demonstrates how Alice Hamilton played with gender boundaries throughout her career—blurring, transgressing, and shifting them—before being recognized as an expert in occupational health by State and Federal governments and international institutions. Influencing the path of technoscientific research has also been in the grasp of some women, considered as research outsiders. This was the case for Margaret Sanger and Katharine McCormick concerning the Pill. Within a eugenicist and feminist American context during the first half of the twentieth century, "they envisioned science as a strategy and reproductive sciences as a crucial resource for women's empowerment" as stated by Angeline Durand Vallot (Chapter 11, p. 199). Their action toward the government of women's own bodies through science allowed them to govern contraceptive research and to change the social issue of birth control. During the same time period in Soviet Russia, when women enrolled in many technoscientific fields, radio engineering remained a masculine field being shaped as a military tool. However, after WWII, the state's heroic narrative about female radio engineer Natalia Freichko—analyzed by Ekaterina Rybkina (Chapter 12)—became part of a new strategy to develop this field. Freichko as an expert embodied the apparent opening of a man's domain to women. However, Rybkina demonstrates that the Soviet state, while trying to feminize the radio field, made it unreachable to ordinary women. In the 1950s, one prominent scientific woman, Vera Varsanofieva, a geologist who reached leadership positions in science and in Moscow City Council, revealed herself as a political force to save a nature reserve. Varsanofieva turned scientific societies into political forces where "science citizens" concerned with ecology drove the state to review its environmental policy (Chapter 13). The last case study proposed in this book retraces the voluntary path of Arlette Leroi-Gourhan in the Musée de l'Homme in Paris during the second half of the 20th century. She first devoted her time to assisting her husband, but quickly dedicated herself to the new archaeological discipline of palaeopalynology that she helped to create. From her apartment to the Museum's lab, she carved a methodology, vocabulary, practices, and norms for which she was later acknowledged by the scientific community. From a counter top in her kitchen, she led a new generation of archaeologists in the use of pollen within archaeology (Chapter 14).

Women's widespread access to education has enabled them to be more active in the technosciences, but much more rarely to reach positions of authority. This book sheds light on the paradox whereby the technosciences have carried on an emancipatory discourse while maintaining forms of gender discrimination within them. The contributions to this volume show the social and political mechanisms that have partially undermined the promise of emancipation for women and female scientists. But they also make clear how these same technosciences can produce tools for counteracting scientific career discrimination within them and in society at large, and they demonstrate that the government of the technosciences can be implemented with a view to greater inclusion and gender equality, from an international to an individual scale.

Notes

1 https://www.timeshighereducation.com/student/best-universities/top-10-universities-led-women.

2 https://www.forbes.com/sites/bizcarson/2019/12/12/most-powerful-women-in-tech-in-2019-beyond-ceoswomen-dominate-the-c-suite/?sh=669362e749e4.
3 https://gender-gap-in-science.org/.
4 See, for example, the pioneer work of Margaret Rossiter (1982) and Evelyn Fox-Keller (1983), and more recent developments: (Abir-Am and Outram 1987; Wajcman 1991; Berner 1997; Rossiter 1998, 2012; Milam and Nye 2015; Jones, Martin and Wolf 2021; Bix 2022; Rentetzi 2023).
5 This understanding of the word "government" in history of science is illustrated in Roy MacLeod and Peter Collins's expression «Parliament of science» to evoke the British Association for the Advancement of Science. We are grateful to Donald L. Opitz for advising us this reading.
6 Many authors have already pointed out that the word governance does not have a settled definition and has become a "suitcase" word, which nevertheless enables politicians and scientists to understand each other. See for example (Fukuyama 2015).
7 The importance of domesticity in historical studies of women and gender in the history of science and technology has already been demonstrated by historians (Opitz et al. 2016).
8 They are also intertwined with discrimination due to racism, validism, classism, transphobia, etc. These intertwining effects are highlighted by the intersectional approach (Crenshaw 1989).
9 Margaret Rossiter (1982) has characterized these limitations as hierarchical and territorial segregation.
10 https://www.nsf.gov/nsb/sei/edTool/data/workforce-07.html.
11 https://www.unesco.org/reports/science/2021/en/dataviz/share-women-researchers-radial.
12 https://nextwavestem.com/stem-resources-news/stem-resources-and-news/breaking-barriers-the-importance-of-women-leaders-in-stem.
13 https://www.universityworldnews.com/post.php?story=2018090509520357.
14 Other metaphors also describe discriminations women scientists endure on their career paths: "leaky pipeline" (Alper 1993), "sticky floor" (Berheide 1992), or "glass cliff" (Ryan and Haslam 2005).
15 About the concept of persona defined by Marcel Mauss and subsequently used in the history of science see (Mauss 1938; Daston and Sibum 2003). Since then, other historians have used the term in a broader sense (Bosch 2016).

References

Alper, Joe. 1993. "The Pipeline is Leaking Women All the Way". *Science* 260(5106): 409–411.

Åsberg, Cecilia and Nina Lykke (ed.). 2010. "Feminist technoscience studies". *European Journal of Women Studies* 17(4): 299–305.

Bensaude-Vincent, Bernadette, Sacha, Loeve, Alfred, Nordmann, *and Astrid, Schwarz.* 2011. "Matters of Interest: the Objects of Research in Science and Technoscience." *Journal for General Philosophy of Science* 42: 365–383.

Berheide, Catherine W. 1992. "Women Still 'Stuck' in Low-Level Jobs." *Women in Public Service: A Bulletin of the Center for Women in Government* 3(1): 1–4.

Berner, Boël. 1997. "L'ingénieur ou le génie du mâle: masculinité et enseignement technique au tournant du xxe siècle". *Cahiers du Gedisst* 19: 7–25.

Bix, Amy Sue. 2022. *Girls Coming to Tech! A History of American Engineering Education for Women.* Cambridge: MIT Press.

Bosch, Mineke. 2016. "Scholarly Personae and Twentieth-Century Historians. Explorations of a Concept". *BMGN. Low Countries Historical Review* 131(4): 33–54.

Brush, Lisa. 2003. *Gender and Governance*. Walnut Creek, CA: Altamira Press.

Chapman, Allan. 1998. *The Victorian Amateur Astronomer: Independent Astronomical Research in Britain 1820–1920*. Hoboken, NJ: Wiley.

Butler, Judith. 1990. *Gender Trouble: Feminism and the Subversion of Identity*. London: Routledge.

Crenshaw, Kimberlé. 1989. "Demarginalizing the Intersection of Race and Sex: A Black Feminist Critique of Antidiscrimination Doctrine, Feminist Theory and Antiracist Politics". *University of Chicago Legal Forum* 1989(1): 139–167.

Daston, Lorraine, and Otto Sibum. 2003. "Scientific Personae and Their Histories". *Science in Context* 1: 1–8.

Des Jardins, Julie. 2010. *The Madame Curie Complex: The Hidden History of Women in Science*. New York: The Feminist Press.

Fages, Volny. 2018. *Savantes nébuleuses: l'origine du monde entre marginalité et autorité scientifique (1860–1920)*. Paris: Editions EHESS.

Foucault, Michel. 1975. *Surveiller et punir*. Paris: Gallimard.

Fauvel, Aude, Jean-Christophe Coffin, and Thibaud Trochu. 2019. "Les carrières de femmes dans les sciences humaines et sociales xixe-xxe siècles: une histoire invisible?" *Revue d'Histoire des Sciences Humaines* 35: 11–24.

Fox Keller, Evelyn. 1982. "Feminism and Science". *Signs* 7(3): 589–602.

Fox Keller, Evelyn. 1983. *A feeling for the organism the life and work of Barbara McClintock*. New York: W.H. Freeman.

Fox Keller, Evelyn. 1985. *Reflections on Gender and Science*. New Haven, CT: Yale University Press.

Fukuyama, Francis. 2016. "Governance: What Do We Know, and How Do We Know It?". *Annual Review of Political Science*, 19 :89–105.

Gardey, Delphine. 2000. "Histoires de pionnières". *Travail, Genre et Sociétés* 4: 29–34.

Graham, Loren, and Irina Dezhina. 2008. *Science in the New Russia: Crisis, Aid, Reform*. Bloomington: Indiana University Press.

Guillemain, Hervé, and Nathalie Richard. 2016. "Introduction. Towards a Contemporary Historiography of Amateurs in Science (18th–20th Century)". *Gesnerus* 73(2): 201–237.

Guillopé, Colette and Françoise Roy. 2020. *A Global Approach to the Gender Gap in Mathematical, Computing, and Natural Sciences: How to Measure It, How to Reduce It?* Berlin: International Mathematical Union.

Haraway, Donna. 1988. "Situated Knowledges: The Science Question in Feminism and the Privilege of Partial Perspective". *Feminist Studies* 14(3): 575–599.

Harding, Sandra. 1986. *The Science Question in Feminism*. Ithaca, NY: Cornell University Press.

Harding, Sandra, and Jean O'Barr. 1987. *Sex and Scientific Inquiry*. Chicago, IL: University of Chicago Press.

Harding, Sandra G. 1991. *Whose Science? Whose Knowledge? Thinking from Women's Lives*. Maidenhead: Open University Press.

Harvey, Joy. 2012. "The Mystery of the Nobel Laureate and His Vanishing Wife." In *For Better or For Worse? Collaborative Couples in the Sciences*, ed. Annette Lykknes, Donald L. Opitz, Brigitte Van Tiggelen, 57–77. Basel: Springer.

Howes, Ruth H., and Caroline L. Herzenberg. 1993. "8 Women in Weapons Development: The Manhattan Project". In *Women and the Use of Military Force*, ed. Ruth H. Howes and Michael R. Stevenson, 95–110. Boulder, CO: Lynne Rienner Publishers.

Huang, Jess, Alexis, Krikovich, Irina, Starikova, Lareina Yee, and Delia Zanoschi. 2019. *Women in the Workplace 2019*. San Francisco: Retrieved from McKinsey & Co. website: https://www.mckinsey.com/featured-insights/genderequality/women-in-the-workplace.

Hunt, Lilian E., and Londa Schiebinger. 2021. "Sex, Gender, and Diversity Analysis in Research Policies of Major Public Granting Agencies: A Global Review." *OSF Preprints*, December 16. doi:10.31219/osf.io/3agxf.

Jasanoff, Sheila. 1990. *The Fifth Branch. Science Advisers as Policymakers*. Cambridge, MA: Harvard University Press.

Jones, Claire G., Alison E. Martin, and Alexis Wolf, eds. 2021. *The Palgrave Handbook of Women and Science Since 1660*. Basingstoke: Palgrave Macmillan.

Jordanova, Ludmilla J. 1993a. "Gender and the Historiography of Science." *The British Journal for the History of Science* 26(4): 469–483.

Jordanova, Ludmilla J. 1993b. *Sexual Visions: Images of Gender in Science and Medicine between the Eighteenth and Twentieth Centuries*. Madison: University of Wisconsin Press.

Kelly, Joan. 1977. "Did Women Have a Renaissance?". In *Becoming Visible, Women in European History*, ed. Renate Bridenthal and Claudia Koonz, 21–47. Boston: Houghton Mifflin.

Kohlstedt, Sally G. 1995. "Women in the History of Science: An Ambiguous Place." *Osiris*, 10: 39–58.

Lamy, Jérôme. 2015. "Des sciences par et pour le gouvernement: sur le régime régulatoire des sciences contemporaines". *Sociologie et sociétés* 47(2): 287–309.

Light, Jennifer S. 1999. "When Computers Were Women." *Technology and Culture*, 40(3): 455–483.

Laslett, Barbara, Sally G. Kohlstedt, Evelynn Hammonds, and Helen Longino, eds. 1996. *Gender and Scientific Authority*. Chicago: University of Chicago Press.

Leggett, Don, and Charlotte, Sleigh. 2016. *Scientific Governance in Britain, 1914–79*. Manchester: Manchester University Press.

Lykknes, Annette, Lise Kvittingen, and Anne Kristine Børresen. 2005. "Ellen Gleditsch: Duty and responsibility in a research and teaching career, 1916–1946". *Historical Studies in the Physical and Biological Sciences* 36(1): 131–188.

Marry, Catherine. 2004. *Les femmes ingénieurs, une révolution respectueuse*. Paris: Belin.

Mauss, Marcel. 1938. *Une catégorie de l'esprit humain: la notion de personne, celle de « moi »: un plan de travail*. Royal Anthropological Institute of Great Britain and Ireland.

Milam, Erika Lorraine, and Robert A. Nye, eds. 2015. "Scientific Masculinities". *Osiris* 30.

Opitz, Donald L., Staffan Bergwik, and Brigitte Van Tiggelen, eds. 2016 *Domesticity in the Making of Modern Science*. London: Palgrave Macmillan.

Pestre, Dominique. 2013. *À contre-science. Politiques et savoirs des sociétés contemporaines*. Paris: Seuil.

Pestre, Dominique. 2014. *Le gouvernement des technosciences: Gouverner le progrès et ses dégâts depuis* 1945. Paris: La Découverte.

Pestre, Dominique, and François Jacq. 1996. "Une recomposition de la recherche académique et industrielle en France dans l'après-guerre, 1945–1970: nouvelles pratiques, formes d'organisation et conceptions politiques". *Sociologie du travail* 38(3): 263–277.

Pitseys, John. 2010, "Le concept de gouvernance". *Revue Interdisciplinaire d'Études Juridiques* 65(2): 207–228.

Pomata, Gianna. 2013. "Amateurs by Choice: Women and the Pursuit of Independent Scholarship in 20th Century Historical Writing". *Centaurus* 55(2): 196–219.

Rentetzi, Maria, éd. 2023. *The Gender of Things: How Epistemic and Technological Objects become Gendered*. London: Routledge.

Rossiter, Margaret W. 1982. *Women Scientists in America: Struggles and Strategies to 1940*. Baltimore, MD: Johns Hopkins University Press.

Rossiter, Margaret W. 1993. The Matthew Matilda Effect in Science. *Social Studies of Science* 23(2): 325–341.

Rossiter, Margaret W. 1998. *Women Scientists in America: Before Affirmative Action, 1940–1972*. Baltimore, MD: Johns Hopkins University Press.

Rossiter, Margaret W. 2012. *Women Scientists in America: Forging a New World Since 1972*. Baltimore, MD: Johns Hopkins University Press.

Ryan, Michelle K., and Alexander S Haslam. 2005. "The Glass Cliff: Evidence that Women Are Over-Represented in Precarious Leadership Positions". *British Journal of Management* 16: 81–89.

Schiebinger, Londa L. 1989. *The Mind Has No Sex? Women in the Origins of Modern Science*. Cambridge, MA: Harvard University Press.

Schiebinger, Londa L. 1999. *Has Feminism Changed Science?* Cambridge, MA: Harvard University Press.

Schiebinger, Londa L. 2003. "The Philosopher's Beard: Women and Gender in Science". In *The Cambridge History of Science. Eighteenth-Century Science 4*, ed. Roy Porter, 184–196. Cambridge: Cambridge University Press.

Scott, Joan W. 1986. "Gender: A Useful Category of Historical Analysis". *The American Historical Review* 91(5): 1053–1075.

Selcer, Perrin. 2015. Fabricating Unity: The FAO-UNESCO Soil Map of the World. *Historical Social Research/Historische Sozialforschung* 40(2(152)): 174–201.

Shapin, Steven. 1989. "The Invisible Technician". *American Scientist* 77(6): 554–563.

Silvertown, Jonathan. 2009. "A new dawn for citizen science". *Trends in Ecology & Evolution* 24(9): 467–471.

Wajcman, Judy. 1991. *Feminism Confronts Technology*. Cambridge: Polity Press.

Part I

Perspectives on Gender and the Government of the Technosciences

1 To "Make a Fuss"

Gender and Governance in the Historiography of Science and Technology

Donald L. Opitz and Brigitte Van Tiggelen

The twenty-first century opened with firestorms precipitated by public invocations of harmful gender stereotypes about women in science. Two instances received significant publicity within the space of a decade. First, there was the resurrection of a biologically deterministic logic for explaining sex differences and the gender gap in science, voiced in a controversial 2005 speech delivered by Harvard President Larry Summers. This was then followed in 2015 by a blatantly sexist "trouble with girls in science" joke made by English biochemist Sir Tim Hunt.[1] These highly publicized cases underlined the durability of gender-based hostilities even after several decades of feminist science studies scholarship and initiatives to promote gender equity in the sciences. As *The Guardian* observed in its report on the Hunt debacle: "Hunt's comments cannot be dismissed as the ramblings of the older generation. Sexism is alive and well in science today" (Sample et al. 2015). For this very reason, Hunt's remarks sent shockwaves through social media, triggering a vigorous response from women scientists, some of whom satirized Hunt's implied reduction of women in laboratories to attractive distractions through the "#DistractinglySexy" Twitter campaign (Tetzlaff et al. 2017). In this particular response, the campaigners fundamentally mocked the idea that women in laboratory attire like "hazmat" (hazardous materials) protective suits could even remotely be perceived as "sexy". Connie St. Louis, director of the science journalism program at City University, London (and author of the tweet that had broadcast Hunt's words), explained the furor over Hunt in this way: "I think people have had enough. There is a problem with sexism in science. It's about the everyday little statements, the little rebuff, the slights that women get, that eventually they think, 'No!' They think, 'No, enough'" (Sample et al. 2015).

As historians who have written about the interplay between women, gender, and science, we cannot help but view these recent public debates through our historiographical lenses and observe the shifts in contexts that have occurred since the emergence of feminist science studies in the 1970s, to the recent, quite public conversations occurring in social media and other forums, indeed in reaction to new instances of sexism and harassment directed at women in STEM (science, technology, engineering, and mathematics). We are reminded of how the "staging" of such discourse is an

DOI: 10.4324/9781003562597-3

inextricable component of the gendering of science and technology in varying contexts, or, to invoke a concept from early gender theory, of how gender is constructed in the very process of its being performed. In this chapter, we wish to explore how revisiting this classic concept is useful for historical studies of gender and governance, the central theme of this volume.

With this in mind, we welcome the opportunity to consider "governance" as an analytical focus in new historical studies of women and gender in the history of science and technology, particularly in a "post-Summers" period of feminist activism (Thompson 2008: 109–110). Our perspective builds on our prior contributions to projects that focused on the dynamics of gendered collaborations across private and public spheres of work/life, and particularly domestic realms (Lykknes et al. 2012; Opitz et al. 2016). In the context of this current volume, we adapt recent theoretical perspectives from studies of governance and gender studies to extend our earlier research agenda, to suggest new, fruitful avenues for historical analyses of women (and other gendered bodies) in STEM. Based on our review of the relevant, recent historiography, we argue that a more nuanced and yet holistic accounting of the interplay between gender and power/politics across private and public spheres of STEM activity—in other words, capturing the full ecology of how governance in science becomes gendered—will not only illuminate more comprehensively the past participation of women in STEM but also reveal critical ways in which that participation is structured according to gender scripts. Moreover, we reassert the criticalness of performance and communication in the constitutions of those gender scripts as they pertain to STEM.

Governance, Communication, and the Gendering of Technoscience

Traditionally, historians of science and technology have interpreted the phenomenon of "governance" in STEM along two fundamental paths. On the one hand, with the rise of the military-industrial complex in various national contexts during the twentieth century, historians have emphasized the entanglement of science and technology with that of government, whether during times of conflict or peace.[2] Interrogated by philosophers and sociologists of science, that entanglement has informed new historical approaches emphasizing social, literary, and material technologies governing the production of knowledge[3] and inspiring revisionist histories. The classic example is Steven Shapin's and Simon Schaffer's (1985) study of the dispute between Robert Boyle and Thomas Hobbes over the constituents and dynamics of air, especially in evacuated chambers, in the seventeenth-century national-political context of the English Restoration. As Shapin and Schaffer analyzed it, the emergent experimental "form of life" (borrowing from Wittgenstein) that has been so fundamental to STEM fields relied on, to paraphrase, the coproduction of norms regarding the nature of knowledge, the polity, and the relationship between the two (1985: 344). Indeed, the literature has emphasized two sides of the entanglement: the role of science and technology in governance, and the role of the state in the governing of science and technology. Shapin and Schaffer concluded with the premonition: "The form of life in which we make our scientific knowledge will stand or fall in the way we order our affairs in the state" (1985: 344). More recently, also regarding Great Britain, Don Leggett and Charlotte Sleigh (2016) noted within historians' explorations of the two-pronged concept of "scientific governance" as in "the governance *of* science, and governance *by* science" (emphasis in original); that the first prong "has been the more common mode to approach the history", whereas the second prong "is a little more elusive to grasp". They also signaled the importance of gender as one of many "often tacit" factors beyond "explicit ideologies in the governance of science", that "shape the production of techno-science" (Leggett and Sleigh 2016: 2).

On the other hand, scholars of science and technology studies (STS) have followed a second major pathway in studying the role of governance, one that we might summarize as that of "demarcation". Most strongly, scholars have associated the concept of demarcation with the processes of defining and regulating disciplinary boundaries, particularly between "science" and "pseudoscience" (Pigliucci and Boudry 2013; Gordin 2021). Historians have additionally attended to the policing of social boundaries that have differentiated "private" from "public" knowledge and "professional" from "popular" or "amateur" expertise, boundaries that constituted the changing status of scientific authority, as well as those who could speak authoritatively in the name of science (and technology) (Barton 2018; Cooter and Pumphrey1991; Eamon 1994; Shapin 1994). To some degree, this agenda has also informed gendered studies of science and technology, particularly in edited volumes attending to geographies of scientific practice that emphasized domestic sites.[4]

In the context of the conversations leading to this present volume, we return to the historiography relating to domestic space and collaborations among scientific partners whose relationships offer windows into the dynamics of work and personal affairs. We are reminded of the centrality of "performance" in explaining the public fate of collaborative outcomes, whether that be equitable credit or the erasure of one partner's contributions in deference to another's.[5] When we earlier revisited (with Annette Lykknes and other colleagues) the topic of "collaborative couples" (Lykknes et al. 2012),[6] we examined a range of cases through the lens of performativity, or how collaborations among intimate partners were produced and then outwardly represented. We now observe how this dimension of our analysis is consistent with the literature on "scientific personae" that emerged in parallel to our work. Advanced by Lorraine Daston, Otto Sibum, and others, this concept emphasizes the mediation between individual persons and social institutions through specified social roles, or personae, as constructed in (and therefore recoverable from) subjects' biographical and autobiographical texts and other publicized commentaries on the social life of science.[7] Persona studies attending to the social influences of gender and other social categories (class, ethnicity, and religion) explicitly built on early gender theory—especially the work of Judith Butler, also an inspiration in our prior research on women and gender in science and to which we now return in the context of this current discussion.[8]

Both Judith Butler (1988) and historian Joan Wallach Scott (1986) emphasized the malleable, performative workings of gender in historical context. In our previous work on collaborative couples in science (Lykknes et al. 2012), we adapted their key concept of gender performance but in terms of "representing" and "projecting" gender. We noted, for example, how a range of our authors' case studies offered "examinations of the dynamics of public representation of collaborative work: the masking of individual contributions within textual and visual imagery projecting the work as undifferentiated labor" (Opitz et al. 2012: 4). Even as we acknowledged the power and politics of social norms that constrained the experiences of partners occupying subaltern positions, both in terms of gender and class, we stressed the agency exercised by individuals in constructing their social roles, even in roles that translated as companionate support vis-à-vis the public "man of science". Those constructions were part and parcel of the public-facing acknowledgments of collaborative scientific work, the professional recognition and awards extended, the scientific iconography, and the biographical memoirs that packaged scientific lives for public reading audiences.[9]

Butler's (1988) classic statement on performativity and gender is useful for us to revisit as we consider the question of gender's interrelationship with governance. We note, in particular, how gender in science, in Butler terms, can be mapped onto the framework of a staged performance: a collective, tacit understanding created through the enactment and reenactment of prescribed scripts, making gender roles in science "natural", even as those roles may shift "onstage" *versus* "offstage": "as a strategy of survival, gender is a performance" and "a construction that regularly conceals its genesis" (522).[10] There is, as we learned from Shapin (1988), in experimental science a process of translating private practice across a "threshold"—at once both physical and epistemological—into the public sphere that establishes reliable knowledge, with communication technologies being critical to that process. Communication, we argue, is in fact core to the question of technoscientific governance, as social hierarchies do get constructed and deconstructed through communication strategies. In agreement with Kristin Roth-Ey and Larissa Zakharova (2015), who considered the political significance of communication and its technologies in Socialist Europe and the Soviet Union, we find that communication can be both "a tool of governance" (para. 29) as well as "a prism through which to evaluate the strength of social ties and understanding" (para. 6). To access this, "we need to take account of what our historical actors tell us and consider their words as forms of engagement, rather than dissimulation" (para. 6). This prompts us to recognize "the emotional experiences at the heart of the communications phenomenon, and thus to a far more complex relationship between the private and the public, the personal and the political than is often considered" (para. 36) particularly in nationalistic contexts such as the Soviet propaganda state. Communication may thus be deployed in efforts to impress social order and uphold authority, but as well it may serve as the means for engagement apart from the agendas of governance.

With this theoretical perspective in mind, we observe the utility for nuancing the concept of governance in gender studies of science and technology, to account for the multiple stages on which governance may be performed, and to give agency to a range of historical actors, particularly women, whose lives and work can too readily be confined to realms offstage and behind the scenes, or onstage roles that are marginal as opposed to leading—confirming a pattern described by Margaret Rossiter (1993) as the "Matilda effect". As such, we advocate for an analysis of the entanglement of governance, gender, and technoscience such that we may trace how scientific governance "is not entirely imposed from above, but powerfully enacted from within" (Leggett and Sleigh 2016: 2). We observed this very enactment in the "#DistractinglySexy" Twitter campaign, which asserted agency indeed from the inner spaces of laboratories, yet translated a narrative of power into the public sphere of discourse, refuting Tim Hunt's joke by leveraging—and satirizing—his masculinist discourse.

The Mask of Gender

The power of social scripts in governing gender behavior and roles is ever more apparent in the context of the recent coronavirus pandemic. Internationally, responses to the pandemic impacted the professional work of men and women including (and especially) in STEM fields, and that impact has once again been disproportionate in terms of gender. With mandates everywhere for families to shelter-in-place, work and family life overlapped in the spaces of private homes on an unprecedented scale. As Deborah Coen (2021: 330–331) recently noted, lockdowns only exacerbated long-established gender-based patterns of household work and tensions in family dynamics, and among the observed impacts were declining rates of journal submissions by women authors, even as overall submission rates increased. In this context, we are reminded again of the offstage and onstage scenes whose enactments portrayed realities differentiated by gender: the labor of care performed overwhelmingly by women, visible only in the private, domestic sphere, while the labor of knowledge-making by men—even when performed at home—proceeding at a quicker pace, made publicly apparent in onstage venues like the publishing industry. We see in this distribution of labor a structural masking that not only demarcates what is seen and unseen, but we are reminded again of the agency of the "mask of gender", or that performance of gender that divides and defines the personal and private in contrast to the political and public.[11] Pandemic masks—and the forms of communication they both enabled (through graphic design) and yet curtailed (for instance, by precluding lip-reading by persons who are deaf or experience hearing loss)—added a poignant material dimension to bodily performances of gender in a world adopting widespread public health precautions. The wearing of masks in this context illustrates again the performance of gender—as suggested in studies demonstrating a correlation between mask-wearing and threats to masculinity[12]—but the whimsy of crossing gender binaries through mask designs may well be interpreted as another kind of crossdressing experiment that exposed the impact of bodily appearance on gender perceptions and relations in STEM fields, as did The Bearded Lady Project (see Figure 1.1).[13]

Scientists who portray themselves through selected styles of dress and cosmetic applications enact gender at the bodily level and yet participate in a social performance that bears all the attributes of theater, as observed by Butler. This dimension to gender in technoscience lays at the core of the aforementioned "#DistractinglySexy" Twitter campaign, and indeed one study of its images suggested the potential for disrupting stereotypes about scientists "by placing the female scientist at the center" with "far-reaching implications" when widely observed by viewers of online media (Tetzlaff et al. 2017: 7). Governance of gender in STEM proves to be as reliant on the micropolitics of bodily presentation and behavior in social settings, as it is on organizational policies and practices that have the effect of privileging men while subverting women, as amply noted in study after study—perhaps most comprehensively, in the recent project, "A Global Approach to the Gender Gap in Mathematical, Computing, and Natural Sciences" ("Gender Gap in Science" 2022). Our point is that gender equity work must attend not only to technoscientific organizational cultures but also to the translation and communication of technoscientific personae publicly, whether through in-person, bodily performances or digital media.

To "Make a Fuss"

We highlight these several dimensions to gender performance in the historiography of technoscience, particularly with current challenges for gender equity in STEM in view, to suggest possibilities for further historical analyses on the entanglement of gender, governance, and science. As a concluding thought, we again urge, as others in the field have recently done (see, e.g., Burgos-Blondelle et al. 2021), the adoption of a *gender* perspective—as opposed to a singularly *woman*-focused perspective—in new studies of technoscience. Indeed, adopting the lens of gender in studies of governance, enables us to observe the fuller scope of knowledge production-as-performance, with particular emphasis on how communication both genders and governs technoscience. Under such a lens, historical actors will reveal a greater complexity of roles beyond the earlier historiographical emphases on exceptionalism, victimhood, sex-based divisions of labor, separate spheres, and the binaries of nature/nurture, reason/emotion, and professional/amateur. New studies will illuminate systems of governance that constitute gender in context-specific ways, and they will

Figure 1.1 Paleontologist Ellen Currano wearing a beard costume. The photograph was taken as part of The Bearded Lady Project. Reproduced by permission. ©Kelsey Vance.

point to the further structural changes and corrections to public stereotypes that are needed to realize full gender equity in STEM. In view of the perpetuation of hostile climates toward women in STEM, as indicated by the recurrence of masculinist performances (in such speeches as Summers's and Hunt's), and the reality of the ongoing "gender gap" in technosciences globally, we recall the activist message articulated so eloquently by academic women across the disciplines (Stengers and Despret 2021)—the ongoing, critical need to "make a fuss"!

Acknowledgements

The authors acknowledge DePaul University Research Council for grant funding that supported this chapter, and Lexi Jamieson Marsh, On Your Feet Entertainment, for the image reproduced as Figure 1.1.

Notes

1 Connie St. Louis quoted Hunt's comments in a post to Twitter: "Let me tell you about my trouble with girls. Three things happen when they are in the lab: you fall in love with them; they fall in love with you; and when you criticize them, they cry" (Ratcliffe 2015). On Summers' controversial remarks, see Jaschik (2005).
2 The classic works, which attend to the US context, are Pursell (1972) and Smith (1985). On the United Kingdom, see especially Leggett and Sleigh (2016). The literature on international contexts has emphasized a Cold War geopolitical framework; see e.g., Krige and Barth (2006) and Krige and Wang (2015).
3 The immense literature builds on landmarks that include Kuhn (1962), Merton (1973), and Latour and Woolgar (1979). For an early overview, see Shapin (1982).
4 The classic anthology emphasizing the gendered impact of the separation of spheres—the public and professional from the private and amateur—is Abir-Am and Outram (1987). Our volume (Opitz et al. 2016), brought this agenda forward in light of developments in the historiography, particularly further collections like Von Oertzen et al. (2013), and the influence of new theoretical perspectives, such as the "geographical turn" as articulated by Livingstone (2003). For an overview of the literature up to 2016, see Opitz (2016). For a recent paper employing the lens of "domestic science", see Coen (2021).
5 The two exemplars of these polar opposites are, respectively, Marie and Pierre Curie (Pycior 1996) and Grace Chisolm and William Henry Young (Wiegand 1996).
6 Our volume added to a rich literature on scientific couples, building on the classic landmark studies collected in Pycior et al. (1996).
7 An outcome of the research program devoted to the subject at the Max Planck Institute for the History of Science, the classic statement on "scientific persona" is Daston and Sibum (2003). This inaugurated a literature that also extended the concept in gender studies; see especially Niskanen et al. (2018).
8 Citing Butler (1990) (among other scholars), Niskanen et al. (2018, 2) write, "Firstly, our investigations are … based on the idea of a dynamic relationship between credibility in scientific assertions and the ways in which *researchers perform and embody their identities* as trusted and credible scientists, and how their personas are influenced and shaped by social categories such as class, gender, ethnicity and/or religious affiliation" (emphasis added). As Butler (2024) notes, the widespread anti-gender backlash against gender studies now uncritically recasts gender as a single, harmful "ideology" within efforts to expunge gender studies from education (20–21, 45).
9 This a strong theme among a range of case studies with an increasing use of the tools provided by gender studies; see particularly Bosch (2013); Burgos-Blondelle et al. (2021); Kohlstedt and Opitz (2002); Opitz (2012).

10 Offering a framework for "staged" behavior in the context of colonialism, see J. C. Scott (1985).
11 The phrase "mask of gender" is widely used in the literature. Suzanne Le-May Sheffield (2001: xiii) adopted the term in the title of her 1997 doctoral dissertation on Victorian women naturalists. Further on this subject in organizational work cultures, see Lewis and Simpson (2010).
12 See, for example, articles appearing in a special issue of *Politics & Gender*, particularly Palmer and Peterson (2020) and Cassino and Besen-Cassino (2020).
13 Schiebinger (1999) analyzed gendered perceptions of scientists in relation to images of men and women in science. The Bearded Lady Project challenged notions of femininity and masculinity in paleontology, and particularly through photographs of women paleontologists donning beard costumes in the field; see Marsh and Currano (2020).

References

Abir-Am, Pnina and Dorinda Outram, eds. 1987. *Uneasy Careers and Intimate Lives: Women in Science, 1789–1979*. New Brunswick: Rutgers University Press.

Barton, Ruth. 2018. *The X Club: Power and Authority in Victorian Science*. Chicago: University of Chicago Press.

Bosch, Mineke. 2013. "Persona and the Performance of Identity: Parallel Developments in the Biographical Historiography of Science and Gender, and the Related Uses of Self Narrative." *L' homme: Zeitschrift für feministische Geschichtswissenschaft*, 24(2): 11–22.

Burgos-Blondelle, Valérie, Juliette Lancel, and Isabelle Lémonon-Waxin, eds. 2021. *Une histoire genrée des sciences est-elle possible? Cahiers François Viète, Série III* (11). https://doi.org/10.4000/cahierscfv.282

Butler, Judith. 1988. "Performative Acts and Gender Constitution: An Essay in Phenomenology and Feminist Theory." *Theatre Journal*, 40(4), 519–531.

Butler, Judith. 1990. *Gender Trouble*. London: Routledge.

Butler, Judith. 2024. *Who's Afraid of Gender?* New York: Farrar, Straus, and Giroux.

Cassino, Dan, and Yasemin Besen-Cassino. 2020. "Of Masks and Men? Gender, Sex, and Protective Measures during COVID-19." *Politics & Gender*, 16(4), 1052–1062.

Coen, Deborah R. 2021. "The Experimental Multispecies Household." *Historical Studies in the Natural Sciences*, 51(3), 330–378.

Cooter, Roger, and Stephen Pumfrey. 1994. "Separate Spheres and Public Places: Reflections on the History of Science Popularization and Science in Popular Culture." *History of Science*, 32(3), 237–267.

Daston, Lorraine, and H. Otto Sibum, eds. 2003. *Scientific Persona. Science in Context*, 16(1–2). https://www.cambridge.org/core/journals/science-in-context/issue/55B8952682FF814C59AAF02D21630D6E

Eamon, William. 1994. *Science and the Secrets of Nature: Books of Secrets in Medieval and Early Modern Culture*. Princeton: Princeton University Press.

"Gender Gap in Science." 2022. "Gender Gap in Science: A Global Approach to the Gender Gap in Mathematical, Computing, and Natural Sciences: How to Measure It, How to Reduce It?" Retrieved 18 February 2022 from https://gender-gap-in-science.org/.

Gordin, Michael D. 2021. *On the Fringe: Where Science Meets Pseudoscience*. Oxford: Oxford University Press.

Jaschik, Scott. 2005. "What Larry Summers Said." *Inside Higher Ed*, 18 February. Retrieved 18 February 2022 from https://www.insidehighered.com/news/2005/02/18/what-larry-summers-said.

Kohlstedt, Sally Gregory, and Donald L. Opitz. 2002. "Re-imag(in)ing Women in Science: Projecting Identity and Negotiating Gender in Science." In *The Changing Image of the Sciences*, ed. Ida H. Stamhuis, Teun Koetsier, Cornelis de Pater, and Albert van Helden, 105–139. Dordrecht: Kluwer.

Krige, John, and Kai-Henrik Barth. 2006. "Introduction: Science, Technology, and International Affairs." *Osiris*, 21(1), 1–21.

Krige, John, and Jessica Wang. 2015. "Nation, Knowledge, and Imagined Futures: Science, Technology, and Nation-Building, Post-1945." *History and Technology*, 31(3), 171–179.

Kuhn, Thomas S. 1962. *The Structure of Scientific Revolutions*. Chicago: University of Chicago Press.

Latour, Bruno, and Steve Woolgar. 1979. *Laboratory Life: The Construction of Scientific Facts*. 2nd edn. Beverly Hills: Sage.

Leggett, Don, and Charlotte Sleigh, eds. 2016. *Scientific Governance in Britain: 1914–79*. Manchester: Manchester University Press.

Lewis, Patricia, and Ruth Simpson. 2010. "Introduction: Theoretical Insights into the Practices of Revealing and Concealing Gender within Organizations." In *Revealing and Concealing Gender: Issues of Visibility in Organizations*, ed. Patricia Lewis and Ruth Simpson, 1–22. London: Palgrave Macmillan.

Livingstone, David N. 2003. *Putting Science in Its Place: Geographies of Scientific Knowledge*. Chicago: University of Chicago Press.

Lykknes, Annette, Donald L. Opitz, and Brigitte Van Tiggelen, eds. 2012. *For Better or for Worse? Collaborative Couples in the Sciences*. Basel: Birkhäuser.

Marsh, Lexi Jamieson, and Ellen Currano, eds. 2020. *The Bearded Lady Project: Challenging the Face of Science*. New York: Columbia University Press.

Merton, Robert K. 1973. *The Sociology of Science: Theoretical and Empirical Investigations*. Chicago: University of Chicago Press.

Mulloor, Jyo John. 2018. "The Care Mask." Retrieved 18 February 2022 from https://www.behance.net/gallery/62813907/The-Care-Mask-.

Niskanen, Kirsti, Mineke Bosch, and Kaat Wils. 2018. "Scientific Personas in Theory and Practice—Ways of Creating Scientific, Scholarly, and Artistic Identities." *Persona Studies,* 4(1), 1–5.

Opitz, Donald L. 2012. "'Not merely wifely devotion': Collaborating in the Construction of Science at Terling Place." In *For Better or For Worse? Collaborative Couples in the Sciences*, ed. Annette Lykknes, Donald L. Opitz, and Brigitte Van Tiggelen, 33–56. Basel: Birkhäuser.

Opitz, Donald L. 2016. "Domestic Space." In *A Companion to the History of Science*, ed. Bernard Lightman, 252–267. Chichester: John Wiley & Sons.

Opitz, Donald L., Staffan Bergwik, and Brigitte Van Tiggelen, eds. 2016. *Domesticity in the Making of Modern Science*. Basingstoke: Palgrave Macmillan.

Opitz, Donald L., Annette Lykknes, and Brigitte Van Tiggelen. 2012. "Introduction." In *For Better or For Worse? Collaborative Couples in the Sciences*, ed. Annette Lykknes, Donald L. Opitz, and Brigitte Van Tiggelen, 1–15. Basel: Birkhäuser.

Palmer, Carl L., and Rolfe D. Peterson. 2020. "Toxic Mask-ulinity: The Link between Masculine Toughness and Affective Reactions to Mask Wearing in the COVID-19 Era." *Politics & Gender,* 16(4), 1044–1051.

Pigliucci, Massimo, and Maarten Boudry, eds. 2013. *Philosophy of Pseudoscience: Reconsidering the Demarcation Problem*. Chicago: University of Chicago Press.

Pursell, Carroll W., Jr. 1972. *The Military-Industrial Complex*. New York City: Harper and Row.

Pycior, Helen M. 1996. "Pierre Curie and 'His Eminent Collaborator Mme Curie': Complementary Partners." In *Creative Couples in the Sciences*, ed. Helen M. Pycior, Nancy G. Slack, and Pnina G. Abir-Am, 39–56. New Brunswick: Rutgers University Press.

Pycior, Helen M., Nancy G. Slack, and Pnina G. Abir-Am, eds. 1996. *Creative Couples in the Sciences*. New Brunswick: Rutgers University Press.

Ratcliffe, Rebecca. 2015. "Nobel Scientist Tim Hunt: Female Scientists Cause Trouble for Men in Labs." *The Guardian*, 10 June. Retrieved 18 February 2022 from https://www.theguardian.com/uk-news/2015/jun/10/nobel-scientist-tim-hunt-female-scientists-cause-trouble-for-men-in-labs.

Rossiter, Margaret. 1993. "The ~~Matthew~~ Matilda Effect in Science." *Social Studies of Science,* 23(2), 325–341.

Roth-Ey, Kristin, and Larissa Zakharova. 2015. "Communications and media in the USSR and Eastern Europe." *Cahiers du monde russe*, 56(2–3), 273–289. Retrieved 18 February 2022 from https://journals.openedition.org/monderusse/8182.

Sample, Ian, Rebecca, Ratcliffe, and Claire Shaw, 2015. "The Trouble with Tim Hunt's 'trouble with girls in science' Comment." *The Guardian*, 12 June. Retrieved 18 February 2022 from https://www.theguardian.com/science/2015/jun/12/tim-hunt-trouble-with-girls-in-science-comment.

Schiebinger, Londa. 1999. *Has Feminism Changed Science?* Cambridge: Harvard University Press.

Scott, James C. 1985. *Weapons of the Weak: Everyday forms of Peasant Resistance*. New Haven: Yale University Press.

Scott, Joan W. 1986. Gender: A Useful Category of Historical Analysis. *The American Historical Review,* 91(5), 1053–1075.

Shapin, Steven. 1982. "History of Science and Its Sociological Reconstructions." *History of Science,* 20(3), 157–211.

Shapin, Steven, and Simon Schaffer. 1985. *Leviathan and the Air Pump*. Princeton University Press.

Shapin, Steven. 1988. "The House of Experiment in Seventeenth-Century England." *Isis*, 79(3), 373–408.

Shapin, Steven. 1994. *A Social History of Truth: Civility and Science in Seventeenth-Century England*. Chicago: University of Chicago Press.

Sheffield, Suzanne Le. May. 2001. *Revealing New Worlds: Three Victorian Naturalists*. New York: Routledge.

Smith, Merritt Roe, ed. 1985. *Military Enterprise and Technological Change: Perspectives on the American Experience*. Cambridge: MIT Press.

Stengers, Isabelle, and Vinciane Despret. 2021. *Women Who Make a Fuss: The Unfaithful Daughters of Virginia Woolf*. 2nd ed. Minneapolis: University of Minnesota Press.

Tetzlaff, Emily, Emily Jago, Ann Pegoraro, and Tammy Eger. 2017. "#Distractingly Sexy: How Social Media was used as a Counter Narrative on Gender in STEM." *#SMSociety17: Proceedings of the 8th International Conference on Social Media & Society*, Toronto, ON, July 28–30, 2017, Article No. 20, 1–10. Retrieved 18 February 2022 from https://dl.acm.org/doi/10.1145/3097286.3097306.

Thompson, Charis. 2008. “Stem Cells, Women, and the New Gender and Science.” In *Gendered Innovations in Science and Engineering*, ed. Londa Schiebinger, 109–30. Stanford: Stanford University Press.

Von Oertzen, Christine, Maria Rentetzi, and Elizabeth Watkins, eds. 2013. *Beyond the Academy: Histories of Gender and Knowledge. Centaurus,* 55(2).

Wiegand, Sylvia. 1996. “Grace Chisholm Young and William Henry Young: A Partnership of Itinerant British Mathematicians.” In *Creative Couples in the Sciences*, ed. Helen M. Pycior, Nancy G. Slack, and Pnina G. Abir-Am, 126–140. New Brunswick: Rutgers University Press.

2 Construction of the Global Surveys of Physicists and Scientists (1999–2020)

Gender and Leadership

Rachel Ivie, Irvy M.A. Gledhill and Silvina Ponce Dawson

The representation of women in physics is among the lowest of the physical sciences in many countries. For example, in the United States, women earn about 20% of physics bachelor's degrees and about the same percentage of physics doctoral degrees. Reliable statistical data for women's representation among degree holders in physics has been difficult to obtain ("Data on Women in Physics" 2019). Many countries lack statistical agencies that collect these data. Countries that have agencies collecting national data on women's participation in science often aggregate the data to the level of "physical science". Given the small size of physics compared to chemistry, these physical science data often obscure the extremely low participation of women in physics. The American Institute of Physics (AIP) collected data on women earning physics degrees in 21 countries for which reliable statistical data were available in 2006. Data with comparable years were difficult to obtain, so AIP had to go back as far as 1998–2000. At that time, women in France earned 33% of physics bachelor's degrees and 24% of physics doctoral degrees. By 2019, when a review of the international situation was published in *Nature* (Skibba 2019), the percentages were almost the same for every country reported as they had been around the turn of the twenty-first century. In other words, 20 years later, despite concerted efforts in many countries, the representation of women in physics had not changed at all.

The reasons for this lack of growth are beyond the scope of this paper. Instead, this paper concentrates on the efforts of international scientific unions, especially the International Union of Pure and Applied Physics (IUPAP),[1] to document not the lack of women in physics, but to measure the inequalities women in physics face during education and employment. These efforts arose out of a very real concern about the opportunities that physics, and science more generally, are missing by their failure to recruit and retain women. Indeed, the result of barriers to women's full participation is a less innovative scientific enterprise. Data are needed about the nature of gender inequalities so that scientific unions and other authoritative bodies can take steps to eliminate them.

DOI: 10.4324/9781003562597-4

The history of IUPAP's efforts to measure gender inequalities reflects the evolving governance structure of an international scientific union. Scientific disciplines have organized themselves into governing bodies that generally advance international cooperation and development of the field. To this end, they convene general assemblies, elect officers, and organize conferences, committees, projects, and working groups. These governance structures often reflect the gendered nature of science itself.

To begin documenting gender inequalities, IUPAP set up a working group that employed several surveys, first of women in physics, and then later of physicists in general (2002–2023), with the aim of documenting gender inequality. The working group also organized eight international conferences on women in physics, and the surveys have been integral parts of the conferences. Gender dynamics have influenced the development, organization, and leadership of the conferences and the surveys. In turn, the conferences and surveys have helped change the leadership structure within the organization sponsoring them.

The global surveys of physicists were first introduced as a component of IUPAP's International Conferences on Women in Physics in 2002. By initiating these surveys, a committee of advocates for women in physics set out to measure the extent of gender inequality in their field. Based on analysis of documents from IUPAP and other organizations, e-mail correspondence, and recollections of our own contributions, the authors find several key drivers of change toward gender equality. These include women acting as change agents within the union by advocating for gender equality. In the 1990s, it was necessary that some of the change was driven by men who had power and influence and used it to advocate for the advancement of women in their discipline. Once these men's influence helped to establish the existence of and rationale for a committee on gender equality, women in the union did the work of setting up the conferences, establishing agendas, and developing surveys designed to measure what was already evident in the structure of the union—gender inequality. Since then, there has been a repositioning of women's power within the union and an increase in collective leadership.

The Beginning of WG5

Recognizing the gendered inequality in physics, IUPAP's governing body, its General Assembly (GA), set up an organizational structure that eventually would have oversight of the Union's activities related to gender. Formed in 1999, Working Group 5 (WG5), for Women in Physics, had a "mandate to survey the situation for women in physics in IUPAP member countries; to analyze and report the data along with suggestions for improvement; to suggest ways that women can become more involved in IUPAP; and to report its findings [at the General Assembly] in 2002" (Hartline et al. 2015: 88). In addition, it was recognized that implementing this charge could include organizing a women's conference. If such a conference occurred, the mandate specified that a report should be made to the next General Assembly (GA) in 2002 (IUPAP n.d.*b*.).

The discussion on forming WG5 started at the IUPAP GA in 1996, where some attendees noted that only three of the more than 80 delegates were women, including Judy Franz, the executive officer of the American Physical Society. In their 2015 article covering the history of efforts to advance women in physics, Franz and her co-authors Beverly Hartline and Millie Dresselhaus note only that the extreme disproportion of women caused "behind the scenes conversations" before the next GA in 1999. Doubtless the role of women in advocating for the creation of WG5 is understated in this article. Franz recalled that she and one of the other women delegates in 1996 made some comments about the small number of women. In her opinion, "this was the beginning of some awareness in IUPAP that some changes were needed".[2] Cecilia Jarlskog,[3] who joined the Swedish delegation in 1999, recalled that Anders Bárány, President of the Swedish Physical Society, who was seated in front of her, introduced the proposal to create a working group on women in physics at her suggestion (Jarlskog 2013: 10). Bárány recalled that

> this suggestion started a heated and somewhat chaotic discussion. Many delegates commented: 'In our country we have no need for this, our women physicists are happy.' Cecilia had arguments that she gave me, and I had to argue for a long time until the meeting decided to retire and sleep on the question. When we gathered again, to my surprise, the decision was accepted.[4]

Others present at the meeting recalled that women delegates also argued strongly for WG5 (Hartline et al. 2015: 88). Burton Richter, the Nobel Prize-winning physicist and incoming IUPAP president, made a strong argument for the creation of WG5. According to one observer:

> some of the men present [at the IUPAP GA] really did not see the need for this WG... Burton was the one who spoke to the importance of this new WG and [said] that it was not a women's issue but a Physics community issue because the community is missing out on a huge number of resources (women).[5]

The resolution passed and WG5 was formed. The implementation of the group was left to the IUPAP Council (IUPAP 1999), whose officers are elected by the GA. Richter asked Council member Franz, then serving as associate secretary general of IUPAP ("IUPAP Elects New Officers" 1999) to form WG5. Franz consulted with physicists, obtained Richter's approval, and recruited nine women members and three men to serve on WG5. Marcia Barbosa of Brazil was the first chair (Ponce Dawson 2018). WG5 held as a principle that men should be included in the conference, the WG, and projects. This principle is based on the observation that exclusion leads to problems and that everyone should look for solutions together. WG5 has always had at least one man on the committee[6] and deliberately includes men in everything.

WG5 formed an international network of women in physics and their allies, led by a team leader from each interested country. The purpose of the network is to keep members "in contact with each other in order to publicize and advocate the resolutions, to monitor their implementation, and to evaluate their impact on the climate for women in physics internationally" (Barbosa 2002). At some point, this network spanned more countries than IUPAP's territorial members.

Background and Previous Research

The formation of WG5 marked the beginning of a change in the governing structure of IUPAP and in the areas with which it would concern itself. At that time, it was relatively common for physicists to express beliefs in "assimilationism" or in "gender blindness". Assimilationism refers to the belief that women in male-dominated fields like physics should adopt masculine characteristics to survive. Gender blindness refers to the belief that there are really no differences between women and men (Banchefsky and Park 2018) and that therefore, the role of gender in organizations, academic disciplines, and scientific research can be ignored or is irrelevant. Related to gender blindness is the supposition that formal governing structures to address gender inequality are unnecessary.[7] These beliefs are likely some of the motivation behind the argument that WG5 was not needed. The beliefs, widely held in physics at the time, were effectively countered by Richter's argument that physics was losing resources by ignoring gender differences and inequality.

In this way, Richter was acting as a change agent and a catalyst for change in the governing structure of IUPAP. According to Dahmen-Adkins and Peterson (2021), change agents are individuals who champion, coordinate, and implement change within organizations. There are two different kinds of change agents for gender equality in organizations: macro and micro. Macro change agents can use status and power to initiate and implement change. They may use a leadership style described as "heroic leadership" (Kelan and Wratil 2018: 8), in which change agents have power and authority and use it by taking charge and directing change. This reflects a masculine[8] style of leading and implementing change (Fletcher 2004: 650). Micro change agents, on the other hand, often have marginalized roles in their fields and work within organizations to implement change or achieve equality. They may use a leadership style known as "tempered radicalism", which derives from the change agent's passion to end gender inequality. In physics, as in other disciplines, women have been aware of and worked to end gender inequality. However, women may have been restricted in their ability to act as macro change agents due to their status as women, which afforded them lower status in the field. In other words, they lacked the authority to act as "heroic" leaders and may have instead used "tempered radicalism". Finally, both of these leadership styles can be contrasted to "post-heroic" leadership, which depends on collaboration and engagement, and may be seen as more feminine (Fletcher 2004: 650).

These categories can serve as guidelines for understanding the role of change agents for gender equality within organizations, but it can be difficult to categorize each change agent as one type using only one leadership style. While the formation of WG5 may have been assisted by macro change agents with governing authority, such as Richter, other change agents with authority certainly propelled the initiative forward. These include Bárány, the president of the Swedish delegation who introduced the resolution, and Franz, who was associate secretary general of IUPAP and who had been empowered to form WG5 by Richter who was implementing the vote of the GA. However, much of the work was done by micro change agents, including the members of WG5 and other contributors to the conferences and surveys. These micro change agents may have been motivated by a passion to end gender equality. However, as the conferences and surveys evolved over the years, a post-heroic, collaborative style of leadership toward gender equality began emerging.

Development of the Global Survey of Physicists

In response to its 1999 mandate to be accountable to the IUPAP's governing General Assembly, WG5 held the first International Conference on Women in Physics (ICWIP) in 2002 in Paris and conducted the first global survey. The purpose was to develop tools to measure and document the situation of women in physics before developing a set of recommendations or resolutions to the GA. Barbosa, chair of WG5, e-mailed WG5's network of country team leaders and asked them to forward the survey to their women colleagues.[9] The survey's goal was to have at least 30 women respondents from at least 10 countries.[10] In the cover letter that accompanied the survey, women were encouraged to respond. Responses from men were not included in the analysis. In the end, there were more than 1,000 women respondents from more than 50 countries, although many of the countries had far fewer than 30 respondents. The distribution method, while methodologically sound, limits the generalizability of the results so that they apply only to the respondents, not to physicists in general.

Midway through the questionnaire dissemination, the Statistical Research Center (SRC) at AIP was asked to revise the survey because of concerns about it. It is likely that AIP was asked to be involved due to its connections with some of the American members of WG5. This edited version of the survey proved to be foundational. In subsequent surveys, some of the questions were asked again so that comparisons could be made over time. One of the key findings from the survey was that having children slowed women's career progress, a finding that has persisted in every global survey.[11]

Impact of the Conferences and Inclusion of Scientific Posters

WG5, as part of the governing structure of IUPAP, has encouraged the development of actions and ideas that advance the situation of women in physics. One such mechanism is the "country posters" that WG5 has asked delegations to present at every conference. Many countries use these posters to document the change activities that they, as leaders in their local physical societies, have been able to enact. For other delegates, these conferences are the first time that they have been able to consider the inequality encountered by women in physics. Ponce Dawson noticed the impact of the first conference on the discussion about women in physics in less developed countries. For example, at the time of the first conference, many women in Argentina argued that there were more pressing inequalities and that gender issues were issues of developed countries. Instead, experience gained over the years has shown that gender is an important issue in developing countries because it intersects with other inequalities to prevent people from developing their full potential.

The second conference held in Rio de Janeiro in 2005 marked the first inclusion of scientific posters from the delegates, thus expanding participation to many more women, since most of the delegates were women. The scientific posters were implemented because the concept of a physicist traveling to a conference at which they would not be presenting a paper on physics did not motivate many leaders at institutions to support travel. The introduction of scientific posters both enabled women physicists to obtain permission to travel from employers and broadened the conference to include scientific interaction.

In connection with the second conference, there was a second global survey. The survey was in English and requested participation from women. While the first survey was in the body of an e-mail that was forwarded, the second survey was web-based but distributed in the same way: by asking women to forward it to other women physicists. There was no plan to discuss the results of the survey during the second conference, so the survey closed about one month after the conference ended. At the time it closed in June 2005, more than 1,350 women physicists from more than 70 countries had answered (Ivie and Guo 2006).

Expansion of the Survey to Include Men and Multiple Languages

Plans for the third survey began almost as soon as the results from the second were published, and its development was influenced by the resolutions sent to the GA from the first and second conferences. The resolutions from the first conference included a charge for professional societies in various countries to "work with other organizations to collect and make available statistical data on the participation of women in physics at all levels" (Barbosa 2002). At this time, the emphasis was still very much on measuring and increasing the participation of women in physics, with less thought to measuring any inequality, or gender gap, between women and men who are already in physics. Also, the resolution focused on an individual country or physical society level, not on global comparisons. This was designed to encourage physical societies to engage in systematic data collection at a national level. In countries that lack reliable statistical agencies, it was often possible only to obtain data for select institutions. However, there is a drawback to this approach: conducting surveys country-by-country would not allow for the comparability of results across countries. Questions that vary even slightly across surveys can produce different results. Therefore, one of the main goals of the third survey was to ensure comparability across countries.

From the second conference, one of the resolutions sent to the IUPAP GA was to "charge the IUPAP Working Group on Women in Physics to oversee a thorough international survey of the status of women in physics in 2007" ("Resolution" 2005). After completing the second survey, WG5 and AIP decided to expand the survey, so it was not put into the field until 2009. One reason for the expansion was that the first two surveys included only women. Thus, the previous survey results could describe women, but no comparisons to men were possible. WG5 and AIP also decided to translate the survey into the United Nations languages (Arabic, Chinese, English, French, Russian, and Spanish), plus German and Japanese, to encourage participation from more physicists.

To update the questionnaire for the third survey, AIP reviewed recent country-level surveys and then included issues that had been identified in its review. AIP asked team leaders from each delegation to the third conference to review the third questionnaire and consider how well it would work in their countries.[12] In this way, the design of the questionnaire included many more physicists, most of whom were women.

Before the survey was available for respondents, the third conference was held in Seoul in 2008. At a workshop, Ivie gave a talk explaining the development of the third survey. One of the third conference resolutions was to charge WG5 with administering "a global survey of physics in 2009" ("Resolution" 2009), a goal that WG5 achieved with the expansion of the survey.

The third survey, funded by AIP (except for the translations),[13] followed a careful dissemination plan. A research analyst at AIP, Casey Tesfaye, contacted each team leader (usually a woman) from every country in the WG5 network and asked each to distribute the questionnaire to her contacts. AIP contacted the team leaders several times to ensure an adequate response. The web-based questionnaire contained instructions to forward to colleagues, both men and women. In addition to this snowball technique, samples of members were drawn by the American and German Physical Societies, so that their results would be representative of their membership. The survey went to the entire membership of the Japanese Physical Society. The survey remained open for a year, from approximately October 2009 until October 2010.

The combination of additional languages and the distribution to men proved effective in increasing the number of respondents. There were almost 15,000 respondents from more than 130 countries. Twenty-two percent of the respondents were women, representing about 3,000 women.[14] Because men were included, the results were used to describe areas of inequality between women and men. In fact, the results showed that having children does not slow the careers of men, but it does slow the career progress of women. At the fourth conference, which was held in Stellenbosch, South Africa in 2011, Ivie gave preliminary results of the third global survey.

Survey Results Document the Gender Gap in Physics

One of the resolutions sent to the GA by the fourth conference was to

> encourage the complete analysis of the Global Survey of Physicists and the wide dissemination through IUPAP Liaison Committees and physical societies of its results and endorse collaborations between physicists and social scientists to understand barriers to women's participation and advancement and to shape interventions.[15]

The first part of this resolution reflected the fact that at the fourth conference, AIP had only just been able to obtain and present the results of a preliminary analysis of the third survey. The resolution as a whole marked something much more important: the beginning of guidelines that would govern the scope of the survey and how it was conducted. Compared to the resolution from the Paris conference in 2002, where the focus was on measuring participation, there was now a recognized awareness that the survey could be used to measure inequality, or as the resolution states "barriers to women's participation and advancement".[16] The value of collaboration between the physics community and the social scientists had also become clear: each was learning a great deal from the other.

Between 2011 and 2014, AIP continued analyzing the results of the third survey by expanding beyond the preliminary results presented in Seoul and including multivariate models that identified differences that were truly due to gender and not to some other identifiable cause. The results showed that compared to men, women had reduced access to career-advancing resources and opportunities. At the fifth conference in Waterloo, Canada in 2014, Tesfaye of AIP gave a talk on some of the findings.

Conference Resolutions Address the Survey

The resolutions from the fifth conference ("Resolution" 2015) were more extensive concerning the survey than in the past and included:

3 There is a need to continue to conduct regular updates of the Global Survey of Physicists presented in 2011 in Stellenbosch.

  a We [the conference] ask IUPAP to take the necessary steps to continue the Global Survey of Physicists periodically, ideally every five years.

  b The continued survey should use the same questions and translations as the first survey.

4 We charge the IUPAP Working Group on Women in Physics with further analysis of country data presented at ICWIP 2014 from the 2010 Global Survey of Physicists...

Resolution 4 reflects the tension between the desire for country-specific findings and the need for international comparisons. Some delegates and WG5 members wished that the third global survey had provided more results to enable local decisions. The main reason country-level analysis was not conducted for most countries was that AIP did not have the capacity to conduct these analyses but could not give respondent data to WG5 without the risk of violating respondent confidentiality. It was not until the development of the Global Survey of Scientists that the problematic nature of respondent confidentiality was addressed by using data suppression software.[17]

The Gender Gap in Science Project

By 2016, it was evident that a follow-up to the third survey would be needed to understand trends and observe developments. The process of merging the International Council for Science, ICSU, with the International Social Science Council, ISSC, opened the way for global multi-union projects under the umbrella of what would become the new International Science Council, ISC. Bruce McKellar, President of IUPAP, forwarded the call for proposals to Irvy (Igle) Gledhill, the WG5 chair. Discussions with Silvina Ponce Dawson and Gillian Butcher resulted in a draft proposal to fund a global survey. Other unions were invited to join the proposal. McKellar contacted the President of the International Union of Pure and Applied Chemistry (IUPAC), Natalia Tarasova. Ponce Dawson contacted Marie-Françoise Roy, champion for women of the International Mathematical Union (IMU). Both enthusiastically joined the leadership of the project. For logistical reasons, IMU and IUPAC took the lead on the project, and IUPAP was a supporting applicant and partner (IUPAP 2022).

Ponce Dawson contacted Silvia Torres-Peimbert, former President of the International Astronomical Union, and drew IAU into the collaboration. With this expansion across unions, the Global Survey of Physicists was on its way to becoming the Global Survey of Scientists. The proposal to ICSU requested funds for AIP to conduct the Global Survey and to maintain consistency with the earlier physics surveys. The project came to be known as the Gender Gap in Science Project and included three parts: the survey, a database of good practices for advancing gender equality, and a study of gender differences in publication patterns. The project was funded by ICSU in February 2017. The number of unions and organizations supporting the project ultimately reached 11 and included:

> IMU, IUPAC, IUPAP, IAU, International Union of the History and Philosophy of Science and Technology (IUHPST), International Union

of Biological Sciences (IUBS), Association for Computing Machinery, International Council for Industrial and Applied Mathematics (ICIAM), the STEM and Gender Advancement project of UNESCO, Organization of Women in Science for the Developing World (OWSD), and Gender in Science, Innovation, Technology and Engineering (GenderInSITE).

The creation of the Gender Gap in Science project depended on a collaborative style of leadership with engagement across many partner unions. There was an executive committee representing the project partners, a smaller coordination group charged with completing the project's three tasks, and an advisory board. As leaders of the partner unions and organizations, many women and a few men took leadership roles, making decisions to join the initiative and sourcing funding for the project. This stands in contrast to the formation of WG5, which was formed within a union and was the result of a formal resolution and voting. In the 1990s, the voice of women was brought to the fore by macro change agents such as Richter and Bárány. The movement was carried forward by other change agents such as Franz, the members of WG5, the country team leaders, and all those collaborating on the conferences and surveys. Once the work of WG5 started, it became more and more collaborative, leading to the creation of the Gender Gap project based on a post-heroic, engaged, and cooperative leadership style.

Development and Implementation of the Global Survey of Scientists

The partner organizations held an initial Gender Gap in Science project workshop in Paris in June 2017. The attendees agreed that once again, the survey would be translated into other languages[18] and focus on measuring gender inequality in career-advancing opportunities and resources.

The Paris meeting was followed by regional workshops in late 2017. At the time of the application, ICSU had three regional offices, one in Asia, one in Africa, and one in Latin America and the Caribbean. Correspondingly, the workshops were held in Cape Town, Taipei, and Bogotá. The workshops were organized by women: Gledhill for Africa, Mei-Hung Chiu for Asia, and Ponce Dawson for Latin American and the Caribbean. Most, but not all, of the participants in the workshops were women. One purpose of holding these workshops in Africa, Asia, and South America was to obtain input into the development of the questionnaire from participants across the world and to ensure that the questionnaire would be appropriate for disciplines other than physics. Workshop participants reviewed a revised questionnaire that was based on the third Global Survey of Physicists. AIP recorded input from the participants then integrated it into a new questionnaire.

One of the most difficult decisions in designing the questionnaire was whether to include a third option other than male/female for gender.[19] As the questionnaire was being developed in 2017/2018, including a choice for non-binary gender was somewhat new in many areas of survey research.

Table 2.1 Number of Respondents to the Global Survey of Scientists

Discipline	*% Women*	N
Astronomy	48	2,597
Biology	69	2,960
Chemistry	51	2,698
Computer Science	55	3,150
History and Philosophy of Science	46	324
Mathematics	43	3,458
Mathematics—Applied	54	2,146
Physics	37	7,570
Overall	50	*

Source: Data are from White and Ivie (2021).

N is the number of respondents.

* Total does not match number of total number of respondents given in the text because there were respondents from other disciplines, and not every respondent provided their gender.

Several members of the coordination group called for a third option. Indeed, this was discussed and recommended by the workshop in Bogotá. However, others on the coordination group were concerned about how this would be received in countries which have less tolerance for gender diversity and whether potential respondents would feel safe answering a questionnaire with this choice on it.[20] Finally, the decision was made by Marie-Françoise Roy, as one of the leaders of the project. The question was worded "male/female/prefer not to respond". It was hoped that "prefer not to respond" would give some option to those who wished not to choose male or female.

The survey was launched 1 May 2018 after approval by the executive committee and translation. The executive committee, rather than AIP, disseminated the questionnaire by asking the project partners to forward to their contacts. The survey closed on 31 December 2018 after having been open eight months. There were more than 32,000 respondents from more than 150 countries. Half of the respondents identified as women, with the percentages varying from a high of 69% for biology to a low of 37% for physics (see Table 2.1). The percentage of women physicists answering is higher than the percentage of women physicists in most countries, perhaps because of the network of country leaders that WG5 founded early on. For unions without similar networks, the survey was a first step in reaching out to wider communities on gender issues. The history of the global physics surveys also may have contributed to the larger number of respondents from physics.

Results of the Global Survey of Scientists

For the initial analysis of the Global Survey of Scientists, AIP used multivariate statistical models to control for factors including gender that could explain various outcomes. In this way, the statistical effects of gender can be pinpointed. In addition to gender, AIP's models controlled for age (as a proxy for career stage), the disciplines represented by the partnering unions, employment sector, geographic region (based on United Nations designations), and the UN human development index.

As with the past surveys, there were gender gaps in (1) access to resources needed to conduct science and (2) opportunities to contribute to the scientific enterprise. Where there were statistically significant differences between men and women, it was always women who had less access to opportunities and resources. Women were also much more likely than men to have had their careers negatively affected by family obligations and were much more likely than men to have experienced sexual harassment. Finally, there was strong evidence indicating that the gender gap persists across all disciplines, all employment sectors, all geographic regions, and all levels of development (Ivie and White 2020).

Ivie presented the results of the Global Survey of Scientists at the seventh IUPAP women's conference, held virtually in 2021 due to the COVID-19 pandemic, and at the 2021 International Conference on the History and Philosophy of Science and Technology. The results of the entire Gender Gap project were published in June 2020 (*A Global Approach to the Gender Gap* 2020).

Formation of the Standing Committee for Gender Equality in Science

At the conclusion of the Gender Gap project, the project partners wished to form a way not only to implement the project's recommendations but also to build on the collaboration that had been constructed between the unions. To this end, representatives of five of the unions that were partners of the Gender Gap in Science Project "worked together to prepare a Memorandum of Understanding on Gender Equality in STEM. These were [four women and one man] Maria J. Esteban (ICIAM), Nathalie Fomproix (IUBS), Marie-Françoise Roy (IMU), Michel Spiro (IUPAP) and... Catherine Jami (IUHPST)".[21] Other partnering organizations were invited to join, so in 2020, nine of the original partners (including IUPAP) formed the Standing Committee for Gender Equality in Science (SCGES) ("Memorandum" 2020). As of the writing of this report, there are now 21 partners (Standing Committee 2022), most of whom are ISC members. In addition to implementing the recommendations of the Gender Gap project, the aims of SCGES include liaising among international scientific unions as they work to foster gender equality (Jami 2022).

Like the collaboration that led to the Gender Gap in Science project, the formation of this committee was the result of collaborative action, engaging many partners from different organizations. The manner in which the SCGES was formed was based on strength and trust across unions that was developed during the Gender Gap project. The SCGES is composed of leaders from different unions who work cooperatively together to effect gender equality. As the invitation letter states: SCGES will

> facilitate communication and cooperation among the signatory partners in future projects and initiatives to promote gender equality. Signatories... commit to keep one another informed of all the new projects and initiatives so that all partners may decide whether they want to take part in them.[22]

How IUPAP Has Changed Since 1999

As previously described, Burton Richter advocated for the 1999 formation of IUPAP's WG5 over some objections. Nevertheless, WG5 has successfully organized eight international women's conferences and implemented four global surveys. It created a very active network of women in physics country-level groups, which has members in more countries than there are IUPAP territorial members. These groups have contributed enormously to bringing gender inequality to the forefront, particularly in communities that may not even have been aware that there was a problem to solve. Every international women's conference has resulted in resolutions, organized and put forth by the conference organizers and delegates, acting as micro change agents. IUPAP received these recommendations and instituted several policy changes designed to support the participation of women in IUPAP. For example, in 2017 the Executive Council recommended that for conferences supported by IUPAP, there should be a 20% target of participation by women on the organizing committees. Further, "conference applications... [that] reflect less than 10% women representation on the three major conference committees (viz International Advisory Committee, Program Committee, and Local Organizing Committee) shall not qualify for IUPAP support" (IUPAP n.d.*a*.). Starting in 2011 there is also a vice president at large who has the duties of Gender Champion (currently Gillian Butcher) who advises on gender and diversity issues, as well as monitors gender data from the conferences and the IUPAP council membership. Finally, each IUPAP-sponsored conference must designate an advisor on sexual harassment.

In 2021, the 30th IUPAP General Assembly approved the Waterloo Charter (IUPAP 2021), named after the fifth ICWIP conference. The Charter was initiated during the fifth conference and finalized during the sixth. Like the other resolutions, it is a product of micro change agents. The Charter and its supporting information outline a commitment to gender inclusion and diversity in physics and make specific recommendations designed to move this commitment forward. The supporting information recommends that physical societies and physics departments continually collect data on "the situation of physicists both locally and internationally", that data be segregated by gender, and that "good gender indicators" be developed (IUPAP 2021).

Conclusions

Concurrent with the efforts of IUPAP to advance gender equality, there also has been a change in the adherence to certain gender ideologies within physics, including the beliefs in "assimilationism" and "gender blindness" (Banchefsky and Park 2018). In their place, there is an increasing level of agreement that gender and other social statuses such as race and ethnicity do influence science and its practitioners, although many are still skeptical that these social constructs are important.

The Global Surveys were created by a network of micro change agents and evolved over time. The first two were surveys of women in physics. The third was a multi-language questionnaire sent to men and women physicists. The fourth was a larger interdisciplinary effort sponsored by multiple international unions. The Global Surveys have sought to document inequality between men and women first in physics specifically, and then in science generally. Even the process of documenting inequality has been shaped by gender. Both types of change agents (macro and micro) have contributed to the development and evolution of the Global Surveys as a product of WG5. Although Burton Richter advocated for the formation of WG5 at the IUPAP General Assembly in 1999, the work of other change agents such as Judy Franz and Cecilia Jarlskog was already in evidence. Richter and the GA made sure that Franz had the funding and resources to implement WG5 and its conferences.

Throughout the history of WG5 and the conferences, there was a shift away from styles of leadership that the literature describes as typical of micro and macro change agents. From the inception of the WG5, the voice of women grew, as did their governing authority within IUPAP through Vice-Presidents, Gender Champions, and the first female IUPAP president, Jarlskog. Currently, Ponce Dawson is the IUPAP President Designate and is likely to become the second woman president of the union (and the first president from Latin America) in 2024. It should be noted that throughout this history, WG5 and IUPAP have upheld the principle that men should be included in any initiative for women based on the values of inclusion and cooperation. There has been a move toward a post-heroic leadership style based on collective action and engagement, culminating in the cooperative formation of the Gender Gap in Science project and the Standing Committee for Gender Equality. By the time SCGES was formed, the leadership roles in these initiatives were largely held by women. What has emerged is a developing picture of fruitful collaboration with an increasing proportion of women, and not a conflict for supremacy. Leadership roles are more equitably distributed, and this has been a result of conscious deliberation and effort.

Notes

1 IUPAP, established in 1922, is an organization of "identified physics communities in countries or regions around the world". There were 13 founding members, and currently, IUPAP has 60 territorial members on every continent except Antarctica.

2 Franz, Judy, e-mail to Ivie, 12 May 2022.
3 Jarlskog later became the first woman president of IUPAP, serving from 2011 to 2014.
4 Bárány, Anders, e-mail to Ponce Dawson, 16 May 2022.
5 Beamon-Kiene, Jackie, e-mail to Butcher, Gledhill, Ivie, and Ponce Dawson, 24 July 2018.
6 Butcher, Gillian, personal communication, 9 May 2022.
7 Butcher, Gillian, personal communication, 9 May 2022.
8 Masculine characteristics can be, and often are, used by women in multiple contexts, and *vice versa*.
9 See Ivie, Czujko, and Stowe (2002: 5).
10 Hartline, Beverly Karplus, "Proposal to Fund US Participation in IUPAP Women's Conference", 2001.
11 See Ivie, Czujko, and Stowe (2002: 1).
12 Kausch et al. (2009: 22).
13 Beverly Hartline obtained funding from the Henry Luce Foundation for using a survey translation service. Hartline also obtained grants that funded not only the translations of the survey, but also for the travel for delegates from less developed countries and for the participation of US delegates in the first four ICWIPs.
14 Ivie and White (2015: 70).
15 "Resolution" (2013: 5).
16 "Resolution" (2013: 5).
17 The coordination group for the Global Survey of Scientists set up a process in which project partners could use the dataset within data suppression software to conduct their own analysis for their specific disciplines or regions.
18 The same languages were used as with the first survey, except that German was excluded to save costs.
19 This chapter also uses a binary construction of gender because during most of the project's history, participants routinely referred to gender only as binary.
20 Although it is not the same concept as gender identity, the International Lesbian, Gay, Bisexual, Trans and Intersex Association found in 2015 that in 33 of the 54 African countries, homosexuality was illegal (Carroll, 2016: 36).
21 Jami, Catherine, e-mail to M-F Roy, 21 April 2020.
22 Roy, Marie-Françoise, letters sent to Jami and the Gender Gap coordination group, 21 April 2020.

References

Banchefsky, Sarah, and Bernadette Park. 2018. "Negative Gender Ideologies and Gender-Science Stereotypes Are More Pervasive in Male-Dominated Academic Disciplines." *Social Sciences* 7(2): 27. doi: 10.3390/socsci7020027.

Barbosa, Marcia. 2002. "IUPAP Women in Physics Working Group Report at the General Assembly, Berlin, Germany." IUPAP website. Retrieved 5 July 2022 from https://archive.iupap.org/wg/wg5/.

Carroll, Aengus. 2016. *State Sponsored Homophobia 2016: A World Survey of Sexual Orientation Laws; Criminalisation, Protection and Recognition*. Geneva: International Lesbian, Gay, Bisexual, Trans and Intersex Association, 11th edn. Retrieved 5 July 2022 from https://ilga.org/downloads/02_ILGA_State_Sponsored_Homophobia_2016_ENG_WEB_150516.pdf.

Dahmen-Adkins, Jennifer, and Helen Peterson. 2021. "Micro Change Agents for Gender Equality: Transforming European Research Performing Organizations." *Frontiers in Sociology* 6: Article 741886. doi: 10.3389/fsoc.2021.741886.

"Data on Women in Physics." 2019. *Nature Reviews Physics* 1: 297. doi: 10.1038/s42254-019-0061-3.

Fletcher, Joyce K. 2004. "The Paradox of Postheroic Leadership: An Essay on Gender, Power, and Transformational Change." *The Leadership Quarterly* 15: 647–661. doi: 10.1016/j.leaqua.2004.07.004.

A Global Approach to the Gender Gap in Mathematical, Computing, and Natural Sciences: How to Measure It, How to Reduce It? 2020. ed. Colette Guillopé and Marie-Françoise Roy. International Mathematical Union, Berlin. Retrieved 12 January 2025 from https://zenodo.org/records/3882609.

Hartline, Beverly Karplus, Millie S. Dresselhaus, and Judy Franz. 2015. "An Eclectic Historical Perspective on Advancing Women in Physics." *Physics in Canada* 71(2): 85–89. Retrieved 20 Feb. 2025 from https://www.cap.ca/wp-content/uploads/2017/03/0f77f67a5b98494792f80009fef20f2a6666086f.pdf.

IUPAP (International Union of Pure and Applied Physics). n.d.*a*. "Conference Policies." Retrieved 15 May 2022 from https://iupap.org/conferences/conference-policies/.

IUPAP. n.d.*b*. "WG5: Women in Physics." Retrieved 15 May 2022 from https://archive.iupap.org/wg/wg5/.

IUPAP. 1999. "Minutes of the 23rd General Assembly." Retrieved 2 July 2022 from https://archive2.iupap.org/wp-content/uploads/2013/12/Minutes23rd-General Assembly.pdf.

IUPAP. 2021. "Waterloo Charter." Retrieved 15 May 2022 from https://iupap.org/strategic-plan/diversity-in-physics-2/waterloo-charter-for-women-in-physics/.

IUPAP. 2022. "Who We Are." Retrieved 21 June 2022 from https://iupap.org/who-we-are/.

"IUPAP Elects New Officers." 1999. *Physics Today* 52(7): 48. doi: 10.1063/1.882721.

Ivie, Rachel, Roman Czujko, and Katie Stowe. 2002. *Women Physicists Speak*. College Park, MD: American Institute of Physics. Retrieved 6 July 2022 from https://ww2.aip.org/statistics/women-physicists-speak.

Ivie, Rachel, and Stacy Guo. 2006. *Women Physicists Speak Again.* College Park, MD: American Institute of Physics. Retrieved 6 July 2022 from https://ww2.aip.org/statistics/women-physicists-speak-again.

Ivie, Rachel, and Susan White. 2015. "Is There a Land of Equality for Physicists? Results from the Global Survey of Physicists." *Physics in Canada* 71(2): 69–73.

Ivie, Rachel, and Susan White. 2020. "Measuring and Analyzing the Gender Gap in Science through the Global Survey of Scientists." In *A Global Approach to the Gender Gap in Mathematical, Computing, and Natural Science,* ed. Colette Guillopé and Marie-Françoise Roy. Retrieved 6 July 2022 from https://zenodo.org/records/3882609#.YnGWtdrMJaR.

Jami, Catherine. 2022. Talk during Standing Committee for Gender Equality in Science webinar. 24 March. Retrieved 15 May 2022 from https://www.youtube.com/watch?v=MEr4qiEa1iQ.

Jarlskog, Cecilia. 2013. "Address to 4th IUPAP International Conference on Women in Physics." In *Women in Physics: Fourth International Conference on Women in Physics,* ed. Beth A. Cunningham, 10–11. Melville, NY: AIP Conference Proceedings. 1517. doi: 10.1063/1.4794207.

Kausch, Corinna, Gillian Butcher, Marcia Barbosa, and Rachel Ivie. 2009. "Organizing Women in Physics Working Groups." In *Women in Physics: Third IUPAP International Conference on Women in Physics,* ed. Beverly Karplus Hartline, K.

Renee Horton, and Catherine M. Kaicher, 20–22. Melville, NY: AIP Conference Proceedings 1119. doi: 10.1063/1.3137802.

Kelan, Elisabeth K. and Patricia Wratil. 2018. "Post-Heroic Leadership, Tempered Radicalism and Senior Leaders as Change Agents for Gender Equality." *European Management Review* 15: 5–18. doi: 10.1111/emre.12117.

"Memorandum of Understanding: Gender Equality in Science." 2020. Retrieved 15 May 2022 from https://indico.cern.ch/event/1065120/contributions/4478638/attachments/2316309/3963306/MoU_SCGES%20IUPAP%20200602signed.pdf.

Ponce Dawson, Silvina. 2018. "Burt Richter and the Working Group on Women in Physics." *IUPAP Newsletter,* Issue 76: 2. Retrieved 6 July 2022 from https://archive2.iupap.org/wp-content/uploads/2018/09/IUPAP_sep_2018_e.pdf.

"Resolution for the IUPAP 25th General Assembly." 2005. In *Women in Physics: 2nd IUPAP International Conference on Women in Physics*, ed. Beverly Karplus Hartline and Ariel Michelman-Ribiero, 3. Melville, NY: AIP Conference Proceedings 795. https://www.researchgate.net/publication/234544190_Second_IUPAP_International_Conference_on_Women_in_Physics.

"Resolution for the IUPAP 26th General Assembly." 2009. In *Women in Physics: Third IUPAP International Conference on Women in Physics,* ed. Beverly Karplus Hartline, K. Renee Horton, and Catherine M. Kaicher, 3. Melville, NY: AIP Conference Proceedings 1119. doi: 10.1063/1.3137905.

"Resolution for the IUPAP 27th General Assembly." 2013. In *Women in Physics: Fourth IUPAP International Conference on Women in Physics*, ed. Beth A. Cunningham, 5. Melville, NY: AIP Conference Proceedings 1517. doi: 10.1063/1.4794205.

"Resolution for the IUPAP 28th General Assembly." 2015. In *Women in Physics: Fifth IUPAP International Conference on Women in Physics*, ed. Beth Cunningham, Catherine O'Riordan, and Shohini Ghose, 040004. Melville, NY: AIP Conference Proceedings 1697. https://doi.org/10.1063/1.4937631.

Skibba, Ramin. 2019. "Women in physics." *Nature Review Physics* 1: 298–300. doi: 10.1038/s42254-019-0059-x

Standing Committee for Gender Equality in Science. 2022. "SCGES Partners." Retrieved 16 August 2022 from https://gender-equality-in-science.org/scges-partners/.

White, Susan and Rachel Ivie. 2021. "The Global Survey of Scientists: encountering sexual harassment". Pure and Applied Chemistry 93(8): 831–837. doi: 10.1515/pac-2021-0304.

Part II

Gender and Technoscientific Policies

3 Women, Science, and Empowerment

The International Federation of University Women (1920s–1930s)

Anna Cabanel

Although transnational networks of intellectual women started to burgeon in the late nineteenth century in North America and the United Kingdom, the foundation of the International Federation of University Women (IFUW) in 1919 marked an important step toward the internationalization and the structuration of the movement of university women (Von Oertzen 2014). The IFUW endeavored simultaneously to promote women, science, and internationalism as it strove, according to the organization's constitution of 1920, "to promote understanding and friendship between the university women of the nations of the world, and thereby to further their interests and develop between their countries sympathy and mutual helpfulness".[1] From the beginning, the IFUW members aligned their work and ambitions with internationalist ideals and general beliefs on the role of education in the peace process.

Perhaps due to its multiple identities, the IFUW had never before been considered to be a full-fledged scientific organization. However, for the first time in history, scientific and academic women from different nations were brought together. By pursuing strong science policies such as the establishment of a research fellowship program for women and organizing large-scale international scientific congresses reflecting on the place of women in science, the ambitions and actions of the IFUW were similar to those of other scientific organizations, but with the specificity of being run by and dedicated for women. In this context, one can wonder to what extent the IFUW functions as a female governing body aiming at addressing and overcoming the underrepresentation of women in science and academia. Bringing together gender history, the history of transnational organizations, and the history of science, this chapter addresses the role of female international networks in promoting and empowering women in science and academia. Using the IFUW as a case study, this chapter reflects on the ways an international organization shapes its own governance to be able to promote women in academia and in (techno)sciences. To do so, it looks at the IFUW ways of governance by exploring the organization's main strategies such as

DOI: 10.4324/9781003562597-6

its funding policies, institutional publications, organization of international congresses, and mentoring.

Although the IFUW still exists today—it has recently been renamed the Graduate Women International (GWI)—the scope of this paper is limited to the interwar period. The two decades following World War I were crucial in the formation of the IFUW and the articulation of its objectives. During this period, the organization follows a clear scientific agenda; and not surprisingly its leaders are all renowned academic and scientific women: the founders, Virginia Gildersleeve and Caroline Spurgeon, are both literary scholars, while the following presidents, Ellen Gleditsch, Johanna Westerdijk, and Winifred Cullis, were internationally recognized for their expertise respectively in radiochemistry, biology, and physiology. If many of the IFUW scientific initiatives endured after the Second World War, such as the fellowship program, the organization priorities evolved toward larger questions regarding education in developing countries, working more closely with international organizations such as UNESCO.

To study the collective and individual strategies elaborated by the IFUW members during the interwar period, and in an attempt to measure the impact of their efforts, this chapter relies on an international corpus of sources of different nature. The organization's institutional archives, located in Amsterdam, offer a good overview of the discussions and decisions taken by the IFUW members regarding the place and role of women in science and give a good insight into the collective strategies they pursued during the 1920s and 1930s. The archives give access both to formal documents, such as the *Bulletins* of the IFUW, whose publication followed the international congresses held every two then three years; and to less formal papers, such as the minutes of the different *Committee* meetings or the correspondence between the IFUW International Office and the national branches. By articulating the collective perspective and individual trajectories this paper also aims to reflect upon the role of the IFUW in empowering other women trying to make their path in the scientific and academic worlds.

In the following pages, I analyze the role of the IFUW members in the creation of new ways of scientific governance, and in the promotion of women in science, by looking at their scientific agenda in the course of the interwar period. After retracing the emergence of the university women's movement from the late nineteenth century, I examine the way the IFUW members justified the need for an all-female organization in a world that they defined as being "man-made".[2] Through the organization's publications and congresses, the IFUW functioned as a public platform from which women could express themselves, and actively promote and legitimize their identity, both as women and scientists. In the last part of this paper, I focus on one of the most important scientific policies carried out by the organization members in the 1920s and 1930s: the foundation of an international fellowship program run by and for women.

Networks of Graduate Women at the Turn of the Twentieth Century

> "We should have", said Miss Spurgeon, "an international federation of university women, so that we at least shall have done all we can to prevent another such catastrophe". Miss Sidgwick and I looked at each other: "Then I guess I must rally the Association of Collegiate Alumnae", I said. Rose Sidgwick added, "and we must go back and talk with the British Federation of University Women".
>
> (Gildersleeve 1954: 129)

The above conversation between Virginia Gildersleeve, Dean of the Barnard College in New York, and two British women, Caroline Spurgeon, Professor of English Literature at Bedford College, and Rose Sidgwick, Lecturer in History at the University of Birmingham; is often regarded as the beginning of the International Federation of University Women. The three women met in 1918 in the United States, as Spurgeon and Sidgwick were commissioned by the British government, along with other university professors, to travel to North America to establish official links between British, Canadian, and US universities. From these initial contacts emerged the project of a federation of university women on an international scale aiming at promoting women's access to higher education and scientific professions to enable them to play a role on the international stage.

Although the IFUW was officially founded in July 1919, scholarly works focusing on women's academic and scientific networks have pointed out the existence of a transatlantic female elite from the late 1880s (Von Oertzen 2014). From the 1880s onward, women in Western countries gradually gained access to university education and created alumnae associations. As early as 1881, following an invitation from Marion Talbot and Ellen Richards, two influential leaders in higher education for women in the United States, former students of American women's colleges met in Boston to discuss the problems faced by women graduates in their country (Talbot and Rosenberry 1931). The following year, these alumnae founded the first association of what would become the movement of "university women": the Association of College Alumnae (ACA), renamed in 1920 the American Association of University Women (AAUW). The association aimed to work "for the advancement of women's opportunities in higher education and the workplace" but also to defend them against "hostility and suspicion, acting as an early pressure group" (Rossiter 1982, 39).

The founding members of the ACA included Ellen Richards, a chemist and initiator of the field of home economics, and Christine Ladd-Franklin, an American mathematician and logician. Both were well aware of the differences in opportunities existing between women and men in the academic and scientific worlds (Spillman 2012). Such inequalities were particularly striking regarding travel opportunities, in a time where studying abroad was regarded as a crucial step in the construction of an academic career. Indeed, while male American students benefited from scholarships to visit foreign universities, women struggled to receive similar funding and travel opportunities (Singer 2003). To remedy this inequality and to provide equal career opportunities for women graduates in the United States, Ladd-Franklin proposed in 1888 to establish a scholarship program to enable women to study abroad (Spillman 2012). Such a program was seen as an "entering wedge" for American women, who hoped to use their experience gained in foreign universities to expand their opportunities in the United States (Rossiter 1982: 39).

Similar initiatives took place in Great Britain. In 1907, women graduates from British universities met in Manchester to discuss "the difficulties in getting money for research and the seeming impossibility of promotion for women in academic life" (Dyhouse 1995a: 472). Participants founded the British Federation of University Women (BFUW). According to its constitution, the aim of the BFUW was "to encourage independent research work by women, to facilitate intercommunication and cooperation between the women of different universities, and to stimulate the interest of women in municipal and public life" (Dyhouse 1995a: 472). Like their US counterparts, the BFUW members quickly raised the issue of the lack of research funding available to British women students, and, in 1909, they launched their postgraduate scholarship program. Of the five women awarded scholarships between 1912 and 1916, three would become professors in British universities. One of them was Caroline Spurgeon, founder and first president of the IFUW (Dyhouse 1995b: 273).

While the American and British organizations laid the foundations for the future IFUW, World War I played an important role in the internationalization of the university women's movement (Fouché 2000). Given the full-scale nature of the conflict, women actively participated in the war effort, seizing the opportunity to take part in the public life in their country. By doing so, they aimed to demonstrate the importance and usefulness of women's education in times of social or political crisis (Von Oertzen 2014: 10–11). In the early 1920s, and in the wake of the internationalist ideal promoted by organizations such as the League of Nations (Laqua 2011), women's movements supported the idea of feminism as a "symbol of peace and friendship between women all over the world" (Fouché 2000: 134). They aspired to prevent the outbreak of another world conflict and hoped to build a new world of peace and equality, particularly between sexes. In this globalized context, new objectives emerged including the promotion, at the international level, of the right to vote and access higher education for women (Goodman 2012). University women, therefore, fully took their place in this post-war movement, promoting, in particular, the exchange of students and teachers to widen women's opportunities and promote an international spirit. By doing so, they developed their own form of governance: through the creation of a female intellectual elite, their aim was, in the long run, to integrate decision-making bodies and to feminize the political class and international leadership.

From the beginning, the IFUW shared other international women's-organizations' ideals about enhancing the place and role of women on the international stage; but its focus was solely on "university women". According to the organization's constitution, only "university women", i.e. "a woman holding a university degree or its equivalent" in either the arts or the sciences, could apply for membership at the national level, as only national organizations could be affiliated to the IFUW.[3] The IFUW's membership conditions were strict in comparison with other international women's organizations that were, at least in theory, open to every woman (Rupp 1997). The specificity of IFUW membership helped to shape a distinct and collective identity for "the university woman".

A Space of their Own in a Man-Made World

In her inaugural speech in 1920, delivered at the occasion of the first IFUW international congress, Martha Carey Thomas, then president of the American Association of University Women, justified the need for a female academic organization, comparing the functioning of the IFUW to the Academies:

> In the United States, we have an Academy, an institution of Arts and Science. It is not a men's Academy, it is a general Academy, and no woman is on the roll. Our American Association of college women has memorialized that Academy and asked to have women's names placed on the roll. It is discouraging to eminent women that they are not, and I am not at all sure that part of the work of this Federation may not be a women's Academy of Arts and Letters and Science, in which we shall honor and praise each other, for praise is a very important thing in success.[4]

Carey Thomas' justification was similar, in many ways, to the idea Virginia Woolf defended a few years later in her famous essay *A Room of One's Own* (1929), in which she argued for a literal and figurative space for women writers. In her essay, Woolf explained the absence of female fiction not as a lack of talent but rather as a lack of opportunity and recognition within a literary tradition dominated by men. Having access to a "room of one's own", to use Woolf's metaphor, would allow women to develop and legitimate their identity as writers. Similarly, by creating an organization of one's own, university women would legitimate their identity and their place.

From the beginning, the IFUW was therefore conceived as an all-female organization. This single-sex structure served university women's ambitions to enhance women's opportunities in science and academia. Historians have shown how women have long been excluded from scientific and academic institutional settings (Stamhuis 2003, 26–27). Many learned societies in the eighteenth and nineteenth centuries limited their membership to males and while the modern universities became the institutional settings for scientific work, connecting education and research, it was not until the end of the nineteenth century that they opened their doors to women, first as students and a few decades later as professors. In this context, one can understand how having a space of their own was one of the founding principles in the organization of women's movements at the turn of the twentieth century. This feminine *entre-soi* was thought of as providing a safe space in which women might fully express themselves without being overshadowed by men.

The IFUW was not the first all-female scientific organization, the Physical Society of the Ladies, established in Middelburg, Netherlands in 1785, is commonly recognized as being the first scientific society exclusively for women (Stamhuis 2003, 21). When the first generations of female students entered the university, they often created their student clubs or association of female graduates as was seen among the former students of women's colleges in the United States and the United Kingdom. In her study of three major women's organizations at the beginning of the twentieth century—namely, the International Council of Women (1888), the International Alliance of Women (1904), and the Women's International League for Peace and Freedom (1915), Leila Rupp notes that the argument for single-sex membership was based on an ideology of difference that relied on either biological or cultural factors (1997: 82–103). Although some individuals questioned the very existence of a difference between sexes, this idea remained marginal at that time. That women have specific qualities–such as the promotion of peace—or that they share common experiences, remains the chief argument of separatist movements whose collective identity relies on a feminine identity.

On several occasions during the first years of the IFUW, founders and presidents had to justify their choice of an all-female organization. The single-sex feature did not meet with unanimous approval, as this separatist vision was considered by some to have been outdated (Skonhoft 1932: 250). At the 1920 International Congress, Virginia Gildersleeve had to explicitly defend the IFUW single-sex structure and she had to reassure participants about the *bien-fondé* of the IFUW: "Because we are forming an organization of university women apparently apart from university men, we do not wish anyone to believe that we are separatists and ultra-feminists in our tendency".[5] The IFUW members were indeed opposed to any form of differentiation between sexes, particularly in education, because they believed that "there are no essential differences between the educational needs of men and women". The aim of the IFUW was not to "play apart" or "duplicate or encroach" on the work already done by various institutions, but to "cooperate", and more importantly "to give women a chance to participate".[6] As Estella Simons, president of the Dutch association, pointed out to the university women gathered in Amsterdam during the Fourth International Conference in 1926:

> We are still living in a "man-made world", and in this world, we must conquer our place and our right.... The general interest is the sum of

> the interest of men and women and it is only then really and truly general when both are taken equally into account.[7]

As long as women were not an integral part of decision-making committees, she concluded, they would not have the same opportunities as men. The single-sex structure of the IFUW was thus justified by the founding members as a strategy to counteract the discrimination—explicit or not—persisting in science and academia against women. Single-sex organizations were seen as a necessary step toward achieving the primary goal of the IFUW, effective gender equality. "The better [a] wom[a]n's position, the less will they feel for the organization", Estella Simons summarized in 1926.[8]

The analysis of the IFUW founding principles, and the justification for the organization's necessity, as shown above, relies above all on discussions from IFUW International Congresses, and the archives these major events generated. The considerable amount of archives left by congresses, as well as the specific nature of such documents, often constitutes a challenge for historians (Rasmussen 1995). The organizers of congresses were conscious of the symbolic dimensions of these events. Thus, the IFUW *Bulletins*, the official archives of their International Congresses, reflect contemporary and cultural intentions, and organizer's efforts toward what might be called "impression management". While it is important for historians to keep a critical distance, as the congress proceedings often directly participated in the self-justification of the events, "impression management" is of interest when one considers the congresses as a stage, as a performance during which a group identity emerges and claims its own right (Peeters et al. 2015). In their chapter in this book, Opitz and van Tiggelen advocate for a better understanding of the issue of governance in a gender perspective by taking into account "the multiple stages on which governance may be performed" (see Chapter 1: 7). In this perspective, the analysis of the IFUW congresses hold during the interwar period might offer new insights on the ways women actively and publicly enacted scientific governance by performing a feminine (scientific) expertise on their own stage.

Staging Feminine Scientific Expertise

Through the organization of its International Congresses, the IFUW created its own (public) stage, from which university women could address and discuss issues regarding women's place in the scientific and academic worlds. Although participation in these conferences was not exclusively limited to IFUW members, such events represented an opportunity for university women to be heard and to be seen. As noted earlier, the participation of women in scientific congresses was not always granted, and the IFUW members were well aware of this. When Johanna Westerdijk, Dutch botanist and president of the IFUW between 1932 and 1936, visited the United States in 1914, her participation in the International Congress of Botanists organized in Saint-Louis was widely commented in the US press. Interestingly, the emphasis was not on the scientific content of her participation, but rather on the fact that she was the only woman present.[9] Such congresses also had a strong social dimension, with dinners and other activities organized in parallel to the more strictly scientific pursuits, but such social gatherings once again often contributed to reinforcing the marginalization of women within organizations and congresses in principle open to both sexes. Until the late 1920s, smoking cigarettes, for instance, constituted an important "and often deliberately intimidating" part of male professional culture, helping to exclude "properly bred" women from such informal gatherings (Rossiter 1982: 92). The IFUW congresses were thus designed, to a certain extent, to remedy this process of marginalization of women in the international realm of science.[10]

While the IFUW congresses played a crucial role in developing and reinforcing an international network beyond national borders, they also functioned as a platform for expression for the university women. Through the choices of the congresses' themes, activities, and the selection of speakers, the university women asserted their identity and expertise. Bu focusing on the reconciliation between private and professional life, or the link between science and femininity, IFUW members sought to promote a new image of female scientists beyond their own circle. In this perspective, the conferences became "places of performance" for university women. Every detail was carefully scrutinized to control the image displayed by IFUW members through the congresses, as shown in photographs taken during such events which were widely publicized in the international press. More than mere illustrations, such photographs can be read as visual performances supporting IFUW communication strategies. As the Norwegian Lilli Skonhoft wrote in a letter to the IFUW general secretary, Theodora Bosanquet, in 1923: "Even the smallest photo says much more than a good many words" (Figure 3.1).[11]

The photograph reproduced below is one of the most famous group portraits of IFUW members, taken during the Third International Congress organized in Oslo (then Christiania) in 1924. In the front row stood the leading figures of the organization, including Caroline Spurgeon and Virginia Gildersleeve, mentioned earlier. While women had been generally absent from group pictures taken during scientific congresses; a process that contributed to women's invisibility and marginalization; this photograph strategically emphasized the numerical importance of university women (and their identity thanks to the wearing of academic robes) demonstrating to the world the existence and legitimacy of the movement.

The IFUW congresses can, in many regards, be compared to the international scientific congresses of the period. The organizers borrow the codes of such scientific events as an attempt to legitimize their congresses and their expertise as academics. Whereas few women were invited as speakers in other scientific events, the IFUW congresses emphasized the role of women in scientific enterprise and research, by inviting world-renowned

Figure 3.1 The IFUW members posing in front of the Oslo University building on the occasion of the 3rd International Congress in 1924 [Riksarkivet, Olso: RA-PA-11 64—Norske Kvinnelige Akademikeres Landsforbund: Db-International Federation of University (I.F.V.W.) 1924–1997—L0018-001 Konferanse i Oslo, 1924].

(female) researchers such as Lise Meitner, the celebrated Austrian physicist, who offered a perfect example of feminine scientific success as envisioned by IFUW members. That being said, the presentation of new research and scientific results, central to the organization of scientific congresses, occupied seemingly less space in those organized by university women. While science is not absent from IFUW congresses, the discussions focused less on scientific content than on the structure and functioning of the scientific field.

The main themes developed at the congresses addressed specific problems women encountered in science and academia, such as the conditions for the contribution and recognition of their work in different scientific fields. Such discussions contributed to identifying and addressing gendered discriminatory practices at stake in science. The main theme selected for the Fourth International Congress organized in Amsterdam in 1926, for instance, addressed the problem of reconciling scientific work and family life. The opinion of Gildersleeve on this topic was ubiquitous. "The problem is this old one", she declared,

> the normal university or professional woman was not born to be celibate, childless and homeless. This is not good for her or for the State. How can she at once achieve and reconcile a career, a home, a husband, and children?[12]

The difficulties of reconciling work and family life were particularly preoccupying for women, as many national legislations did not allow married women to continue their work, especially in academic settings. Even in US women's colleges, female faculty members very often had to leave their positions when they married (Rossiter 1982). The situation was similar in Great Britain where, according to Carol Dyhouse, between 79% and 85% of academic women remained single throughout their lives (1995b: 161). The term "marriage mortality", used by Johanna Westerdijk in a petition to the League of Nations in 1935 concerning the legal status of women, captures the impact of marriage on women lives.[13] For Westerdijk, this marital bar helps to explain the very low proportion of female university professors.

Interestingly, university women, but also feminists in the first half of the twentieth century, very often mobilized statistical surveys to support and legitimize their claims. This choice of methodology was adopted by the IFUW members as a strategy to change the governance of academia. By using statistics, the IFUW members contributed to developing a new language that could approach science through the prism of gender. With the introduction of the category of marriage or children, they underlined the influence of social categories in the structure of the scientific field, pointing out the profound social dimension of science. By paying attention to marriage as an important break in women's scientific careers, they challenged the idea that women were incapable (physically or mentally) of conducting scientific research and producing scholarly work of quality. Because of their marginalized position in the scientific field, the IFUW members could adopt a critical and reflexive stance on science and the conditions of knowledge production, in some ways being in the vanguard of science studies and critical feminist thinking of the 1970s. Identifying discriminatory practices but also the conditions of success, notably the importance of research funding, they were among the first to reveal the impact of social conditions on scientific production and academic recognition.

Empowering Women through Funding and Travel Opportunities: The IFUW International Fellowship Program

The lack of funding and travel opportunities was identified as one of the main discriminatory practices affecting women's careers in science and academia. Since the foundation of the ACA in 1882, and the foundation of the BFUW in 1907, graduate women were well aware of the importance of a period of research abroad in the construction of a scientific career; and both national associations organized fellowship programs exclusively dedicated to women. While such programs were limited in scope, and open only to national candidates, the creation of the IFUW fellowship program in 1924 pursued the same objectives on an international scale. The IFUW funding system was specially designed to enable women to pursue research abroad for a year, regardless of nationality or discipline. While the International Fellowship Fund Appeal

Committee took care of organizing and promoting fundraising to finance the program, the Committee for the Award of International Fellowships was in charge of selecting the IFUW fellowship holders.

Since the beginning, the establishment of such research fellowships was considered "the most vital and necessary" work of the organization.[14] As the members of the IFUW Council stated in 1924:

> Such Fellowships are urgently needed for encouraging scholarship and advancing the status of university women, for spreading knowledge, [for] improving educational methods and ideals, and [for] actively promoting friendship and sympathy between the nations through the medium of these chosen representatives.[15]

The IFUW founders and first presidents were particularly aware of the limited opportunities for women to study or conduct research abroad, even though they had experienced the importance of such visits in building their careers. This was emphasized by Ellen Gleditsch in a speech entitled "The Need and Value of International Fellowships for Research":

> If we want women to do research work we must give young women the opportunity of advanced study in their own country and in other countries. In fact, research in science will nearly always, and research in arts very often, necessitate years of study at foreign universities. I need not dwell on this; it is known and recognized by us all.[16]

The Norwegian chemist saw international fellowships as a necessary stepping stone to allow women to devote a year to continuing or completing their research in a foreign country. She had spent time in Marie Curie's laboratory in Paris between 1907 and 1912, before moving to the United States on a grant from the American-Scandinavian Foundation in 1914– 1915. Her international experience proved to be of crucial importance when she applied to a professorship position at the University of Oslo in 1929 (Lykknes et al. 2004).

The emphasis on the establishment of this fellowship program responded to the structural changes affecting the scientific and academic worlds from the turn of the twentieth century through the first half of that century (Tournès and Scott-Smith 2018). During that period, funding bodies came to play an increasingly important role in shaping both scientific practices and scientific identities and ideals through the selection of what they considered being the best fellows. The emergence of research funding programs supported in particular by new philanthropic organizations such as the Rockefeller or Carnegie Foundations accelerated the internationalization of science, the promotion of researchers' mobility, and knowledge circulation; but too few women benefited from them (Rossiter 1982: 267). In addition to the practical and financial aspects, obtaining grants assumed a more symbolic dimension as it functioned as a proof of quality, legitimacy, and scientific authority (Paul 2016, Huistra & Wils 2016, Cabanel 2019).

As Robert Kohler wrote in his book addressing the relation between scientific researchers and sponsors in the twentieth century, "ideas are often made good by the process of securing resources… Fundability has become, for better or for worse, an important element in the system by which credibility is allocated to researchers and their work" (1991: 2). Looking at the situation of American women in the 1920s and 1930s, Margaret Rossiter demonstrated how the functioning of the scientific world, particularly the system of prizes and awards, resulted in the marginalization of certain groups, including women, who were relegated to the rank of amateurs (1982: 267), while at the same it reinforced the masculine ideal associated with the image of the scientist. In this context, the development of fellowship programs by and for women functioned as a "system of compensatory recognition" (Rossiter 1982: 297). By giving women the chance to be active as experts or as fellowship holders, or more generally within leadership positions denied to them in male-dominated societies, such single-sex programs contributed to establishing women's visibility and legitimacy in the scientific and intellectual communities.

As Rossiter noted, such parallel systems were never presented as an attack on the existing system (1982). By achieving "work of absolutely first-rate quality, of distinction", women would prove their capacity to produce quality scientific work, and progressively the issue of gender would disappear as a marker or guarantee of scientific quality (or of lesser quality and lesser expectations).[17] The strategy advocated by Gildersleeve was a non-transgressive and "pragmatic" feminism in opposition to the militant feminist movement. Not by ambitions but by means of action, Gildersleeve in particular, and university women in general advocated for a type of infiltrating feminism allowing women to work from within and thus to avoid antagonism with their male colleagues (Von Oertzen 2014, Dilley 2017).

In the course of the interwar period, the IFUW awarded about 48 fellowships and grants to women from different countries and various disciplines. It remains difficult, of course, to evaluate the role of the IFUW in effectively empowering women through its fellowship program. The prosopographic study of the IFUW fellows, however, tends to show the crucial role the fellowship played in the fellows' career as 15 of them became professors (Cabanel 2019). The trajectories of Margarete Bieber, Margaret Mes, or Erszébet Kol, three of the IFUW fellows in the interwar period, illustrate this dynamic. After conducting research abroad, the three women succeeded in finding a position in a university in their native countries. Bieber became Extraordinary Professor and Director of the Archaeology Department at the University of Giessen, in Germany. Compelled by the Nazi government to leave her position, due to discriminatory policies against Jewish intellectuals, she later became a Professor at Columbia University. Margaret Mes became the first woman to obtain a teaching position at the University of Pretoria,

South Africa. After defending her dissertation on tobacco plantation diseases under the supervision of Johanna Westerdijk at the University of Utrecht, she was awarded an IFUW scholarship to study at the University of California Berkeley in the United States. Thanks to her research on plant physiology with Professor Hoagland, a professor specializing in plant morphology, she became a recognized expert in South Africa and took charge of developing a research laboratory in this field of study. The trajectory of Erzsébet Kol followed a similar path. A Hungarian botanist who specialized in snow and ice algae, she benefited greatly from her trip to North America in 1935 (Cabanel 2021). Upon her return to Hungary, Kol's career took a new turn as she obtained a permanent position as an Assistant Professor at the University of Szeged in 1937, and as a full Professor in 1940. In a letter she sent to the IFUW fellowship committee in 1937, she dedicated a part of her academic and scientific recognition to the IFUW:

> I am really glad to have received this fellowship, not only because it gave me an opportunity… to carry on my studies on another continent, but because the results of these studies have aided me to receive a permanent post in the botanical department of Ferenc József University in Szeged from the Hungarian government.[18]

Such testimonies are not rare in the fellows' archives. Although this rhetoric must be understood in the light of unwritten conventions inherent to any funding system, the success stories of many of the IFUW fellows in the interwar period validate, to a certain extent, the IFUW collective strategies of scientific governance to empower scientific and scholarly women.

Conclusion

During the Interwar period, IFUW leaders pursued clear scientific governance strategies. They aimed to strengthen the place of women in universities and the scientific world, and through the promotion of higher education and the development of an international policy, to empower women and integrate them into the governance of academia and science. The IFUW served both as a laboratory and platform of expression for the shaping of a new international, gender-free scientific identity—one that men, as well as women, could claim as their own and be identified with by the scientific community and the larger public.

The existence of an international organization such as the IFUW demonstrates the dynamism of female networks and the importance of their collective strategies in their engagement for equal chances between sexes in the scientific and academic worlds of the first half of the twentieth century. Through their scientific initiatives such as the creation of a funding system dedicated to women and the organization of international congresses, the IFUW members (publicly) contributed to support and promote a new vision of women scientists. They demonstrated, playing on long-lasting and negative clichés on women and science, and using statistical surveys, that the low proportion of women in science and academia was not due to their lack of talent or their inabilities to conduct research, but rather to the structure of science itself and the gendered discriminatory practices at work in the scientific field. This strategy shows the IFUW members and more largely university women of the interwar period not only aspired to take part in the governance of science but also to use science as a way of governance.

Notes

1 Records of the International Federation of University Women (IFUW), inv.no 256: IFUW complete set of Constitutions and By-laws since 1920. 1920–1992. Article 1, Collection International Archives for the Women's Movement (IAV), in Atria, Institute on Gender Equality and Women's History.
2 Records IFUW, inv.no 71: Bulletins (Bluebooks), 4th Conference, Amsterdam, The Netherlands. 1926: 6.
3 Records IFUW, inv.no 109: Minutes of Conferences. 1920.
4 Records IFUW, inv.no 67: *Bulletins* (Bluebooks), 1st Conference, London, Great Britain. 1920. Report 1920–1921: 71.
5 Records IFUW, inv.no 67: *Bulletins* (Bluebooks): 19.
6 Records IFUW, inv.no 67: *Bulletins* (Bluebooks): 19.
7 Records IFUW, inv.no 71: *Bulletins*, 4th Conference, Amsterdam. 1926: 6 and 8.
8 Records IFUW, inv.no 71: *Bulletins*, 4th Conference: 8.
9 Archief Johanna Westerdijk, collectie Internationaal Archief voor de Vrouwenbeweging (IAV), in Atria, kennisinstituut voor emancipatie en vrouwengeschiedenis, inv.no 211: "Shaw's Garden Honors Woman: Botanists to Hear Her on Phytopathology in Tropics" *St Louis Republic*, 16 October 1914.
10 In the course of the Interwar period, the IFUW organized eight congresses in different European capitals: London (1920), Paris (1922), Christiania (1924), Amsterdam (1926), Geneva (1929), Edinburgh (1932), Cracow (1936), and Stockholm (1939).
11 Norske Kvinnelige Akademikeres Landsforbund, Riksarkivet, Oslo [PA-1164], inv.no DB/L0018-0001, Konferance i Oslo, 1924: letter from L. Skonhoft to T. Bosanquet, 30 December 1923.
12 Records IFUW, inv.no 71: *Bulletins*, 4th Conference, Amsterdam. 1926: 28.
13 Records of the British Federation of University Women (BFUW), 1907–1997. London: London School of Economics, The Women's Library, inv.no 5BFW/04/20: "Notes of the marriage bar", 17 January 1933.
14 Records IFUW, inv.no 69: *Bulletins* (Bluebooks), 3rd Conference, Christiania, Norway. 1924: 29.
15 Records IFUW, inv.no 494: Committee for the Award of International Fellowships. Minutes (1925–1962). 1924.
16 Gleditsch, Ellen. "The Need and Value of International Fellowships for Research". Records IFUW, inv.no 71: Bulletins (Bluebooks), 4th Conference, Amsterdam, The Netherlands. 1926: 112–114.
17 Records IFUW, inv.no 69: *Bulletins*, 3rd Conference, Christiania, Norway. 1924: 28–29.
18 Archives of the American Association of University Women. AAUW headquarters, Washington, DC. Box 441: Kol, Erzsébet. "Final Report, Fellowship Crusade International". 1938: 7. Ferenc József was the Hungarian name of Franz Joseph University.

References

Bard, Christine. 1995. *Les filles de Marianne. Histoire des féminismes 1914–1940.* Paris: Fayard.

Bosch, Mineke. 2016. "Scholarly Personae and Twentieth-Century Historians. Explorations of a Concept," *BMGN – Low Countries Historical Review* 131 (4): 33–54.

Cabanel, Anna. 2021. "A Woman in a 'Man Made World': Erzsébet Kol (1897–1980)." In *Gender, Embodiment, and the History of the Scholarly Persona: Incarnations and Contestations* ed. Kirsti Niskanen and Michael Barany, 113–146. London: Palgrave Macmillan.

Cabanel, Anna. 2025. *Femmes de sciences. Le mouvement des university women dans l'entre-deux-guerres.* Rennes: Presses Universitaires de Rennes.

Cabanel, Anna. 2018. "'How Excellent… for a Woman?' The Fellowship Programme of the International Federation of University Women in the Interwar Period," *Persona Studies* 4 (1): 88–102.

Daston, Lorraine, and Otto Sibum. 2003. "Introduction: Scientific Personae and Their Histories," *Science in Context* 16 (2):1–8.

Dilley, Patrick. 2017. *The Transformation of Women's Collegiate Education: The Legacy of Virginia Gildersleeve.* London: Palgrave Macmillan.

Dyhouse, Carol. 1995a. "The British Federation of University Women and the Status of Women in Universities, 1907–1939" *Women's History Review* 4 (4): 469–470.

Dyhouse, Carol. 1995b. *No Distinction of Sex? Women in British Universities* 1870– *1939*. London: UCL Press.

Fouché, Nicole. 2000. "Des américaines protestantes à l'origine des 'University Women' françaises 1919–1964" In *Femmes protestantes au XIXe et au XXe siècles,* ed. Gabrielle Cadier-Rey. *Bulletin de la Société de l'Histoire du Protestantisme Français* 1: 133–152.

Gildersleeve, Virginia. 1954. *Many A Good Crusade.* New York: Macmillan.

Gleditsch, Ellen. 1926. "The Need and Value of International Fellowship for Research" *Occasional Papers* 5: 20–21.

Goodman, Joyce. 2012. "Women and the International Intellectual Co-operation," *Pedagogica Historica* 48 (3): 357–368.

Huistra, Pieter, and Kaat Wils. 2016. "Fit to Travel': The Exchange Programme of the Belgian American Educational Foundation: An Institutional Perspective on Scientific Persona Formation (1920–1940)," *BMGN - Low Countries Historical Review* 131 (4): 112–134.

Kohler, Robert. 1991. *Partners in Science. Foundations and Natural Scientists, 1900– 1945.* University of Chicago Press.

Laqua, Daniel. 2011. "Transnational Intellectual Cooperation, the League of Nations, and the Problem of Order," *Journal of Global History* 6 (2): 223–247.

Lykknes, Annette, Lise Kvittingen, and Anne Kristine Børresen. 2004. "Appreciated abroad, depreciated at home. The career of a radiochemist in Norway: Ellen Gleditsch (1879–1968)," *Isis* 95 (4): 576–609.

Paul, Herman. 2016. "Sources of the Self. Scholarly Personae as Repertoires of Scholarly Selfhood," *BMGN - Low Countries Historical Review* 131 (4):135–154.

Peeters, Evert, Joris Vandendriessche, and Kaat Wils. 2015. *Scientists' Expertise as Performance: Between State and Society, 1860–1960.* London: Pickering & Chatto.

Rasmussen, Anne. 1995. "L'internationale scientifique 1890–1914" PhD dissertation, EHESS.

Rossiter, Margaret W. 1982. *Women Scientists in America, Struggles and Strategies to 1940*. Baltimore: The Johns Hopkins University Press.
Rupp, Leila. 1997. *Worlds of Women: The Making of an International Women Movement*. Princeton University Press.
Singer, Sandra L. 2003. *Adventures Abroad. North American Women at German-Speaking Universities, 1868–1915*. Westport: Praeger.
Skonhoft, Lilli. 1932. "Norske Kvinnelige Akademikeres Landsforbund." In *Kvinnelige studenter 1882–1932*, ed. Norske Kvinnelige Akademikeres Landsforbund, 249–263. Oslo: Gyldendal Norsk Forla.
Spillman, Scott. 2012. "Institutional Limits: Christine Ladd-Franklin, Fellowships, and American Women's Academic Careers, 1880–1920" *History of Education Quarterly* 52 (2): 196–221.
Stamhuis, Ida H. 2004. "Historical Considerations on 'Women Scholars and Institutions." In *Women Scholars and Institutions. Proceedings of the International Conference, vol. 13* B, *Women pioneers in Radioactivity Research*, ed. Sona Strbanova, Ida Stamhuis, and Katerina Mojsejova, 17–48. Prague: Research Center for the History of Sciences and Humanity.
Talbot, Marion, and Lois K. M. Rosenberry. 1931. *The History of the American Association of University Women 1881–1931*. Boston: Riverside Press.
Tournès, Ludovic, and Gilles Scott-Smith. 2018. *Global Exchanges. Scholarships and Transnational Circulations in the Modern World*. New York: Berghahn Books.
Von Oertzen, Christine. 2014. *Science, Gender, and Internationalism. Women's Academic Networks, 1917–1955*. New York: Palgrave Macmillan.
Woolf, Virginia. 1929. *A Room of One's Own* [first edition]. Richmond: Hogarth Press.

4 Abortions, Eugenics, and Artificial Reproduction in the Soviet Union, 1920–1936

Alexei Kojevnikov and Kirill Rossiianov

A mass movement for women's rights arose in the Russian Empire ca. 1860, with one of its primary demands for access to higher education. At the time when universities accepted only male students, hundreds, then thousands of young women pushed for desegregation by auditing classes, obtaining academic degrees, traveling to other countries in Europe, and asking for special permissions to enroll (Koblitz 1988; Stites 1978). By the century's end, special colleges for women and/or individual precedents of female students studying at predominantly male universities became common enough, paving the way toward the official adoption of co-education in most European countries following WWI. One of the first acts of the Russian Revolution in 1917 was to allow equal rights for women to study at every educational institution, any level and field. Ending legal discrimination did not, of course, mean factual equality, as post-revolutionary generations of students continued advancing through educational and academic ranks facing informal prejudices, inequitable governance, and glass ceilings. It took two additional decades until at least some female scientists reached the very top echelons of Soviet research establishment.

In the meantime, new reproductive technologies and practices that had been considered ideologically unacceptable prior to the 1917 Revolution—abortion, birth control, and artificial insemination—became openly available during the 1920s. Together with the early Soviet legislation on family and marriage, libertarian even by today's standards, and decriminalization of homosexuality, they created a social and intellectual environment conducive to experimentation with sensitive issues of human reproduction. Ilya Ivanov, a pioneer in artificial insemination of mammals, attempted to confirm Darwin's theory of human evolution by crossbreeding humans and apes (Rossiianov 2002). Antonina Shorokhova developed and applied the techniques of artificial insemination for human patients in gynecological clinics in Tashkent. Vera Danchakova returned to revolutionary Russia in 1926 to undertake a program of experimental research with human embryos provided by an abortion clinic, with the goal to grow and develop fetal tissues outside maternal bodies.

DOI: 10.4324/9781003562597-7

While extending technological options to govern the human bodies, these new scientific approaches also allowed some opportunities for women's own government of and through science. This paper focuses primarily on the spectrum of relationships between new reproductive practices and the discussion of women's liberation in the revolutionary Russian society and government of the 1920s, including women's control over their own bodies and reproductive future. They were part of a more general, leftist, and modernist project that envisioned plasticity and flexibility of human nature challenging traditional gender roles and bio-social boundaries. While pioneering many women's rights in the social and public sphere, the Soviet society continued to be more resistant toward radical proposals for women's sexual liberation. The new techniques of human reproduction, however, enabled discussions of new, emerging "women's rights": for birth control, independence from men in pregnancy and motherhood, choosing genetic material for reproduction, and hormonal sex change.

Legalization of Abortions, Act I

November 2020 marked the hundredth anniversary of the first decision by a modern state to legalize abortions and to make the procedure freely available in medical clinics, upon a woman's demand. We provide below the full translation of the pathbreaking decree that is often referenced but only briefly analyzed in historical literature (Goldman 1993: 255) (see Figure 4.1).

> **"Decision of the People's Commissariats of Public Health and Justice No. 471: On Protection of Women's Health.**
>
> In recent decades, the number of women who resort to aborting their pregnancies has been increasing both here and in the West. In all countries, legislation is fighting this evil by punishing women who decide to induce a miscarriage as well as physicians who perform it. Far from producing positive results, this method of fighting has pushed the operation underground and turned women into victims of egoistic and often ignorant abortionists, profiteers from covert operations. As a result, up to 50% of women become infected, and up to 4% die.
>
> The Workers' and Peasants' Government takes into consideration the harmful effects of this phenomenon on the collective. It is fighting this evil by strengthening socialist society and by agitating against abortions among the working female population. By broader realization of the principles of Protection of Maternity and Infancy, it envisions gradual disappearance of this practice. In the meantime, moral vestiges of the past and difficult economic conditions of the present are still forcing some women to decide upon such a procedure. The People's Commissariat of Public Health and the People's Commissariat of Justice, acting to protect women's health and the wellbeing of race from ignorant

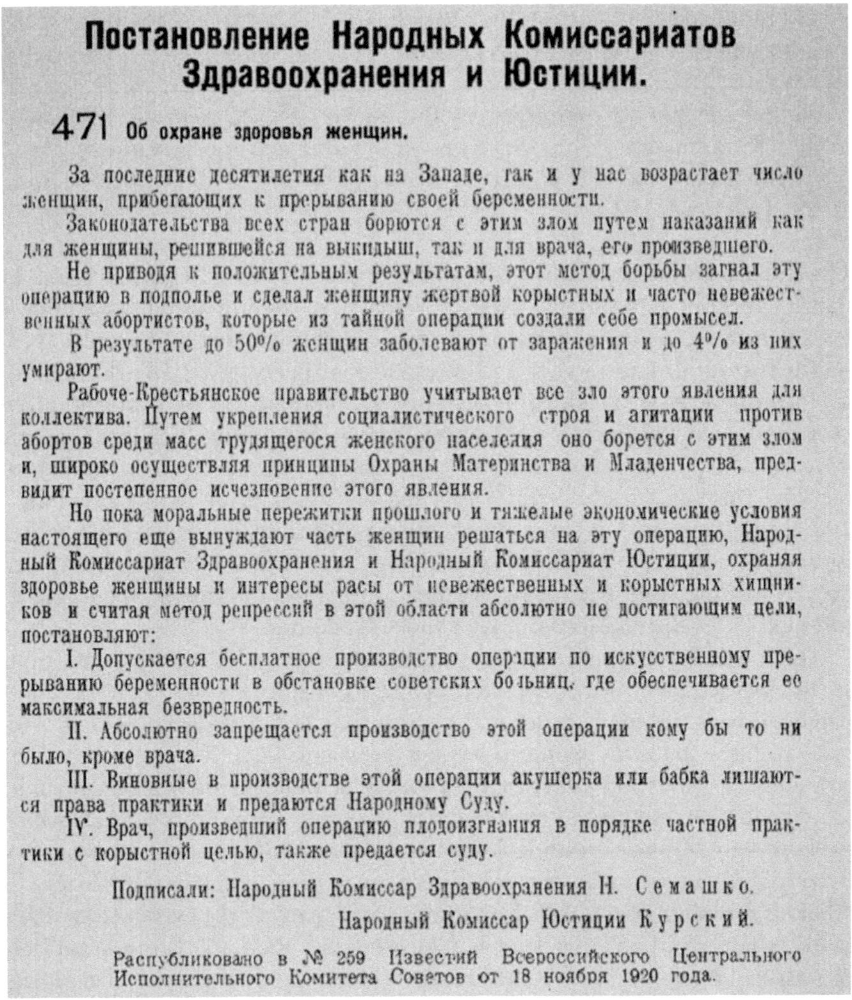

Постановление Народных Комиссариатов Здравоохранения и Юстиции.

471 Об охране здоровья женщин.

За последние десятилетия как на Западе, так и у нас возрастает число женщин, прибегающих к прерыванию своей беременности.

Законодательства всех стран борются с этим злом путем наказаний как для женщины, решившейся на выкидыш, так и для врача, его произведшего.

Не приводя к положительным результатам, этот метод борьбы загнал эту операцию в подполье и сделал женщину жертвой корыстных и часто невежественных абортистов, которые из тайной операции создали себе промысел.

В результате до 50% женщин заболевают от заражения и до 4% из них умирают.

Рабоче-Крестьянское правительство учитывает все зло этого явления для коллектива. Путем укрепления социалистического строя и агитации против абортов среди масс трудящегося женского населения оно борется с этим злом и, широко осуществляя принципы Охраны Материнства и Младенчества, предвидит постепенное исчезновение этого явления.

Но пока моральные пережитки прошлого и тяжелые экономические условия настоящего еще вынуждают часть женщин решаться на эту операцию, Народный Комиссариат Здравоохранения и Народный Комиссариат Юстиции, охраняя здоровье женщины и интересы расы от невежественных и корыстных хищников и считая метод репрессий в этой области абсолютно не достигающим цели, постановляют:

I. Допускается бесплатное производство операции по искусственному прерыванию беременности в обстановке советских больниц, где обеспечивается ее максимальная безвредность.

II. Абсолютно запрещается производство этой операции кому бы то ни было, кроме врача.

III. Виновные в производстве этой операции акушерка или бабка лишаются права практики и предаются Народному Суду.

IV. Врач, произведший операцию плодоизгнания в порядке частной практики с корыстной целью, также предается суду.

Подписали: Народный Комиссар Здравоохранения Н. Семашко.

Народный Комиссар Юстиции Курский.

Распубликовано в № 259 Известий Всероссийского Центрального Исполнительного Комитета Советов от 18 ноября 1920 года.

Figure 4.1 Newspaper announcement of the November 1920 decree. Public Domain.

and mercenary predators, and considering the repressive method in this field absolutely incapable of attaining its goal, have decided:

1 To allow performing the procedure of the artificial interruption of pregnancy free of charge and in the conditions of Soviet clinics, where its maximal harmlessness is ensured;
2 To completely ban the performance of this procedure by anyone except a professional physician;
3 Midwives or birth helpers guilty of performing this procedure will be prosecuted in People's Court and lose their right to provide midwifery;

4 Physicians who perform the procedure of miscarriage in their private practice for profit will also be prosecuted in court.

Signed: People's Commissar of Public Health, *N. Semashko*, People's Commissar of Justice, *Kurskiy.*" Published in *Izvestiia of the All-Russian Central Executive Committee of the Soviets*, # 259, on 18 November 1920.

The brevity and tone of this groundbreaking governmental decision convey that its authors, in particular Nikolai Semashko, did not attach strong ideological meanings and political importance that their decision would eventually acquire in later Soviet and international perspectives. At the time of its introduction, legal abortion was not yet understood or propagandized as a fundamental principle of women's rights, but primarily as a pragmatic if regrettable, temporary medical necessity. A hypothetical possibility of relaxing the ban on abortions had been discussed prior to the revolution among some medical doctors. It was not a political priority either for Marxist revolutionaries, the Bolshevik government that had just barely survived the devastating Civil War, or for women's liberation movement, including its foremost radical spokesperson, Alexandra Kollontai.[1]

The initiative thus came from medical authorities who legalized abortions as an emergency measure to protect women's health, a lesser "evil" as compared to the widespread, unrecognized, unprofessional, and often lethal practice which the state was unable to prevent (see Figure 4.2). Though presented as a pragmatic concession rather than a proudly proclaimed ideological principle, the decision certainly relied on revolutionary politics as a precondition, which explains why Bolshevik authorities were able to take this unconventional step. By revolutionary instincts, the government was strongly inclined to disrespect religious prohibitions and beliefs that motivated many of the existing legal bans in the sphere of sex and marriage. It thus adopted a libertarian attitude and easily canceled many restrictions, including those imposed on divorce, illegitimate children, and homosexuals—and also on abortions, despite disapproving the latter practice.[2] Their openly anti-capitalist argument shifted the blame from poor women in precarious situations to private providers and profit-makers, hence the decision to offer abortions in state clinics without requiring any payment. As Marxists, Soviet officials saw the primary cause of abortions in economic conditions that made it difficult for women to provide for their future babies and were optimistic that once liberated from such fears by the improvement of economy and social welfare, women would feel much less compelled to seek deliberate termination of pregnancy.

"The wellbeing of race" is an untypical for Bolsheviks turn of phrase reflecting an input from Russian eugenicists, in their culturally specific meaning referring not to any particular racial group or ethnicity of the country's diverse demographics, but to the social hygiene of all its population in entirety, in a meaning closer to that of "the human race" in English. Meanwhile,

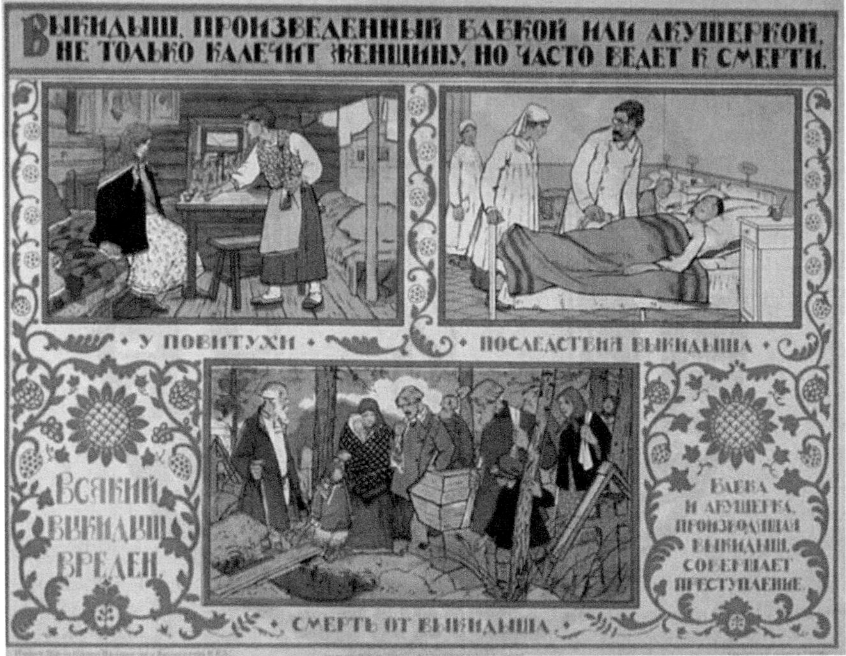

Figure 4.2 Early Soviet visual propaganda against unprofessional abortions, 1920 (Il'ina 2019: 58). Public domain.

activists of women's emancipation, in particular Kollontai, focused primarily on developing gender equality in public and social life, "equal pay for equal work", and equal access to education and professions (Kollontai 1977: 39). In the sphere of gender relations, Kollontai pushed for a new, libertarian marriage code, social welfare, support for maternity and childcare, women's liberation from housework burdens, and greater sexual freedoms. Ahead of time in many other aspects of radical feminism, Kollontai refrained from extending the concept of "women's rights" to abortions—the latter did not figure prominently in her discourse—but accepted the logic of the medical establishment and generally supported the legalization decree. Her goal was not to emancipate women from maternity, which, in her time, in the absence of any effective birth control, would have practically implied religiously motivated sexual abstinence. Instead, she strove for recognition of specifically women's rights to safe and protected "childbearing and motherhood", which, in her opinion, was also a moral and "social obligation" for any woman as "the reproducer of the race" (Kollontai 1977: 145–146). The term "race" here also reflected Kollontai's awareness of eugenics in its Soviet version and, as in the text of the 1920 decree, referred to the overall population in demographic sense. Strongly pro-natalist in general, as almost everyone in the Russian society at the time, Kollontai deviated from this standard stance at least once, when she defended birth control: "Let there be fewer children born but let them be of better 'quality'. Every child should be wanted by its mother" (Kollontai 1977: 309).

The gradual development of attitudes toward abortion as a woman's right followed from, rather than preceded the established practice. Already in the 1920s, abortions were effectively understood as a patient's right, since government regulations obliged doctors to provide this service to women upon request. With growing numbers of cases, also due to a partial return to capitalism under the New Economic Policy, this right was also formalized bureaucratically and grounded in class. Special commissions that included doctors and representatives of women's organizations reviewed applications and granted the priority access to free abortions to women from the working class and in social need, while requiring economically better-off female patients to pay for the procedure. Although disagreements continued in the medical literature (Gens 1925), the demand for legal abortions quickly grew to become a widespread practice and new normality for younger generations of Soviet women well before the reversal of the decision and a renewed ban on abortions in 1936. During the 15 years when this social norm remained unique to the Soviet Union, it enabled an intellectual environment conducive to innovative biological and medical experimentation that undermined traditional bio-social boundaries and gender roles, while also transgressing cultural, and sometimes traditional moral, taboos.

Vera Danchakova: From Stem Cells to the Developmental Plasticity of Sex

In March 1931 the new director of the Institute of Experimental Morphogenesis in Moscow, Rafail Belkin, summarized the research plans designed by his predecessor, Vera Danchakova:

> The main goal of the Institute's research was to give humans a possibility to change their nature… through a biological method, in order to achieve a higher development of the nervous system, especially the brain, in the process of embryological development and at the expense of some other tissues.
>
> (Belkin 1931: 19)

Though Belkin may have rhetorically exaggerated the intentions, Danchakova (born Vera Mikhailovna Grigorevskaia in 1879) conducted groundbreaking investigations on blood stem cells and the development of embryonic tissues without, however, receiving a well-deserved academic recognition (see Figure 4.3). Her entire scientific career developed in transit, from one country to another and from one short-term, insecure position to another one, repeatedly being pushed to the margins. Like many women, she endured discrimination

THE FIRST LADY UNIVERSITY TEACHER IN RUSSIA
This proud distinction has befallen Dr. Vera Dantschakoff, who, after a hard fight, has been officially recognised as a Professor at Moscow University. She lectured at the latest Congress of the Anatomical Society in Berlin. Photograph by Bolak.

Figure 4.3 Vera Danchakova at her desk at the Imperial Moscow University, 1908. *The Graphic*, Saturday 20 June 1908 Vol. LXXVII, No. 2012, 28. Public domain.

in academia, suffered from political instabilities of the time, and also because of exceptional originality, unconventionality of her research.

Danchakova's autobiography (1931a) describes her studying first at the Pedagogical Institute in St. Petersburg and then as medicine and biology student at the University of Lausanne, Switzerland, the country that since the 1860s provided educational opportunities for many Russian women. She then worked in Russia as a medical pathologist and in 1907 defended her MD in histology at the St. Petersburg Military-Medical Academy (Danchakoff 1907). Shortly thereafter, at the start of her academic career at Moscow University, Danchakova made her first major discovery by describing the existence of what are now called blood stem cells (Danchakoff 1908, 1909). Simultaneously and independently, Alexander Maksimov, professor of histology in St. Petersburg, made a similar breakthrough (Maximov 1908/1909). In the literature, Danchakova was often referred to as a student of Maksimov, which downplayed the originality and independence of her own investigations.[3] In the atmosphere of political reaction following the defeated 1905 revolution, the conservative Ministry of

Education refused to confirm Danchakova's appointment to a regular university position. Discrimination that she experienced within the Russian academic system motivated her to look for an academic career abroad, first in Germany, and later in the United States, at Woods Hole, at the Rockefeller Institute for Medical Research, at the Wistar Institute, and eventually, in the early 1920s, at Columbia University.

Danchakova's interpretation of stem cells anticipated today's views in some important respects. She understood them as undifferentiated precursors that continue to exist in adult tissues and, in certain situations when needed, or under the influence of certain stimuli, have the capacity of developing into a variety of directions, producing differently specialized blood cells. Somewhat later she also proposed the existence of similar stem cells for other tissues. Her subsequent lifelong research primarily focused on investigating experimental possibilities to influence, modify, direct, or channel the development of stem cells and tissues, and on the importance of stem cells for embryology. While in the United States, Danchakova started experiments with "heteroplastic transplantations". She used the allantoic cavity of chicken eggs as a medium to grow embryos, transplanted stem cells, and tissues of other species (Danchakoff 1924). Economic and logistical constraints on these investigations came from the limited availability of embryos, especially from large organisms. It appears that the existence of abortion clinics, and thus logistical possibilities to use human embryos for biological experimentation, provided the main motivation for Danchakova's transfer of her experimental program to the Soviet Union. In 1926 she returned to Moscow and was elected a full member to the Timiryazev Scientific-Research Institute, one of the main experimental institutions supported by the People's Commissariat of Enlightenment (Fando 2020: 251). There she reached technoscientific leadership by founding and directing for five years the Laboratory of Experimental Morphogenesis which in 1931 was further reorganized into a separate Institute.[4]

By early 1929, Danchakova succeeded in growing various human tissues in chicken embryos and submitted a paper on the successful *in vitro* cultivation of an embryonic human heart (Danchakoff and Gagarin 1929). Her laboratory received human embryos from Moscow maternity homes and negotiated with the Commissariat of Public Health for a formalized permanent arrangement with an abortion clinic. It was somewhat more difficult to ensure a year-round supply of appropriately fertilized chicken eggs, but by the end of 1927, this problem was also solved by establishing a farm of Leghorn chickens imported from the United States. Danchakova's main problems appeared to be bureaucratic. She either lost initial patrons in Soviet government or did not manage to maintain good relations with them[5] and complained about the lack of support for her laboratory from other colleagues in the Institute (Danchakoff 1928), who saw her as an arrogant and demanding foreigner (she did have a US passport).

By 1931, Danchakova lost administrative control of her Moscow laboratory and with it, the possibility to experiment on human embryos. Her new projects focused on sex transformations in the embryo and the possibility of manipulating it toward the development of either male or female sexual organs. She could still work in the USSR until 1933, when she was refused a Soviet visa (Fando 2020), and then at several universities in Eastern and Central Europe, survived the war in countries that mostly came under the control of Nazi Germany, and after another brief postwar stay in the USSR, by 1950 ended up back in Lausanne, in neutral Switzerland. Despite all these political troubles and the lack of a permanent, secure position, Danchakova remained a productive researcher throughout.

Her pathbreaking investigations involved new possibilities for experimentation with human embryos resulting from political decisions such as the removal of the ban on abortions and other restrictions. No less important were the cultural changes in understanding gender relations encouraged by the new technologies of reproduction. Danchakova's interest in the developmental plasticity of embryonic tissues came from an analogy with the multi-potentiality of stem cells and also reflected widespread hopes for the transformability of human nature popularized internationally within the leftist modernist project of the 1920s (Squier 1994). Before or concurrently with her research in the USSR, during the debate over the possibility of socialist eugenics, other Soviet biologists were entertaining various alternatives to the constraints of "hard" Mendelian heredity. The first secretary of the Russian Eugenic Society, Mikhail Volotskoy, envisioned a Lamarckian "proletarian eugenics" (Gaissinovitch 1980: 21), whereas the geneticist Alexander Serebrovsky published a proposal to educate women on how to differentiate love life from procreation and encourage them to choose better genetic material than their husbands for artificial insemination (Babkov 2008). In the theoretical conflict between neo-Lamarckism and genetic determinism, Danchakova subscribed to neither side, but she did oppose the idea of "genetic preformism" popular in the twentieth century, according to which, in Jane Maienschein's words, "development brings differentiation that is unidirectional" (Maienschein 2005). Danchakova's interest in the plasticity of the process of embryonic development positioned her research in the broad tradition of developmental biology, which Evelyn Fox Keller (1997) described as a specifically feminist approach.

Politically, Danchakova does not appear to have supported the Bolsheviks, but she strongly endorsed their goals of gender equality, in particular the attempts to adopt new legislation on marriage that would not only proclaim equality between men and women in words but also de-facto compensate for the "actual conditions of economic inequality" and for women carrying "the physiological burden" (Danchakoff 1927: 188). Her idea of the embryo's pluripotency, or the possibility of transforming the direction of embryonic development, materialized further in her 1930s research on the lability of expression of sexual characteristics. By injecting newly available "histogenetic substances" (chemically purified male and female hormones) she was able to form rudimentary sexual organs of the opposite sex in the embryos of mammals and birds and to modify the sexual behavior of grownup organisms (Danchakoff 1938a, 1938b). In her interpretation, sex-changing experiments contradicted the then-prevalent chromosomal theory of sex determination.

They also contradicted the dominant, much more conservative, ideological trends of the 1930s. The window of opportunity for taboo-breaking biological experimentation that allowed her research in the Soviet Union narrowed significantly after 1930.[6] Her successor in the Moscow laboratory, Belkin, referred with disapproval, as "risky", to her plan to biologically modify human nature. German Nazism was much more hostile to any blurring of the biological distinction between the sexes, and Danchakova had to deny rhetorically having such an "impudent thought" when in 1941 she published in German her book on development of sex (Danchakoff 1941: IV–V). Politically dogmatic support for the hard-wired chromosomal theory of heredity in the West after WWII, likewise, made her views appear heretical and led to their marginalization. An alternative discourse on women's rights in reproduction, emancipation from male domination, improvement of the human stock, and treatment of infertility came from another post-revolutionary technology, artificial insemination.

Antonina Shorokhova: Artificial Insemination and Women's Rights

Shorokhova's case also reveals a combination of possibilities that became available following the revolution thanks to the new government's scientific policy: the use of new artificial technologies in human reproduction, the removal of earlier, religiously inspired prohibitions, and the inspiring discourse of women's emancipation. Born Antonina Alekseevna Vasil'eva in Saratov in 1881, she enrolled in 1907 at the newly opened St. Petersburg's Women's Medical Institute. Ever since the inception of the powerful movement for women's higher education in Russia ca. 1860, medicine was the subject of prioritized demand among female students. By 1900, despite existing restrictions, about 1,000 women practiced as certified medical doctors in the Russian Empire. Typically, they were not allowed to study alongside male students at state-run medical schools and universities; but could still receive education at some specially established private or community colleges, or abroad (Koblitz 1988). Female doctors were not accepted into state civil service and thus, typically, were not employed at major state hospitals, but they could open private medical practice or work at community-run *zemstvo* clinics. Sometimes, additional exceptions were allowed, which in Shorokhova's case, helped her personal medical career: the field of gynecology, in particular, was traditionally more open for female practitioners (see Figure 4.4). Also, in a few cases, the Russian Imperial state agreed to employ female doctors in areas with indigenous Muslim populations in order to provide service to female patients who could otherwise be reluctant or not allowed by their families to see a male medical official.

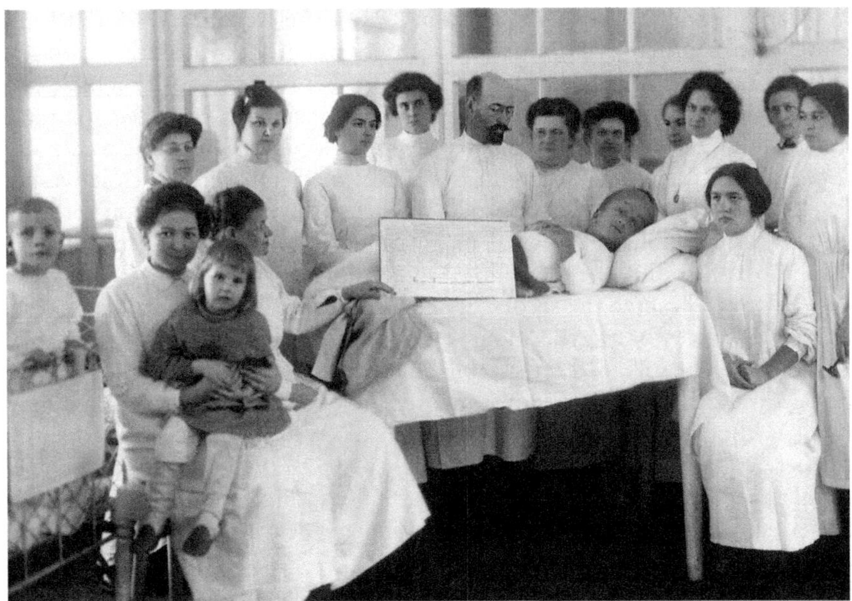

Figure 4.4 Professor N.I. Rachinsky with students at the Department of Midwifery and Gynecology, Women's Medical Institute, St. Petersburg, ca. 1904–1908. Central State Archive of Film and Photo Documents in St. Petersburg. Public domain.

Having worked for several years at a provincial *zemstvo* clinic in Tula, in 1916 Shorokhova moved to Tashkent, the Empire's main outpost in Central Asia, where her husband, military doctor and pathologist Stefan Shorokhov, had received a new appointment. The same year she organized the first gynecological service at the city's hospital, located in the "new town" area (Shorokhova 1970: 7). Typical for a colonial city, Tashkent built a modern European-style settlement for its Christian population, adjacent to the existing "old town" center (Sahadeo 2010). The gynecological department quickly filled up with Russian patients, but it took some time and effort, recalled Shorokhova, to earn the trust of local families so they would bring their young women to a "European" clinic for help in childbirth. If a Muslim family was reluctant, Shorokhova visited patients in Tashkent's "old town", where in December 1916 she performed her first Cesarean section on an Uzbek woman and saved the mother's and the child's lives. Horrified by the lack of gynecological services and "barbaric" traditional practices, she saw her main goal as establishing a scientific, "European" gynecology for women in Tashkent of all faiths and ethnicities (Shorokhova 1970; Shadmanova 2017).

These personal desires of hers coincided with the program of the new, post-revolutionary Soviet state toward a modernizing, anti-racist, and anti-colonial nation-building process in Central Asia, especially after the establishment of the Uzbek Soviet Socialist Republic in 1924. The Revolution of 1917 also removed all formal discriminatory restrictions on women's education and careers. Within two decades, Soviet medicine became a majority-female profession. The glass ceiling and other informal inequities persisted for much longer, but by the late 1930s, some individual women were rising to the very top of the academic hierarchy in the Soviet medical sciences (Dolgova and Streltsova 2019). In 1923, Shorokhova defended her doctoral dissertation at the newly established Central Asian University in Tashkent. In 1930, she helped organize the first specialized gynecological hospital in Central Asia and, in 1933, became the second woman in the Soviet Union to be awarded the title of professor of gynecology. Until she retired in 1951 at the age of 70, Shorokhova spearheaded the establishment of the field of professional gynecology in Uzbekistan. Altogether, in 54 years of service, she performed more than 32 thousand serious operations and received high official recognition for her work. Shorokhova died in Tashkent in 1979 aged 98, and her personal papers are preserved in the Central State Archive of Scientific, Technical and Medical Documentation of the now-independent Republic of Uzbekistan (Shadmanova 2015).

As part of her job, and in accordance with official Soviet policies, Shorokhova administered abortions. Her main ethical and research interests, however, were directed toward curing infertility and establishing women's rights to motherhood. Already in 1917, she commenced experiments with artificial insemination in the hopes of alleviating "women's misfortunes" of not being able to have children (Shorokhova 1923: 55). By that time, artificial insemination had been developed as a reproductive technology and used primarily for large, domesticated animals, especially horses (Ivanov 1922; Rossiianov 2002). One of the top international experts and inventors in the field, Professor Ilya Ivanov, headed a laboratory in St. Petersburg, where Shorokhova learned the technique in 1911. Obviously, she was already then thinking about applying it to women, despite the existing restrictions. Shorokhova cited Ivanov's and her own experiments to argue that the method of artificial insemination was safe and efficient also for humans, capable of producing normal and healthy offspring and helping the cause of women's liberation, not only as a way to treat infertility but as a possible substitute for sexual intercourse as such.

In her first publication (1923: 68), Shorokhova called "the desire to have children a natural, inalienable right of every woman", which could be ensured with the sperm of an unrelated, anonymous donor, outside of marriage, and without any need to enter into a relationship with male partner. The following year, speaking at the sixth All-Union Congress of Gynecology and Midwifery, she again discussed this right and asked rhetorically, "is there any reason to refuse artificial insemination to women who want to have babies without intercourse?" (Shorokhova 1925: 420). She mentioned female patients who wanted her professional help in getting pregnant while avoiding intimacy with men. "The happiness of motherhood", she complained, "is often acquired by women at a very high price, the loss of personal freedom". In revolutionary Russia, the first marriage code of 1918 radically liberalized both marriage and divorce, which could now be granted easily and on request of either partner. It also abolished any legal distinction between babies conceived in and outside of a formal marriage, eliminating the category of "illegitimate children" altogether. These legal innovations provided the context and possibility for Shorokhova's discourse on extending the concept of women's emancipation further into the domain of reproductive rights.

A summary of her research published in French described 50 successful cases of human birth achieved via artificial insemination (Shorokhova 1927). A German commentator underscored as an unusual fact that three of her cases were reported to have used the sperm of an anonymous donor (Geppert 1928). But Shorokhova could have additional reasons to resort to a donor's sperm to help her female patients, Russian as well as Uzbek. As a gynecologist, she knew from her research that a couple's infertility was at least as often caused by the male as by the female partner, but that in traditional patriarchal families the blame invariably would be laid on the woman. Her desire to help women avoid such personal tragedies strengthened her resolve to view artificial insemination as a woman's right. She also referred to cases when local men resisted and prevented their spouses, sometimes violently, from attending the Tashkent women's club where they could hear lectures related to female hygiene and health (Shorokhova 1970: 8).[7]

In her publications, Shorokhova mentioned a possible eugenic utility of artificial insemination. Indeed, insemination by a donor's sperm was the key element of proposals for positive eugenics, compatible with socialism, formulated by Serebrovsky in 1929 and by Hermann J. Muller in the 1930s.[8] Socialist eugenicists could use neither the privilege of class nor racial differences as indicators for the quality of genetic material. Serebrovsky did not specify what would qualify as "recommended sperm", whereas Muller metaphorically alluded to intellectual capacities, exclaiming that "many future mothers, liberated from the shackles of religious superstitions, would be proud to mix their own plasm with the plasm of Lenin or Darwin, to give the society a child with their inherited biological qualities" (Babkov 2008: 532, 684). Shorokhova could not be aware of the latter suggestion, but in 1929, she designed a physiological, rather than racial or social, criterion for the eugenic quality of sperm: the agility of spermatozoids, which she measured by observing the speed of movement in a specially designed capillary device (Shorokhova 1929). Even after the 1930 ban on eugenics in the Soviet Union, the simple test on agility, deprived of its eugenic connotations, remained the physiological standard for deciding and choosing which donor sperm could ensure a higher probability of successful conception in artificial insemination.

Legalization of Abortions: The First Hundred Years

An entitlement is perceived most acutely when it is lost, which happened to Soviet women in 1936 with another government decree that re-criminalized abortions except when recommended by doctors on medical grounds. After 15 years of legalized practice, many women took it for granted as an actual right to decide and choose the most appropriate time for conceiving babies, as revealed by many letters they sent to newspapers during the discussion of the proposed new law (Chatterjee 1999; Lapidus 1978: 113). Although the editors preferentially published opinions that supported the government initiative, correspondence received by them reflected strong opposition to the proposed ban and defense of the existing practice. These responses chronologically coincided with high-profile public discussions on the fundamental rights and obligations of men and women as proclaimed in the draft of the new Soviet constitution of 1936. The official and publicly used justification for banning abortions once again invoked the need to protect women's health but reversed the logic of the 1920 decree and lamented the residual medical harm from clinical abortions, rather than from underground, unprofessional practice. The government reverted from earlier libertarian ideals to the promotion of more traditional family values and proclaimed that economic hardships that made abortions unavoidable in the 1920s had been resolved by the successes in economic construction of socialism, social welfare, and support for the protection of motherhood and infancy. Later historians have usually interpreted the measure as resulting from concerns about declining fertility and a slower than anticipated population growth, although the demographic crisis in the 1930s, caused by Stalinist collectivization and urbanization, was not as catastrophic as in 1920.

The window of opportunity during which abortions were legal defined the period of weakening of many traditional restrictions in family and gender practices and of experimental openness for some previously unimaginable investigations in reproductive biology. Our analysis of these developments helped reveal the changing roles, possibilities, and limitations for women in the cultural transformations of the 1920s, both as objects of social and technological control and also increasingly as agents of science and of their own emancipation. For example, in 1929, when male biologists sought volunteers for artificial insemination with the sperm of an orangutan, they looked for ideologically committed and sexually liberated women, without considering, however, what possible moral burden and consequences their experimental subjects would have to struggle with (Rossiianov 2002). In Danchakova's experiments, female donors of embryos remained practically invisible. In any case, her research did not exert much influence on the already widespread practice of abortion. Her publications, while generally supportive of women's social and sexual emancipation and experimenting with the transformability of sex, took abortions for granted, as an existing norm. Combined with interest in "positive eugenics" (Adams 1990), this encouraged her scientific investigations aimed at a radical biological transformation of human beings, such as the use of fetal organs and tissues in experiments on directional modification of embryological development.

Shorokhova's experiments were designed to develop a new practice of artificial insemination, complementary to existing abortions, which she saw as a liberating technology for women to have babies independently or outside of sexual relations with men. These ideas were possibly inspired by early Russian feminists' speculations regarding the separation of procreation from sex and the possibilities of asexual reproduction (Kochetkova 1915). When male eugenicists, such as Serebrovsky or Muller, discussed experiments on artificial insemination, they envisioned a plan in which the experts—geneticists—would have the main authority to select the eugenically appropriate sperm. To Shorokhova, the relatively simple and available technique of artificial insemination opened a way toward empowering women with control over their own reproductive options and criteria, including the choice of donor's sperm.

By 1931, eugenics was declared ideologically unacceptable and effectively censored in the Soviet Union. Even before the Nazis came to power in Germany and enacted their eugenic legislation, Soviet authors had diagnosed the hegemonic American and European eugenical movements as pseudoscientific, irredeemably built upon racist or class prejudices, while pretending to naturalize and legitimize the latter as biological, hereditary hierarchies. Thereafter, the official Soviet discourse proclaimed that the main problems of modern societies had social rather than biological causes and rejected discriminatory racial hygiene in favor of an inclusive social hygiene. But the 1930s international swing to the right in cultural norms and policies had its counterpart in Soviet conditions, too, usually referred to as the "Great Retreat", the extent and causes of which are still a subject of historical debate. A comparison and rivalry with the fascist movement, whose international appeal and ability to attract and mobilize masses were then growing much faster than for its communist enemy, can help to sort out the complex, layered structure of Soviet ideological adaptations during that decade. On some issues of principled ideological and political importance for their project, Soviet spokesmen continued expressing, openly and vehemently, their absolute and categorical opposition to fascism. With regard to women and reproduction, this concerned the wholesale rejection of eugenics, racial hierarchies, and biological reductionism, and insistence on women's equal rights in education, social and public spheres. On other issues deemed less ideologically crucial, the USSR tacitly or not so tacitly retreated from revolutionary radicalism toward what could be characterized as the traditional cultural mainstream of the era. This category included various aspects of gender relations and family life, which had played a much less central role for Soviet ideology. The ensuing return to more conservative patterns and international norms of the time brought about a return to stricter rules for marriage and divorce, endorsement of "family values", parental authority, sexual Puritanism, and recriminalization of male homosexuality and of abortions.

These trends largely continued for the rest of Stalin's years (Nakachi 2008) and meant a major reduction of expectations for the earlier project of gender equality formulated by Kollontai and her followers. Kollontai became mostly silent in public throughout these years, retreating to her quiet diplomatic duties. The window also closed, for the most part, on pioneering and often risky lines of experimentation which were redefining and violating the traditional demarcation between sexes, between animals and humans, and between what was considered "natural", and therefore normal and moral, and "artificial", and possibly dangerous in the area of human reproduction. Yet, some of the practices did not completely disappear and remained preserved in memory, even if no longer allowed to be exercised or discussed openly. It is hard to tell whether Shorokhova continued to help her patients with artificial insemination, as the Soviet media kept a decades-long silence about the method.[9] Abortions continued to be performed illegally and semi-legally but with much less information available publicly. Many couples who had not bothered to register their relationships during the libertarian period continued to live in de-facto unregistered or open partnerships even after the legal marriage code became much stricter. And Kollontai's earlier publications were not forgotten either.

Abortions were legalized for the second, and final, time in the Soviet Union in 1955. Maria Kovrigina, the minister of public health, and other women who made possible this correction understood the right to abortion as a socialist practice, one of the core Soviet values which she as a young student internalized ca. 1930. For her, restoring this right was part of the general "return to revolutionary and Leninist norms" following the corrupting excesses of Stalin's period (Nakachi 2008; Talaver 2020). This time around, Soviet experience was generally attracting much more attention from the rest of the world. The case of abortions, too, helped trigger a much stronger international following. Several socialist countries in Europe instituted similar legislation almost immediately, and during the rebellious 1960s, the movement for legalizing abortions also spread to Western Europe and North America (Roemer 1967). The second wave of legalization did not bring back eugenics, which had been thoroughly discredited by the Nazi abuses and remained effectively banned in the Soviet Union and elsewhere. But as the abortion reform widened internationally, it was accompanied by a renewed and much stronger discourse on women's rights and by possibilities for more advanced, if not necessarily bolder, experimentation with human reproduction.

Notes

1 A generation later, Kollontai's main ideas would form the core of the "second wave" of Western feminism. In her own time, however, she did not call herself a "feminist", but used the term dismissively as applying only to the much less radical "bourgeois feminism" of the early twentieth century. She viewed her own communist project of women's emancipation as a much bolder revolution in gender norms and also an inseparable part of class struggle, realizable only together

with—not in competition with—men, as a mutual liberation in the process of building socialism (Kollontai 1977; Stites 1978).

2 The term "evil" in the text by medical (atheistic) authorities should be understood as anti-social, i.e., "social evil" damaging the interests of society. See a commentary on the logic of the decree by Vera Lebedeva, head of the Department on the Protection of Maternity and Infancy (Lebedeva 1927: 88–89).

3 Danchakova (or Danchakoff as she spelled her name in international scientific publications) denied that she was a student of Maksimov. The complicated history of the discovery and rediscovery, and reinterpretations of stem cells, and the contributions from several researchers, are still awaiting detailed investigation.

4 Gender equality in higher education advanced rapidly in 1920's revolutionary Russia, including complete co-education at the undergraduate level and efforts to increase the number of female graduate students. At professorial positions, women were still rare, but in 1925, the physiologist Lina Shtern also returned to Russia and by 1939 would become the first woman elected to the highest academic rank of full member of the Soviet Academy of Sciences.

5 Her main patron appears to have been the mathematician and astronomer Vladimir Kostitsyn, who directed the Scientific Directorate at the People's Commissariat of Enlightenment and had invited her to the Soviet Union but left the country in 1928 (Kostitsyn 2017).

6 See also a similar reinstatement of cultural taboos in the case of Ilya Ivanov's cross-breeding experiments (Rossiianov 2002).

7 She alluded to cases related or similar to those described in (Northrop 2003; Kamp 2011) as violent resistance to the unveiling campaign initiated by radical female activists in the late 1920s.

8 On the history of Soviet eugenics see (Adams 1990; Babkov 2008).

9 There seem to be no Soviet publications on artificial insemination for families between (Imerlishvili 1933) and (Vozvrashchenie 1969). The USSR health ministry officially approved the practice only in 1981 (Parashchuk 1987).

References

Adams, Mark B., ed. 1990. *The Wellborn Science: Eugenics in Germany, France, Brazil, and Russia.* Oxford: Oxford University Press.

Babkov, Vasilii V. 2008. *Zaria genetiki cheloveka: Russkoe evgenichskoe dvizhenie i nachalo meditsinskoi genetiki.* Moscow: Progress-Traditsiia.

Belkin, Rafail A. 1931. "Vystuplenie. Stenogramma soveshchaniia direktorov nauchno-issledovatelskikh institutov Glavnauki Narkomprosa RSFSR." *Central State Archive of Russian Federation (GARF).* Collection. A-2307. Dossier 16. Folder 7. P. 19–20 ob.

Chatterjee, Choi. 1999. *Soviet Heroines and Public Identity, 1930–1939.* Pittsburgh: University of Pittsburgh.

Danchakoff, Vera. 1907. *K voprosu o neirofibrilliarnom apparate nervnykh kletok i ego izmeneniiakh pri beshenstve.* Saint Petersburg: Stasiulevich.

Danchakoff, Vera. 1908. "Untersuchungen über die Entwicklung von Blut und Bindegewebe bei Vögeln." *Archiv für mikroskopische Anatomie.* 73: 117–181.

Danchakoff, Vera. 1909. "Über die Entwicklung des Knochenmarks bei den Vögeln und über dessen Veränderungen bei Blutentziehungen und Ernährungsstörungen." *Archiv für mikroskopische Anatomie.* 74: 855–926.

Danchakoff, Vera. 1924. "Wachstum transplantierter embryonaler Gewebe in der Allantois." *Zeitschrift für Anatomie und Entwicklungsgeschichte.* 74: 401–431.

Danchakoff, Vera. 1927. "Russia's New Marriage Code." *Current History* (May 1927), 187–189.

Danchakoff, Vera. 1928. "Annual Reports of the Laboratory of Experimental Morphogenesis," *Archive of the Russian Academy of Sciences.* Fond 356. Opis' 1. Dela 83, 89.

Danchakoff, Vera. 1931. "Autobiography." *Central State Archive of Russian Federation (GARF).* Collection. A-2307. Dossier 23. Folder 95. P. 23–24 ob.

Danchakoff, Vera. 1938a. "Das Hormon im Aufbau der Geschlechter." *Biologisches Zentralblatt.* 58: 302–328.

Danchakoff, Vera. 1938b. "Gewebeplastizität, Hormone und Geschlecht." *Ergebnisse der Physiologie, biologischen Chemie und experimentellen Pharmakologie.* 40: 101–163.

Danchakoff, Vera. 1941. *Der Aufbau des Geschlechts beim höheren Wirbeltier.* Jena: Fisher.

Danchakoff, Vera and Agafangel Gagarin. 1929. "Embryoherz in der Chorio-Allantois des Hühnchens." *Zeitschrift für Anatomie und Entwicklungsgeschichte.* 89: 754–762.

Dolgova, Evgenia A. and Ekaterina A. Streltsova. 2019. "'Dobro pozhalovat' v klub': Polozhenie zhenshchin v sovetskoi nauke 1920-kh gg." *Sotsiologicheskie Issledovaniia.* 2: 97–107.

Fando, Roman A. 2020. "'Delo professora V.M. Danchakovoi,' ili neprostye gody russkoi amerikanki v strane Sovetov." *Voprosy istorii estestvoznaniia i tekhniki.* 41: 244–279.

Gaissinovitch, Abba E. 1980. "The origins of Soviet genetics and the struggle with Lamarckism, 1922–1929." *Journal of the History of Biology.* 13: 1–51.

Gens, Abram. 1925. *Problema aborta v SSSR.* Moscow: Gosmedizdat.

Geppert. 1928. "La fécondation artificielle dans l'espèce humaine." *Deutsche Zeitschrift für die gesamte gerichtliche Medizin. Referate.* 11: 122.

Goldman, Wendy. 1993. *Women, the State and Revolution: Soviet Family Policy and Social Life, 1917–1936.* Cambridge: Cambridge University Press.

Il'ina, Viktoriia V., 2019. "Osobennosti sovremennoi rossiiskoi sotsial'noi reklamy" *Kommunikologiia* 4: 46–64.

Imerlishvili, Ya. 1933. "Sluchai iskustvennogo oplodotvoreniia u zhenshchin." *Zhurnal akusherstva i zhenskikh boleznei.* 44: 380–384.

Ivanov, Ilya I. 1922. "On the Use of Artificial Insemination for Zootechnical Purposes in Russia." *Journal of Agricultural Science (London).* 12: 244–256.

Kamp, Marianne. 2011. *The New Woman in Uzbekistan: Islam, Modernity, and Unveiling under Communism.* Seattle: University of Washington Press.

Keller, Evelyn Fox. 1997. "Developmental Biology as a Feminist Cause?" *Osiris.* 12: 16–28.

Koblitz, Ann Hibner. 1988. "Science, Women, and the Russian Intelligentsia: The Generation of the 1860s." *Isis.* 79 (2): 208–226.

Kochetkova, Lidiya P. 1915. *Vymiranie muzhskogo pola v mire rastenii, zhivotnykh i liudei.* Moscow: V.M. Sablin.

Kollontai, Alexandra. 1977. *Selected Writings.* New York: Norton.

Kostitsyn, Vladimir A. 2017. *"Moe utrachennoe schast'e...": Vospominaniia, Dnevniki.* Moscow: Novoe literaturnoe obozrenie.

Lapidus, Gail. 1978. *Women in Soviet Society: Equality, Development, and Social Change.* Berkeley: University of California Press.

Lebedeva, Vera P. 1927. *Proidennye etapy. Stat'i i rechi.* Moscow: Okhrana materinstva i mladenchestva.

Maienschein, Jane. 2005. "Epigenesis and Preformationism." *Stanford Encyclopedia of Philosophy* (online). https://plato.stanford.edu/entries/theories-biological-development/

Maximow, Alexander. 1908/1909. "Untersuchungen über Blut und Bindegewebe." *Archiv für mikroskopische Anatomie.* 73: 444–561; 74: 525–625.

Nakachi, Mie. 2008. *Replacing the Dead: The Politics of Reproduction in the Postwar Soviet Union, 1944–1955.* PhD Dissertation. University of Chicago.

Northrop, Douglas. 2003. *Veiled Empire: Gender and Power in Stalinist Central Asia.* Ithaca: Cornell University Press.

Parashchuk, Yury S. 1987. *Iskusstvennaia inseminatsiia spermoi donora pri lechenii besplodiia.* PhD Dissertation. Kiev.

Roemer, Ruth. 1967. "Abortion Law: The Approaches of Different Nations." *American Journal of Public Health and the Nation's Health.* 57: 1906–1922.

Rohleder, Hermann O. 1911. *Die Zeugung beim Menschen. Eine sexualphysiologische Studie aus der Praxis. Mit Anhang: Die künstliche Zeugung (Befruchtung) beim Menschen.* Leipzig: Georg Thieme.

Rossiianov, Kirill. 2002. "Beyond Species: Ilya Ivanov and His Experiments on Cross-Breeding Humans with Anthropoid Apes." *Science in Context.* 15: 277–316.

Sahadeo, Jeff. 2010. *Russian Colonial Society in Tashkent, 1865–1923.* Bloomington: Indiana University Press.

Shadmanova, Sanobar B. 2015. "Professional and Personal Life of Doctor of Medical Sciences, Professor A.A. Shorokhova." *IX World Congress of the International Council for Central and East European Studies.* Makuhari, Japan.

Shadmanova, Sanobar B. 2017. "Meditsina i naselenie Turkestana: Traditsii i Novatsii (konets XIX – nachalo XX vv.)." *Istoricheskaia Etnologiia.* 2: 119–139.

Shorokhova, Antonina A. 1923. "Iskustvennoe oplodotvorenie u liudei." *Turkestanskii meditsinskii zhurnal.* No. 1–2: 55–68.

Shorokhova, Antonina A. 1925. "Iskustvennoe oplodotvorenie u liudei." *Trudy VI s'ezda Obshchestva vsesoiuznykh akusherov i ginekologov.* Moskva 1–6 Iiunia 1924. 420–428.

Shorokhova, Antonina A. 1927. "La fécondation artificielle dans l'espèce humaine" *Gynécologie* et *Obstétrique.* 5: 132–139.

Shorokhova, Antonina A. 1929. "Novye puti v selektsii cheloveka i mlekopitaiushchikh. Doklad na Vsesoiuznom s'ezde po genetike i selektsii 15 ianv. 1929 g." *Vrachebnaia gazeta.* 3–4: 180–184.

Shorokhova, Antonina A. 1970. *Tak Bylo.* Tashkent: Meditsina.

Squier, Susan. 1994. *Babies in Bottles: Twentieth-Century Visions of Reproductive Technology.* New Brunswick, N.J.: Rutgers University Press.

Stites, Richard. 1978. *The Women's Liberation Movement in Russia: Feminism, Nihilism, and Bolshevism, 1860–1930.* Princetown: Princeton University Press.

Talaver, Sasha. 2020. "When Soviet Women Won the Right to Abortion (For the Second Time)" *Jacobin* (8 March 2020) (online). https://jacobin.com/2020/03/soviet-women-abortion-ussr-history-health-care

"Vozvrashchenie materinstva." *Literaturnaia gazeta*, 21 May 1969.

5 Governing Psychiatry

The Importance of Networks for Brazilian Women Psychiatrists, 1941–1970

Valentine Mercier and Ygor Martins

During the 1920s and 1930s, seven women practiced psychiatry in all of Brazil. Because they represented the first generation of women psychiatrists, studying their professional paths shines a light on a new configuration of gender diversity within the profession. Inspired by prosopography, as a collective study of their careers (Stone 1971), we emphasize their common characteristics; understanding them as a group sharing salient features of *governance*, defined as an act of managing science through financial, organizational, practical, theoretical, and human resources (Brooks 1968). As they directed hospitals and clinics and obtained university chairs, these women psychiatrists were the first to assume positions of power and responsibility that permitted them to make crucial decisions in their field and gain empowerment for themselves and others.[1] *Governance* is an interesting tool of analysis for tracking their career paths. The way these women psychiatrists got to their positions, and stayed there, was as much a stake for women in the *governance* of psychiatry, as the choices they made once they had the power to decide. Throughout this process, the mobilization of networks intersected with different facets of *governance* as part of each woman's agency (Butler 1990).

From this first cohort of women psychiatrist's need to gain legitimacy, to their efforts to implement specific policies and to choose who to work with,the archives we explored[2] revealed a great range of professional, personal, and political contacts that served their careers, weighed in their positions, and testified about their preferences; giving evidence of their strategies and constraints in the act of *governance*. In line with concepts from the histories of science and of institutions, this paper portrays these Brazilian women psychiatrists as situated at the intersection between institutions that inculcate a "thought style" (Fleck 2005), collectives that embody these values while constituting solidarity networks, and the dynamics of societal transformation to which they contribute (Douglas et al. 2004).

In Brazil, the professionalization of women was shaped by important political changes regarding health policies and women's place in society. In 1941, Getúlio Vargas, *Estado Novo* movement President of Brazil, created the Serviço Nacional de Doenças Mentais (SNDM, National Service for Mental Diseases), dedicated to psychiatry among other health administrations, which

DOI: 10.4324/9781003562597-8

increased the number of beds and professionals. This policy was retained after the end of the *Estado Novo* dictatorship in 1945, until the establishment of a military government in 1964, which relegated public health to the background. Overall, SNDM policies created opportunities for women. Some of them became official doctors at SNDM institutions. The SNDM also impacted women psychiatrist's careers by encouraging the development of psychoanalysis, praxis therapy, and outpatient services. Women psychiatrists were generally more familiar than men with these practical and theoretical options, not only in Brazil (Martins and Mercier 2021; de Alameida Moura 2019; Ferreira Abrão 2018) but also in the United States and Europe (Borgos 2019; McGovern 1984). These changes occurred simultaneously with changes in women's social position. Thus it is in the tension between emancipation and the persistence of rigid gender norms that the professional careers of psychiatrists and their public reputations must be considered.

Very little has been written about women psychiatrists in Brazil (Facchinetti and Venancio 2018), and in international histories of institutions in general, which have focused on the relation between psychiatry and the construction of nation-states (Huertas García-Alejo 2012; Ablard 2008; Ríos Molina and Giraldo Granada 2017; Agostini 2008; Zulawski 2007; Lambe 2017; Meyer 2017; Favier 2021). There are exceptions like Felicia Gordon's portraits of French psychiatrists Constance Pascal and Madeleine Pelletier (Gordon 1990, 2013). In an effort to fill this gap, we will analyze how scientific networks formed in the 1930s around recognized male professionals proved indispensable to the advancement of women psychiatrists' careers between the 1940s and 1960s. These scientific networks were intertwined with political ones, fundamental in understanding psychiatrists' presence in the public arena. Finally, specifically female associative networks represented a source of considerable support for psychiatrists, who in turn tended to encourage the presence of other women in their professional circle.

Essential Male Scientific Networks

In the medical field in general, and in the psychiatric field in particular, all psychiatrists stood side by side with individuals who composed their "thought collective" (Fleck 2005). Women psychiatrists were often associated with men who already held positions of high professional prestige. By connecting with these prestigious names, women accumulated the political impetus necessary for them to act effectively and implement *governance* projects of their own. Networks with recognized male scientists were a part of Brazilian women psychiatrists' agency (Butler 1990). By forging links with recognized figures, women psychiatrists could gain their trust, assert their skills, and benefit from the influence that their recognition could generate. More concretely, these relationships could also open doors to specialized training courses, medical societies, or scientific journals in which to publish. For example, Antônio Austregésilo (1876–1960), chair of Neurology at the

Faculdade de Medicina do Rio de Janiero (FMRJ), founder of the periodicals *Arquivos Brasileiros de Medicina* and *Arquivos Brasileiros de Neurologia e de Psiquiatria*, author, member of the Academia Brasileira de Letras, and politician (Teive et al. 1999), was indispensable for articulating interactions that some of these women psychiatrists implemented during the 1940s. His importance is particularly interesting as he was close to the biologist, feminist, and president of the Federação Brasiliera pelo Progresso Feminino, Bertha Lutz. As such, he introduced psychiatrist Nise da Silveira to her.[3] He played this role as a part of his political convictions regarding women's place in society and science.

> Nise da Silveira (1905, Maceió—1999, Rio de Janeiro). Graduated in 1926, internship with Austregésilo in 1933, director of the Occupational and Rehabilitation Therapy Service of the National Psychiatric Center between 1946 and 1975, and director of *Casa das Palmeiras* from 1956 to the 1970s.

In particular, for three of the psychiatrists under analysis, Dr. Austregésilo was the mentor responsible for advancing the first steps of their careers. He assisted in channeling Alice Marques dos Santos's, Nise da Silveira's, and Eurydice de Magalhães Borges Fortes's interests into neurology and psychiatry. Dr. Silveira's case has already been explored to some extent by historians (Sales Magaldi 2018; Melo 2005; Fernandes 2015). After she completed an internship with Austregésilo in 1933 (Mello 2014), Dr. Silveira relied once again on collaboration with her mentor when she was reinstated as a civil servant in 1944. In 1936, on charges of subversion associated with communism, Dr Silveira was imprisoned for 18 months and, when released, she remained in partial hiding with her husband until 1944. With Austregésilo's encouragement, Dr. Silveira published two articles in the *Revista de Medicina, Cirurgia e Farmácia*, in which Dr. Austregésilo also published an original article, and in *Cultura Médica*; both journals of the Brazilian medical scientific establishment at that time. Thus, Nise da Silveira mobilized the medical networks she had composed in previous moments to re-establish herself professionally and legitimize her performance in the new social context in which she found herself.

> Alice Marques dos Santos (1911, São Gonçalo—1996, Rio de Janeiro). Graduated in 1933, assistant of Austregésilo in the 1930s, psychiatrist at the Centro Psiquiátrico Nacional (CPN) from the 1940s to the 1960s, director of Odilon Galloti Hospital from 1964 to the 1970s.

More recently, one of us analyzed Alice Marques dos Santos's interactions with Dr. Austregésilo (Martins 2022). Dr. Santos was the first woman in Brazil to direct a psychiatric hospital, in the 1960s, and had already been described in the newspapers in the second half of the 1930s as

> "the illustrious assistant to Professor Austregésilo".[4] Thus, the beginning of her professional career was in line with the intellectual field in which her mentor stood out: neuropsychiatry. Therefore, the first years of her socio-professional career were linked to the networks provided by the "the father of Brazilian neurology.
>
> (Teive et al. 1999)

The cross-references about these women psychiatrist's relations with Dr. Austregésilo have allowed researchers to assess his role in advancing their careers. In Eurydice de Magalhães Borges Fortes's case, there is still more research to be done. However, we know that Dr. Borges remained very close to Dr. Austregésilo's group when she was his assistant, as an item in *Jornal do Brasil* indicated.[5] Her husband[6] and long-time professional partner, the psychiatrist Dr. Ary Borges Fortes, was also part of that circle[7] and her success as a doctor may have been linked to these connections. Thus, from the point of view of the future career that these three women would accomplish, Austregésilo stood out as the driving force.

> Eurydice de Magalhães Borges Fortes (1911, State of Rio de Janeiro—?). Graduated in 1934, student of Austregésilo in the 1930s and his assistant in the 1940s, assistant professor of clinical neurology at the FMRJ from 1938, professional at the Merchants' Retirement and Pension Institute, and private practice in the 1960s.

It is also necessary to consider the socio-political background in which these interactions took place. In Brazil, women from the urban middle class began to have greater opportunities in the 1940s. This was particularly true in Rio de Janiero. Women entered the tertiary sector, some of them as scientists (Azevedo and Ferreira 2006; Rossiter 1984). A women's press developed (Bassanezi Pinsky, and Pedro 2013; Zuckerman 1998). Women's politicization widened, and a mass movement was formed (Mercier 2020). Although sometimes repressed, despite a more democratic period that stretched from 1945 to 1964, this mobilization was still influential. Thus, among other social changes underway, women could work in careers such as mental health.

This movement toward increased career opportunities for women in mental health also took place in other national contexts, as it had earlier in France with Dr Constance Pascal (Bourgeois 2015; Gordon 2013). On an international level, it was also evident that interactions with men as mentors were central to the establishment of women within mental health. When compared to the history of psychiatry, the history of psychoanalysis was very illuminating. Around Sigmund Freud, for example, women like Melanie Klein (Grosskurth 1986) and Anna Freud (Young-Bruehl 2008) organized their professional careers. Similarly, psychoanalysts such as Antônio Austregésilo in Brazil surrounded themselves with women psychiatrists who became part of their social networks. From this connection, these women psychiatrists were able to build, develop, and establish their clinical projects and their *governance* formats.

Partisan and Nationalist Networks

For some of these women psychiatrists, direct connections with politicians and members of political parties were also part of their social networks; which may have made a difference in the projects they pursued. Recent research shows how political matters influenced the *government of science* by encouraging certain domains of research (Lamy 2018; Pestre 2014; Rindzeviciute 2016; Nieto-Galan 2019). As already noted, psychiatry in particular was deemed necessary for the construction of nation-states. Thus a part of Brazilian women psychiatrists' agency involved broad connections with political groups. All seven of the women psychiatrists studied here thought of their work in connection with a political project. This allowed them to integrate into networks that legitimated them, or which they themselves legitimized once they had acquired a position of power.

Two types of integration into the networks that would propel their careers can be distinguished among these seven women psychiatrists: partisan integration versus more broadly ideological integration. Nise da Silveira was a good example of the first case. Her trajectory was initially linked to her militancy in the Partido Comunista do Brasil (PCB). Other psychiatrists were also members of the PCB and could support Silveira's career. In the late 1930s, communism suffered repression and the police noticed Silveira was in correspondence with Osório César,[8] a communist psychiatrist who had worked at Juqueri Hospital in São Paulo since 1925. Osório César was already famous in the 1930s. He had published his principal work *A expressão artistica nos alienados: Contribuição para o estudo dos símbolos na arte* (*Artistic Expression of the Insane – Studies of Symbols in Art)* (Cesar 1929), when Nise da Silveira was just starting her career as a psychiatrist. His importance became evident when Nise da Silveira returned to public service in 1944, and she became famous for developing artistic expression among patients, which had also been part of Cesar's work.

Second, some political connections were more broadly ideological. During the first half of the twentieth century, public health policies in Brazil involved significant hygienic discourse, as was also the case in Europe and the United States. This hygienic discourse influenced mental health and temperance movements in particular (Bourdelais 2001; Hochman 2012; Rorabaugh 2018; Tyrrell 2010). The regulation of public health policies involved the support of scientific intellectuals who produced eugenic theories applicable to the Brazilian nation. At the same time, the political utility of hygienic, mental health, temperance, and eugenic discourses thrust psychiatrists into the limelight and legitimated their discipline (Meyer 2017). For women psychiatrists, involvement in Brazil's social and political development was a way to integrate themselves into the profession.

The *governance* of psychiatry was thus closely associated with the politics of the nation. For example, Juana Mancusi de Lopes found in the fight against alcoholism a fundamental framework for the social development of Brazil. She was part of the Liga Brasiliera de Higiene Mental (LBHM), a civic association recognized through federal funding (Anderson Seixas et al. 2009). The LBMH was composed of the most important Brazilian psychiatrists: among them, its founder Gustavo Riedel, Henrique Roxo, and Ernani Lopes, Juana Mancusi's husband, all of whom directed public psychiatric institutions. Thus, during LBMH meetings, Juana Mancusi rubbed shoulders with renowned psychiatrists and made a place for herself among them. At the same time she forged links with politicians. For example, the Organizing Committee of the Anti-Alcohol Week, scheduled for September 1932,[9] of which Juana Mancusi was a member, noted that the Minister of Education and Public Health, Washington Pires, had given his support to the event, and had promised to attend the inaugural session to be held at the Academia Nacional de Medicina.

Juana Mancusi de Lopes (1894, Buenos Aires—?). Graduated in 1917, gynecologist at the Settlement Hospital for Insane Women from 1918 to the 1940s, Chief Gynecologist in the same hospital from 1933, SNDM psychiatrist from 1943 on.

Thus, Juana Mancusi became influential in the fight against alcoholism, first as vice-president[10] and then as president of the pro-temperance league[11]; and as organizer of and participant in yearly anti-alcohol weeks on many occasions.[12] Internationally, she represented Brazil at the 1956 Pro-Temperance Congress[13] and thus made the issue a main point of action in her political agenda. The *governance* of psychiatry for Juana Mancusi de Lopes was inseparable from integration into networks that went beyond this discipline alone to interact with broader political issues.

The political dimension of these women psychiatrists' trajectories was manifested in the support they gave to certain projects, and from the position of power established by their careers. While the dynamics of this network are harder to describe due to a lack of sources, it seems important to mention this because it is a part of the *government of science*: the use of scientific legitimacy and social status by psychiatrists to defend their political convictions.

Ursulina Penteado Bueno (before ca. 1915, Campinas—after ca. 1960). Graduated in 1937, private practice, owner and director of the Physiotherapeutic Institute of São Paulo in the 1940s, Chief Psychiatrist of the Juvenile Ward of Juquery Hospital of São Paulo from the 1940s to the 1950s.

For example, in São Paulo since the 1940s, Ursulina Penteado Bueno had been involved in political issues. In a letter addressed to General Maurício Cardoso, she offered to make available the services of her clinic, the Instituto Fisioterápico de São Paulo, for military service coordinated by him.[14] Dr. Penteado Bueno's words expressed a nationalist tone behind her offer: "I would feel proud if, as a doctor and a Brazilian, I could render some service in favor of my country". Similarly, when Hercília Rocha Pitta ran for city councilor in Rio de Janeiro in 1947 for the Partido dos Trabalhadores do Brasil, her doctorate and her profession as a practicing physician were the main arguments for her candidacy (see Figure 5.1). Status as a scientist could serve political ambitions as a part of the stakes in the *government of science*.

Political networks were an important element in the careers of the psychiatrists presented here. Perhaps the first generation of Brazilian women psychiatrists all share this common characteristic. In comparison, the trajectory of the first two French women psychiatrists was not the same. Madeleine Pelletier (1874–1939) had a very strong political commitment to socialism, communism, anarchism, and feminism; but this commitment did not serve her professional career (Gordon 1990). As for Constance Pascal (1877–1937), Felicia Gordon has noted that, if Pascal's psychiatric activity was subversive, she herself was a model of social conformity and political discretion (Gordon 2013). Regarding feminism, Pascal preferred to become an example as a physician rather than to participate in activist campaigns. How might women's associative networks differ then, in the case of Brazilian women psychiatrists?

Figure 5.1 Hercilia Rocha Pitta. *O Cruzeiro*, edition 0013, 01/18/1947. Public Domain.

Hercília Rocha Pitta (ca. 1905—after ca. 1965). Graduated in 1931, psychiatrist at the Botafogo Polyclinic, Rio de Janeiro, responsible for the Pediatrics and Childcare Library; In-house Physician of the Economy Ministry from 1943 on.

Women's Associative Networks

It is worth noticing the importance of all-female organizations among the different forms of commitments of these psychiatrists. For Brazilian women, the *governance* of psychiatry also meant being part of all-female groups, whether in their direct professional environment or in broader networks. Because women were thought to have special, innate, and common skills; joining women-only spaces at work was expected, and granted women a certain legitimacy, from which they could implement their own practices and defend ideas. In social contexts, all-female organizations also became spaces where women could claim their merits, meet other women who could help them, and where they could become influential references. These links provided support for gaining and staying in positions of power; and these spaces became a focus for women psychiatrists putting their skills, their names, and their professions to the service of women's causes.

One of the most important women's organizations for female psychiatrists was the União Universitária Feminina (UUF). This university association was not only dedicated to women scientists; it was open to all students and graduates. Most often, the women who belonged to the UUF were licensed professionals: lawyers or doctors, engineers or biologists. Acting through moral and financial support, the UUF aimed at "coordinating and directing the efforts of women graduates of Universities and Faculties"[15] and it provided significant human and material resources for women to achieve professional success.

Being part of this network was, for some of the psychiatrists, a way to be recognized. For example, Juana Mancusi and Iracy Doyle met Bertha Lutz, the engineer Carmen Portinho, and the lawyer Maria Rita Soares at the UUF.[16] The activities carried out by the organization gave visibility to a member's work and a major professional boost. In May 1933, the UUF invited Juana Mancusi to give a conference on premarital examinations.[17] In November 1934, the UUF demonstrated its "spirit of solidarity" by celebrating "a friendship tea" with Iracy Doyle upon her successful competition to become an intern at the Psychiatry Clinic of the Rio Medical School.[18]

Iracy Doyle (1911, Rio de Janeiro—1956, Rio de Janeiro). Graduated in 1935, owner and director of the Tijuca Clinic from the late 1930s to 1956, chair of the Psychiatric Clinic at the FMRJ from 1942 to 1956.

Moreover, politically, the UUF was also a place to forge relationships. The Minister of Health and Education was present in 1935 when the new board of the UUF took office, with Iracy Doyle as vice-president.[19] Several UUF members were to be found alongside psychiatrists in other contexts, which testifies to the importance of the UUF as a node in women's professional networks. It is no coincidence that Maria Rita Soares authored several articles about Iracy Doyle[20]! In one of them, Soares defended the financial independence of women and used Iracy Doyle as a successful example of a licensed professional woman.

Carmen Portinho and Bertha Lutz, together with Iracy Doyle, were on the list of women signatories to the manifesto against the Conferência Latino-Americana de Mulheres in 1954 (Latin-American Women's Conference), a document that was read in a session of the Senate.[21] Under a presumed defense against communism, and criticism of its supposed anti-democratic bias, the signatories positioned themselves against the event. In addition, Iracy Doyle and Juana Lopes, alongside other women, represented the Conselho Nacional de Mulheres (National Council of Women) in a meeting with the President of the Republic, Café Filho in 1954.[22] Finally, after Iracy Doyle had died, the UUF set up a scholarship named after her,[23] which simultaneously showed her importance in the university women's movement, and that movement's willingness to perpetuate her importance.

Even those women psychiatrists who were not formal members of the UUF had relations with their members. Nise da Silveira had been introduced to Bertha Lutz at the UUF. Lutz supported some of Silveira's causes, such as animal protection, even when they disagreed about other political subjects: Nise da Silveira supported the Latin-American Women's Conference against which Bertha Lutz had signed a manifesto as mentioned above. On the other hand, these women's networks could be strictly professional and confined to the field of psychiatry: in the 1960s, Ursulina Penteado Bueno, for instance, was a member of a women's group set up to assist the patients of Franco da Rocha Hospital, where she worked.[24]

Despite some differences in women psychiatrists' values, these network links not only show the acute awareness they had of the significance of female solidarity but also of the ways in which networks contributed to supporting them professionally and increasing their social visibility. Conversely, psychiatrists added their status to the causes they supported. Their social visibility was sufficiently important for their names to carry weight in the legitimation of manifestos and communiqués that they signed on behalf of women's movements. Female associative networks helped these women psychiatrists in the *government of science*; which for them could also mean putting their scientific legitimacy at the service of political causes that went beyond the strict framework of their professional activity. This type of sociability probably was not specific to Brazilian psychiatrists. It might have been part of the global dynamic suggested by Anna Cabanel's work on the International Federation of University Women (IFUW), to which the UUF had been affiliated since 1931.[25]

Including Women in Professional Networks

While the networks they developed proved essential to building their careers, women psychiatrists also helped to integrate others, including many women, into their community. They did not surround themselves exclusively with recognized experts (often male when it came to psychiatry), but also included less famous women in their network, whose skills they recognized and who supported them in their projects. Networks of professional women tended to favor the presence of other women in their profession. Women used their positions to empower other women.

One important example of this phenomenon was the case of Nise da Silveira. When she created the *Casa das Palmeiras* in 1956, she surrounded herself with the following founding members: psychiatrist Maria Stela Braga, sculptor Belah Paes Leme, educator Alzira Lopes Cortes, and social worker Ligia Loureiro. Soon after, Alice Marques dos Santos joined the initiative. The all-female composition of this team can be explained by their professional profile: since women were the majority among educators and social workers at that time (Volpato 2011), the choice to integrate these professions into the *Casa* team made women's presence increasingly probable. In addition, this choice recognized the skills women would bring to the proper functioning of the project.

On the other hand, since men were the majority among psychiatrists, the presence of two women in a majority-male environment was significant. For example, Maria Stela Braga and Alice Marques dos Santos worked with Nise da Silveira at Engenho de Dentro Hospital. Their subsequent cooperation at the *Casa das Palmeiras*, outside of the place where they were first professionally socialized, testified to the strength of their relationship and the mutual trust they had in each other. Maria Stela Braga was then in her thirties and just starting her career. Nise da Silveira and Alice Marques dos Santos were older and had more experience. They may have supported Braga in a male environment such as the hospital, thus encouraging a rapprochement between them.

Moreover, the first premises for the *Casa das Palmeiras* had been given to Nise da Silveira by a friend, Alzira Lafayette, a private school principal. The initiative was exclusively female, and thus it revealed undeniable gender affinities between the participants. However, female solidarity in itself was not enough to explain this partnership between women. Other structural and cultural elements can help us now to understand it. On the one hand, the male psychiatrists in Nise da Silveira's professional network already enjoyed prestigious positions and recognition. They may have had little interest in an alternative project developed outside of official networks. Osório César, interested in patients' art and close to Nise da Silveira, was already thriving through his own activities: in the same year as *Casa das Palmeiras*, he opened the Escola Livre de Artes Plásticas at Juquery Hospital in São Paulo. On the other hand, for Nise da Silveira and Alice Marques dos Santos, *Casa das Palmeiras* was probably a way to build their own space, to extend the scope of their personal research, and to implement it autonomously without hierarchical support. For Maria Stela Braga, *Casa das Palmeiras* may have been a way of acquiring new skills, in connection with the development of dynamic psychiatry. Then again, the psychiatrists who practiced at the *Casa das Palmeiras* did so on a voluntary basis. As it was a non-profit establishment, open to as many people as possible, it depended on financial donations and on the volunteered time of healthcare professionals. Only the monitors, who ran the workshops, were paid. Volunteer work seems to have been a key factor in the female pre-eminence at *Casa das Palmeiras*. Charity was an important part of women's culture in the 1920s and 1930s, as a means for women from upper-class backgrounds to engage outside of their homes (Bassanezi Pinsky and Pedro 2013; Perrot 2007, 2005). It was not certain that only women would have accepted to work for free in an idealized vision of care, but it seemed that women's social status predisposed them to do so, making the very existence of the institution possible.

Other examples show women psychiatrist's cooperation and mutual support for each other. Alice Marques dos Santos referred her patients to the Occupational Therapy Section run by Nise da Silveira at *Casa das Palmeiras*, thus contributing to its maintenance and continuity (Mello 2014). Reciprocally, after having stayed at the Carl Jung Institute in Zurich in 1957, Nise da Silveira recommended, in a letter from 29 July 1959 to the Jung Institute's secretary, Aniéla Jaffé, that her colleague Alice Marques dos Santos, "one of the most enthusiastic members of our study group" be sent to Zurich "to attend courses and for personal analysis".[26] If the case of Nise da Silveira and Alice Marques dos Santos was particularly significant,[27] it could be generalized through the friendship and solidarity it demonstrated to the whole cohort, defining it as common practice among the women psychiatrists we have studied.

Indeed, there was a similar occurrence with Iracy Doyle. In 1953, she founded the Instituto de Medicina Psicológica (Institute of Psychological Medicine) with the psychologist Margarida Reno; and with two doctors: her brother Américo Doyle, and Henrique Novaes Filho. An equally male-female team was unusual for the management of a higher educational institution at that time. Moreover, in 1955, as a member of the Brazilian Society of Neurology, Psychiatry, and Mental Health; Iracy Doyle invited psychoanalyst Clara Thompson (1893–1958), with whom she had worked in the United States, to give a lecture during the plenary session of the association, and Margarida Reno simultaneously translated it. This, again, testified to well-functioning all-female collaborations, and to the propensity of psychiatrists to promote the circulation of women they knew in their professional environment.

Regarding Eurydice de Magalhães Borges Fortes, although her professional activity did not depend exclusively on her husband's career, she still cooperated with him closely, and no close female colleague seemed to be known to her, with whom she could have provided encouragement, or who could have valued her own work in return. Juana Mancusi's case was similar. Thus, when several women psychiatrists met in the same institution, there might have been bonds of solidarity and mutual recognition between them. However, if they did not share the same day-to-day experience in the workplace, the women did not act particularly in favor of other women. Female solidarity seemed to arise first from rare encounters in a male environment, then from the maintenance of interpersonal relationships resulting in reciprocal valuing. This distinction coincides with the private lives of these psychiatrists. Those who married another psychiatrist prioritized professional cooperation with them. Their possible female involvement was then played out in other spheres.

Conclusion

The expansion of social networks was central to the processes structuring the careers of these Brazilian women psychiatrists. In a way, this sedimentation of professional layers culminated in their arrival at the positions and functions they began to perform from the 1940s onwards. These women psychiatrists were able to recruit male and female allies within several fields of medicine, politics, and society. By bringing figures from different circles into their own areas of action, these psychiatrists surrounded themselves with relationships that allowed them to amalgamate elements into professional choices while promoting the visibility and presence of other women in their community. Women tended to fill as many scientific positions around them as possible with other women: to demonstrate their commitment to their profession; to demonstrate to colleagues their ability to advance in their careers; and to establish their own creative and innovative projects within the *governance* of psychiatry. These seven women psychiatrists reached coordination positions and thus accumulated new responsibilities and more power. Their professional trajectories were successful, and they became influential within the thought collectives and social networks they built. The composition of social networks was a crucial element for these women psychiatrists to reach the positions they occupied. Scientific, political, and women-only support for publications; and symbolic and material resources played their parts. At the same time, influential positions allowed women to help other women. Other themes—such as these women psychiatrist's intense dedication to their work and individual merits; and their investment in theoretical formulations under construction also warrant further analysis, since they are part of the strategies these women psychiatrists activated.

Notes

1 Understood as "the right to influence the direction of change, through the ability to gain control over crucial material and nonmaterial resources" (Moser 1989).
2 Brazilian newspapers *Correio da Manhã*, *Jornal do Brasil* and *Jornal do Comércio*, the magazines *O Cruzeiro* and *A Casa*, the newspaper *Imprensa Popular*, the archives of the DOPS, and those of the ETH-Bibliothek.
3 Letter from Austregésilo to Bertha Lutz, Brazilian National Archives, Reference code: BR AN, RIO Q0.ADM, COR.A928.39.
4 *Diário Carioca*, edition 02417, 03/06/1936.
5 *Jornal do Brasil*, edition 00150, 28/06/1941.
6 *O Jornal*, edition 04561, 29/08/1934.
7 *O Jornal*, edition 04529, 22/07/1934.
8 DOPS, prontuário 13–990.

9 *Correio da Manhã*, edition: 11593, 09/23/1932.
10 *Correio da Manhã*, editions 11210, 07/03/1931; 11642, 11/18/1932; 11962, 11/28/1933; 12599, 12/13/1935.
11 *Correio da Manhã*, edition 18898, 11/02/1954.
12 *Correio da Manhã*, editions 18898, 11/02/1954 and 21043, 10/29/1961; *Jornal do Brasil*, edition 00250, 10/25/1958.
13 *Correio da Manhã*, edition 19388, 06/09/1956.
14 *Correio Paulistano*, edition 26556, 10/04/1942.
15 *Correio da Manhã*, edition 10439, 01/15/1929.
16 *Correio da Manhã*, edition 20185, 14/01/1959.
17 *Correio da Manhã*, edition 11786, 05/06/1933.
18 *Correio da Manhã*, edition 12269, 11/22/1934.
19 *Jornal do Commercio*, edition 00209, 06/05/1935.
20 *Jornal do Brasil*, editions 00197, 24/08/1956 e 00257, 4/11/1956.
21 *Correio da Manhã*, edition 18824, 07/08/1954.
22 *Correio da Manhã*, edition 18894, 28/10/1954.
23 *Correio da Manhã*, editions 20270, 26/04/1959, 20403, 30/09/1959 e 20539, 12/03/1966.
24 *Correio Paulistano*, edition 32796, 16/04/1963.
25 Atria Archives, IFUW, Collection IIAV00000533.
26 ETH-Bibliothek—1056: 21567.
27 A set of letters in the custody of the Museum of Images of the Unconscious in Rio testify of their deep professional bonds and intimate friendship.

References

Ablard, Jonathan. 2008. *Madness in Buenos Aires: Patients, Psychiatrists, and the Argentine State, 1880–1983*. Calgary: University of Calgary Press.

Agostini, Claudia. 2008. *Curar, sanar y educar: enfermedad y sociedad en México. Siglos XIX y XX*. México: Universidad Nacional Autónoma de México.

Anderson Seixas, André Augusto, André Mota, and Monica L. Zilbreman. 2009. "A origem da Liga Brasileira de Higiene Mental e seu contexto histórico". *Revista de Psiquiatria do Rio Grande do Sul* 31: 82–82. https://doi.org/10.1590/S0101-81082009000100015.

Azevedo, Nara, and Luiz Otavio Ferreira. 2006. 'Modernização, Políticas Públicas e Sistema de Gênero No Brasil'. *Cadernos Pagu* 27: 213–254.

Bassanezi Pinsky, Carla, and Joana Maria Pedro, eds. 2013. *Nova História Das Mulheres No Brasil*. São Paulo: Contexto.

Borgos, Anna. 2019. "Alice Bálint at the Intersection of the Personal, the Professional, and the Political". In *Psychology and Politics*, edited by Anna Borgos, Ferenc Erős, and Júlia Gyimesi, 53–78. Intersections of Science and Ideology in the History of Psy-Sciences. Budapest: Central European University Press. https://www.jstor.org/stable/10.7829/j.ctvs1g9j3.7.

Bourdelais, Patrice. 2001. *Les hygiénistes: enjeux, modèles et pratiques, XVIIIe-XXe siècles*. Paris: Belin." 'Constance Pascal (1877–1937), Première Femme Psychiatre et Médecin-Chef Des Hôpitaux Psychiatriques En France". *Annales Médico-Psychologiques, Revue Psychiatrique* 173 (9): 815–816.

Brooks, Harvey. 1968. *The Government of Science*. Cambridge, MA: MIT Press.

Butler, Judith. 1990. *Gender Trouble: Feminism and the Subversion of Identity*. New York: Routledge.

Cesar, Osorio. 1929. *A expressão artistica nos alienados: (contribuição para o estudo dos symbols na arte).* Officinas Graphicas do Hospital de Juquery.

de Alameida Moura, Vanessa. 2019. *Marialzira Perestrello: a trajetória profissional de uma médica e psicanalista carioca (1934–1962).* Rio de Janeiro: Fiocruz. https://www.arca.fiocruz.br/handle/icict/37029.

Douglas, Mary, Alain Caillé, Georges Balandier, and Anne Abeillé. 2004. *Comment pensent les institutions.* Paris: La Découverte.

Facchinetti, Cristiana, and Ana Teresa Venancio. 2018. "Da psiquiatria e de suas instituições: um balanço historiográfico". In *História da Saúde no Brasil*, edited by Luiz Antonio Teixeira, Tânia Salgado Pimenta, and Gilberto Hochman, 356–402. São Paulo: Hucitec. https://www.arca.fiocruz.br/handle/icict/38638.

Favier, Irène. 2021. "De pioneros y hombres. Una mirada historiadora sobre el Hospital Víctor Larco Herrera a lo largo del siglo 20". Presented at the Jornada Académica por el 103 Aniversario del Hospital Víctor Larco Herrera, 22 January 2021. Online: https://www.youtube.com/watch?v=-SgPO8FwVdg.

Fernandes, Sandra Michelle. 2015. "Nise Da Silveira e a Saúde Mental No Brasil: Um Intinerário de Resistência". PhD dissertation. Universidade Federal do Rio Grande do Norte.

Ferreira Abrão, Jorge Luis. 2018. *Marialzira Perestrello. Mulher de Vanguarda e Pioneira da Psicanálise.* São Paulo: Zagodoni.

Fleck, Ludwik. 2005. *Genèse et Développement d'un Fait Scientifique*, trans. Nathalie Jas. Paris: Les Belles Lettres.

Gordon, Felicia. 1990. *The Integral Feminist: Madeleine Pelletier, 1874–1939.* Cambridge: Polity Press.

Gordon, Felicia. 2013. *Constance Pascal (1877–1937): Authority, Femininity and Feminism in French Psychiatry.* London: Institute of Germanic & Romance Studies.

Grosskurth, Phyllis. 1986. *Melanie Klein: Her World and Her Work.* New York: Random House.

Hochman, Gilberto. 2012. *A era do saneamento: As bases da política de saúde pública no brasil.* 3a edição. São Paulo: Hucitec.

Huertas García-Alejo, Rafael. 2012. *Historia cultural de la psiquiatría: (re)pensar la locura.* Madrid: Catarata.

Lambe, Jennifer L. 2017. *Madhouse: Psychiatry and Politics in Cuban history.* Chapel Hill: University of North Carolina Press.

Lamy, Jérôme. 2018. *Faire de la sociologie historique des sciences et des techniques.* Paris: Hermann.

Sales Magaldi, Felipe. 2018. "A Unidade Das Coisas: Nise Da Silveira e a Genealogia de Uma Psiquiatra Rebelde No Rio de Janeiro, Brasil". PhD dissertation. Museu Nacional/Universidade Federal do Rio de Janeiro.

Martins, Ygor. 2022. "Presença e Participação Feminina Na Psiquiatria Brasileira: A Trajetória Da Dra. Alice Marques Dos Santos (Rio de Janeiro, 1928–1964)". Masters Thesis. Rio de Janeiro: COC/FIOCRUZ.

Martins, Ygor, and Valentine Mercier. 2021. "Estratégias Das Primeiras Psiquiatras Brasileiras Na Consolidação de Suas Carreiras: A Centralidade Da Psicanálise Para o Avanço Profissional (1941–1970)". *Faces Da História* 8 (2): 165–186.

McGovern, Constance M. 1984. 'Psychiatry, Psychoanalysis, and Women in America: An Historical Note'. *Psychoanalytic Review* 71 (4): 541–552.

Mello, Luiz Carlos. 2014. *Nise Da Silveira: Caminhos de Uma Psiquiatra Rebelde.* Rio de Janeiro: Automática Edições.

Melo, Walter. 2005. “Ninguém Vai Sozinho Ao Paraíso: O Percurso de Nise Da Silveira Na Psiquiatria Do Brasil”. PhD dissertation. Rio de Janeiro: UERJ.
Mercier, Valentine. 2020. *Mouvements féminins et Parti communiste au Brésil, Des sociabilités militantes à la politisation des femmes (1945–1961)*. France: Éditions de l’Institut des Hautes Études de l’Amérique Latine. Chrysalides.
Meyer, Manuella. 2017. *Reasoning against Madness: Psychiatry and the State in Rio de Janeiro, 1830–1944*. Rochester, NY: University of Rochester Press.
Moser, Caroline O. N. 1989. “Gender Planning in the Third World: Meeting Practical and Strategic Gender Needs” *World Development* 17 (11): 1799–1825. https://doi.org/10.1016/0305-750X(89)90201-5.
Nieto-Galan, Agusti. 2019. *The Politics of Chemistry: Science and Power in Twentieth-Century Spain*. New York: Cambridge University Press.
Perrot, Michelle. 2005. *As Mulheres Ou Os Silêncios Da História*. Bauru: EDUSC.
Perrot, Michelle. 2007. *Minha História Das Mulheres*. São Paulo: Contexto.
Pestre, Dominique, ed. 2014. *Le gouvernement des technosciences. Gouverner le progrès et ses dégâts depuis* 1945. Paris: La Découverte.
Ríos Molina, Andrés, and Alejandro Giraldo Granada, eds. 2017. *Los pacientes del Manicomio La Castañeda y sus diagnósticos: una historia de la clínica psiquiátrica en México, 1910–1968*. Serie Historia moderna y contemporánea 72. Ciudad de México: Universidad Nacional Autónoma de México,
Rindzeviciute, Egle. 2016. *The Power of Systems: How Policy Sciences Opened Up the Cold War World*. Ithaca: Cornell University Press.
Rorabaugh, William Joseph. 2018. *Prohibition: A Concise History*. New York: Oxford University Press.
Rossiter, Margaret. 1984. *Women Scientists in America: Struggles and Strategies to 1940*. 3rd edn. Baltimore: Johns Hopkins University Press.
Stone, Lawrence. 1971. “Prosopography”. *Daedalus* 100 (1): 46–79.
Teive, Hélio, Daniel Sá, Octavio Silveira Neto, Octavio Da Silveira, and Lineu Werneck. 1999. “Professor Antonio Austregésilo: o pioneiro da neurologia e do estudo dos distúrbios do movimento no Brasil”. *Arquivos de Neuro-Psiquiatria* 57 (3B): 898–902. https://doi.org/10.1590/S0004-282X1999000500030.
Tyrrell, Ian R. 2010. *Reforming the World: The Creation of America’s Moral Empire*. Princeton and Oxford: Princeton University Press.
Volpato, Adriéli. 2011. “A predominância do sexo feminino na profissão do serviço social: uma discussão em torno desta questão.” In *Anais II Simpósio Gênero e Políticas Públicas*, ed Cássia Maria Carloto, 11. Universidade Estadual de Londrina. https://www.uel.br/eventos/gpp/pages/arquivos/jessica.pdf
Young-Bruehl, Elizabeth. 2008. *Anna Freud*. New Haven, CT: Yale University Press.
Zuckerman, Mary Ellen. 1998. *A History of Popular Women’s Magazines in the United States, 1792–1995*. Westport: Greenwood Press.
Zulawski, Ann. 2007. *Unequal Cures: public health and political change in Bolivia, 1900–1950*. Durham, NC: Duke University Press.

6 Women in Soviet Meteorology and Climatology

Governing Blindness toward Gender, 1919–1991

Katja Doose

There are many narratives and projections about Soviet women in science. They oscillate between the ideas that Soviet women were forced into science and that there was absolute gender equality among scientists in the socialist countries. The latter is in particular often stressed for the usually very male-dominated STEM subjects (science, technology, engineering, mathematics) and noted for the past and present time with a certain admiration and astonishment (Eveleth 2013). Moreover, as a recent study reveals, female scientists working in these disciplines in Russia today likewise seem to perceive their work environment as gender-neutral and free of gender discrimination, which the authors argue is influenced by the Soviet legacy (Khanukaeva and Di Puppo 2021). But although the Soviet academic workforce has been made up of almost 40% women since the 1960s, very few women held positions of authority in Soviet science. By 1980, for instance, only 3 out of 247 full members of the Soviet Academy of Science—the most prestigious scientific institution of the country—in all disciplines were women (Kneen 1984: 30).

Taking these contradictions as a starting point, this chapter asks how the Soviet state (1922–1991) has created this perception of gender neutrality while at the same time continuing inequality—an inequality even greater at the top of the scientific hierarchy. It explores how the official discourse on gender impacted the way female scientists thought about their academic abilities. While there is much research on female scientists and women's education in tsarist Russia (Valkova 2019a, 2019b, 2008; Creese and Creese 2015; Hibner 2000), the Soviet period still suffers from lacunae and Soviet science has to date, with a couple exceptions, not been thoroughly analyzed through the category of gender (Kalemeneva and Lajus 2018). Much has been said about the double standard and burdens of Soviet women that point to life conditions far from equal to men (Buckley 1989; Goldman 2002). But little is still known about the circumstances of female scientists in Soviet society, which by adopting the authoritative discourse of the State regarded a woman in science as the symbol of gender equality while the reality looked very different (Yurchak 2005).

DOI: 10.4324/9781003562597-9

This chapter takes meteorology and climatology as a lens to examine the impact of gendered state discourses. These disciplines were institutionalized in the 1840s and were the first in Russia to officially allow women to work as scientific staff. The analysis is based on published materials such as scientific journals and autobiographies as well as on interviews with six women scientists. It focuses on the ways the Soviet gender policy of equality shaped the careers and their self-perception of female scientists. Personal accounts are problematic sources in many respects since they rely on memories shaped by forgetfulness, hindsight, motives, and society's collective memory. Nonetheless, such accounts are no more misleading than archival documents that were written and archived for particular reasons and with certain motives. In addition, subjectivity can also be of great value to a historical study. Interviews can illustrate the power of discourse in the Foucauldian sense by showing how state power produced a discourse about women's participation in Soviet work life which in turn not only shaped women's ways of thinking about themselves and their particular roles in Soviet society but also informed women's understandings of existing possibilities that could help fight for improvements. Using such a small sample seems legitimate as there is little data available that gives an insight into the experience of being a female scientist in these disciplines in this particular historical situation. The advanced age of some of the interviewees makes their subjective memories all the more valuable.[1]

This chapter argues that politically claimed gender equality led to blindness toward gender among the general public and scientific women alike. State-proclaimed parity made it difficult to demand changes and better access to the highest and most prestigious positions and to broach the issue in public and work environments. A perception of gender equality thus coexisted with inequities and hidden biases among the academic workforce. As a result, gender only plays a role in scientists' narratives when it discredits women and describes female deficiencies in their professional lives. Such narratives do not speak about gender inequalities that could be attributed to the Soviet system. After introducing the general situation of Soviet women in science, I explore how this situation mirrors the biographies of female meteorologists and climatologists who were active throughout the Soviet period.

Soviet Women at Work and in Science

By now it is common knowledge that women in the Soviet Union for the most part were involved in science, much more so than in any Western country during the same period. The Bolsheviks had decided early on to promise equal rights for all citizens, and that women would be able to study and be allowed to work in traditionally male-dominated occupations. According to Lenin, women had to be freed from domestic servitude so that they can participate in the social production process. By 1936, the Soviet constitution proclaimed that a "problem of enormous magnitude had been resolved" and that "for the first time in history, authentic equality for women" was assured (Aivazova 2003: 628).

Although this was far from the truth, as many studies have already shown, women's participation in science and education was nevertheless remarkable in international comparison. Between the 1950s and the 1980s, the proportion of women among scientific workers was between 36% and 40% in the USSR (Kneen 1984: 28), compared to less than 3% in 1947, and 30% in 1980 in the United States (Kohlstedt 2004: 16), despite the high numbers of women who entered engineering colleges following the launch of the Soviet satellite Sputnik in 1957 (Crane 2003: 7). In Russia now, the percentage of women in science remains equally high, which can partly be explained through the way scientists were recruited during Soviet times, but also by the naturalization of gender differences in Soviet ideology. For the Soviet case, it has been suggested that the significant interest of women for science since the 1960s can be attributed to the USSR's standard school curriculum which focused very much on the study of math and hard sciences to generate future generations of scientific workers for the State and in particular for the industrial and defense ministerial system. It was considered socially important to receive a high-level education, and afterwards to work in a university or research institute (Hargittai 2015: 216). Since educators at schools had to identify talented students in general to increase the numbers, little difference was made about gender (Sylvester 2013). The Soviet State employed large amounts of workers in all sorts of fields to compensate for the lack of technological infrastructure, such as in agriculture, construction, and in science. Indeed, by the 1980s, the Soviet scientific establishment had become the largest in the world, counting 1,520,000 people, which amounted approximately to between 10% and 30% more scientists and engineers than in the United States, depending on the definition of degrees and fields (Graham 1992: 50). But of these, only very few women reached the stage of finishing their doctoral dissertation, the prerequisite for professorial appointment and a requirement to enter any higher-ranking administrative post in the Soviet scientific establishment, including the ranks in the Soviet Academy of Science (Kneen 1984: 28).

However, the numbers are less important and show at worst how historians have been guided by Soviet ideology and repeat the same moments of success within Soviet history. It is therefore more interesting to show the consequences of Soviet official gender discourse. One way to explore this is to examine cultural perceptions of female scientists over time. One of the most popular Soviet journals, *Ogonek,* regularly published articles during the 1920s about female scientists or women as pilots and ship captains (Valkova 2008: 159). Women's magazines of the 1940s and 1950s would stress the absolute devotion of female scientists in the service to their country and their readiness to sacrifice family life for it.[2] As such, the participation of women in science became part of the narrative of Soviet heroism (Kalemeneva and

Lajus 2018: 267). In the beginning of the 1970s and throughout the 1980s, questions about the ability of women to combine family life and a scientific career became increasingly prominent in state propaganda, as articles in *Soviet Woman* show. In contrast to the 1940s and 1950s, the role of women for household and family duties became increasingly taken as given in the 1980s. This was naturalistically stated: "a woman is a woman" and she "cannot get away from these [woman's] obligations" such as family, children, and home (Ilyina 1980: 36). It is likely that this naturalization discourse was used because the Soviet state needed women to contribute to the scientific work only as long as they could still ensure Soviet demography. Indeed, population decline has served as a main political calculation for Soviet policies concerning women: for example, with the changing laws about abortion (Randall 2011; Evans 1981). The compromise between allowing women to participate in research, but to still contribute to increasing Soviet population was possible through armies of female lab technicians who helped research projects to advance, but who also left academia after the birth of their first child. This gendered hierarchy received support among female scientists. Biologist Raisa Butenko, for example, herself a very successful scientist, put it in an interview for *Sovetskaya Rossiya* in 1980 as: "Not every woman has to become an academician... Exceptional personalities aside, modern experimental science also needs the service of a large number of people... We really need these 'masters of their own art'" (Butenko 1980: 4). This statement came at a time when Soviet population was shrinking and State propaganda tried to encourage female scientists to have children; thus to settle for lesser ambitions in order to avoid jeopardizing their private life (Kosheleva 1983: 38–39).

Women's underrepresentation in the higher-ranking academic positions was part of public culture. For instance, Soviet movies on science showed individual and single female researchers among large groups of male researchers (Novikova 2019)—an imbalance, that was never discussed officially. If anything, female scientists were sometimes used in Soviet satire to depict the usefulness or idleness of scientists; as shown in examples from the satirical magazine *Krokodil* (see Figure 6.1) (Grin'ko and Shevtsova 2018: 283). Mostly, however, female scientists only appeared in special women's magazines. The caricature "In the Researcher's Column" (see Figure 6.2), in *Krokodil,* for example, illustrates this fact pointedly. It is a humorous allusion to the typical communist demonstrations on 1 May of different professional groups, but the striking detail here is that only male scientists are marching, despite the Soviet academic workforce being between 30% and 40% women. Though the magazine frequently ridiculed Soviet working conditions in academia, there were no caricatures touching upon, for example, additional barriers for women in reaching high-ranking academic posts (Grin'ko and Shevtsova 2019: 88).

Even though criticism of gender equality in science was not visible in Soviet public communication, it did not mean it was not happening, but thorough research is still pending. What becomes evident already, is that the public

Figure 6.1 Caricature by Galina and Valentin Karavaev, *Krokodil* (1970s).

— How much time is left until the end of the work day?

— Half sleeve and back.

discourse did not question gender dynamic, because it universalized gender by merging men and women into one sex. In an article on gender equality in education published 1975 in *Soviet Woman* the reader is told that gender equality is "so natural for Soviet women that we do not even think of any other way… When we talk of… scientists, we mean men and women also. Only statistics, with their impartial accuracy, make the difference between men and women" (Dorokhova 1975: 12). A discourse that had a profound impact on how female scientists felt about themselves and their achievements as the following case studies on female meteorologists and climatologists will

Figure 6.2 No author, "In the Researcher's Column", *Krokodil* (1971, 30: 3).

show: their modest position in the academic hierarchy seemed to them a reflection of the natural state of affairs.

Women in Meteorology and Climatology

Meteorology was the first discipline within the Russian scientific establishment that allowed women officially to participate in science. The reform movements of the 1860s that affected most institutions of Russian society enabled women to informally participate in courses, but they were not yet able to receive degrees. In 1878, the tsarist government created so-called Higher Courses for Women, in order to avoid having to enroll women in traditional universities with men. However, the content of the education and the acquired degree only became equal to men in a few disciplines after 1914 (Koblitz 2000: 23–24).

Women's entry into the science of meteorology was in part owed to the beginning of World War I that required a large personal infrastructure to monitor the weather and to produce forecasts useful at the front. In 1913, the head of the Main Geophysical Observatory (GGO) in Saint Petersburg Boris Golitsyn (1862–1916), released a charter according to which women were given the possibility to join the Observatory as a research associate, provided they had the required educational qualification, a condition equal to men. Women were to receive the same salaries as their male colleagues doing the same job, but had no right to ranks and awarding orders (Selezneva 1989: 57). Simply put, the GGO needed scientific staff to support the increasing research work on the climate of the Russian empire (Budyko et al. 1967: 4). Within the first three years, between 1914 and 1916 the Observatory hired 15 women, that had mostly come from the Higher Courses (Valkova 2019b: 584). Evgeniya Rubinstein (1891–1981), who held a diploma in mathematics and astronomy from the Higher Courses, entered the Observatory as a scientific assistant in the Department for station networks and climatological works. By 1919, she had moved on to the newly founded Department for Climatology, where she was employed as a "fizik", a senior research associate. Despite this prestigious-sounding job title, her main tasks were establishing weather and climate statistics and doing data processing. At the time all women in tsarist academia occupied similar subordinate positions.

This continued also when the Soviets took power. Women would come from universities and be employed as scientific-technical staff, and do work as "human calculators" (Selezneva 1998: 115). Another typical job for these newly acquired female scientists was to look after instruments. Berta P. Karol (1892–1993) for instance, who held a diploma in history from the Higher Courses for Women, began at the GGO as an assistant for instrument maintenance and helped to prepare large polar and glacier expeditions. Later she also worked on her own research and eventually wrote her dissertation (Selezneva 1989).

As the above examples show, the tasks assigned to female scientists were often low-skilled, regarded as "routine" work, and served mainly to help with the research men were doing. The fundamental research was still done only by men as reflected in the underrepresentation of final research outputs in scientific articles. For instance, contributions by female scientists in the main meteorological journal *Meteorologia i Gidrologia* for the 1970s and 1980s made up 15% of the overall contributions, most of which are second authorships.[3] Moreover, the sections, in which the milestone birthdays of important scientists in climatology and meteorology were marked, were almost exclusively dedicated to male scientists. Similarly, book reviews during that period were never on a book written by a female scientist. Of course, a more thorough study would also reveal the percentage of male-authored books not being reviewed to put things into perspective. Meteorology and climatology are no different from the conditions for most natural sciences during the Soviet era, but they can serve as a powerful lens focusing on those circumstances, and therefore, on the general social situation in the Soviet Union.

Female Success Despite Obstacles

While it is not difficult to single out the few female scientists who excelled in their field, it is incredibly difficult to find information on them either in archives or through published material. Partly this is due to the fact that information pertaining to the Soviet Hydrometeorological Services are well guarded, and partly it can be attributed to the general lack of information on female individuals which we can observe elsewhere. For Ekaterina Blinova (1906–1982) for example, who developed an important long-range forecasting method in the 1950s, and who became one of the few women with full membership to the prestigious Soviet Academy of Science (1953); a file seems to be available in the archives of the Russian Academy of Sciences, but it is inaccessible to the public. Similarly, Evgenija Rubenstein (1891–1981), who was one of the first women to obtain an academic position in meteorology at the GGO in 1913, and who is considered the first female climatologist in Russia, was given her first milestone birthday announcement in the observatory's journal *Trudy GGO* in 2016, 35 years after her death. The only female meteorologist who left something in writing seems to be Maria Gavrilova (1928–2010). She became famous for her permafrost studies in Yakutia during the 1960s and 1970s, for which she received a medal from the Geographical Society of the USSR, the first woman ever to receive this award, in 1985. Generally, in anniversary publications on the history of the GGO or the Hydrometeorological Services the role of women is ignored, even the fact that it was their institutions where the first female scientists in Russia were employed. The blindness toward and the invisibility of gender affected all spheres.

Left with little written data like most scholars working on women's archives, I tried to understand how Soviet female meteorologists and climatologists did succeed to make a name for themselves in an area that at least in its upper ranks was male-dominated. Studies have pointed to the influence of parents, teachers, and peers; stereotypes permeating at home or through mass culture; and the quality of science teaching in schools. The overall image of science and the presence of role models currently are decisive factors for women to pursue career opportunities in science (OECD 2008; Dasgupta and Stout 2014). Based on five semi-structured interviews with female scientists notable for their academic performance and reputation, and on one autobiography by Maria Gavrilova, this section focuses only on the role of parental influence and the larger network.[4] All six women were educated during Soviet times in meteorology or climatology. Their life paths and family backgrounds are all quite different depending on the time period during which they were educated; and yet they all became scientists in the field of meteorology or climatology. The reasons why they followed the academic path after their studies in comparison to many of their other female fellow students are similarly diverse. While, as Roshanna Sylvester has argued, the 1963 space flight of Valentina Tereshkova strongly impacted girl's decision to enter science-related professions, this section shows that State efforts were practically absent and that parents and networks played a large role in shaping career paths (Sylvester 2013). Although, the sample size is too small to draw general conclusions or statements, a few cases still shed some light on the social profile of a successful female scientist throughout the Soviet period.[5]

One very important path to academic success, as most studies on male and female career paths show, was the social background and the parent's level of education. For the early Soviet period, for women born in the 1920s or 1930s, a noble background seemed to have played some role in being able to pursue an academic career. Although the nobility was persecuted during the 1920s and 1930s and fell victim to the Stalinist purges, education continued to play an important role in noble families as it secured them jobs under more comfortable working conditions such as for example as teachers, in local government offices in the health and land departments or in banks, but also because they could build on existing educational skills and cultural interests (Rendle 2008). In the case of Nina Kobysheva (b.1925), who had been working as a lecturer in meteorology at the Military Space Academy A.F. Mozhaisky, and became a climatologist at the GGO in 1974; her noble background and the social habitus of aristocracy, which, included surrounding oneself with servants, led to her having nannies, household aids, and the financial means to pursue the studies she liked. In the interview, it became clear, that she took for granted that all female scientists had nannies at the time for them to be able to pursue their academic passion. Moreover, her parents were not only from a noble origin that placed importance on education, but both her mother and her father were professors, in surgery and pediatrics respectively, and thus most likely transmitted their values of education to their children. Moreover, their professional background made them qualified specialists worthy of being evacuated to Ashkhabad during World War II (Zharkynbayeva, Abdiryaimova, and Dadabaeva 2020). Here, Nina Kobysheva could begin her studies at the evacuated Kharkiv Institute of Hydrometeorological Engineering. For her, she said, becoming a scientist was the natural path thanks to her family background: "I tried [to become a scientist], but I, you understand, had such a family. I am from such a family… I really wanted to go further".[6] In comparison with fellow female students, she considered herself as harder working and less focused on her child. In her opinion, "[t]hat's why women are bad, because they are always thinking about children. They have children in their heads". Asked how she explains her own academic success she pointed to different priorities. "Well, here I am different [from others], my work has always come first… I like to think, to invent".[7]

Anna Mesherskaya (b. 1932), who had worked as a climatologist at the GGO since 1955, is another example of a female climatologist with noble origin who advanced in Soviet academic ranks, and that even despite her father's arrest in 1931 based on his social background. During tsarist times, he had been a high-ranking, well-educated military officer. His fate did not seem to have an impact on her academic career as she herself claimed. She began her studies in 1950 and received her PhD in 1955 at the GGO three years prior to her father's rehabilitation in 1958. It is likely that family networks helped her to overcome the obstacle of having a close family member that was not yet rehabilitated—a fact that generally excluded descendants from studying and pursuing an academic career in the early 1950s.

In addition to noble origin, the parent's level of education—both stand in close correlation to each other—played a crucial role in that generation. For example, Maria Gavrilova (b. 1928), who was considered the country's permafrost specialist for decades, benefited from her father's education and networks. Kuzma Gavrilov, the famous intellectual and writer in Yakutia, had died in 1938, when Maria Gavrilova was only 10 years old. She did not directly refer to him as source of intellectual inspiration in her memoirs, suggesting her uncle instead (Gavrilova 2007). Kuzma Gavrilov's fame in Yakutia and his works still must have influenced her path. Her father had been politically well connected in Yakutsk, as her memoirs suggested, and his social and intellectual credit helped her, among other things, to get accepted to college in Moscow in 1949, despite the fact that Kuzma Gavrilov's reputation, was only officially rehabilitated in 1957, similar to the case of Anna Mescherskaya's father. However, she was still denied a PhD at the Moscow State University and finally did her doctorate at the GGO (Gavrilova 2005: 231). These few cases illustrate that the father's level of education, but also connections through either an aristocratic background or political ties may have helped women of this generation educated during the 1940s and early 1950s to pursue an academic career. Moreover, Gavrilova's autobiography shows that she brought along a high sense of herself enabling her to push further against authorities at moments when she was rejected. According to her own narrative, she reached out to the Council of Ministers in Yakutsk, when the Ph.D. commission decided that she was to work as a school teacher instead of doing her Ph.D. and when after her Ph.D. the commission of the Soviet Hydrological Services sent her to work in Vladivostok, she complained until she was sent to Yakutsk, where she fundamentally shaped Soviet permafrost studies for half a century.

The 1960s brought a range of changes that eased women's access to higher-ranking positions in the USSR (and worldwide). The network of Soviet research institutions widened significantly during those years in order to create an intellectual workforce for the incessant Cold War competition with the West, and to insure economic development. This logic required the increased employment of women as well, although their numbers grew only in the absolute sense, overall, between 1960 and 1988 women continued to make up between 37% and 40% of the overall Soviet scientific workers (Allakhverdyan 2014: 84). In the Soviet Union, as in China and in the West, it was also a time, when "science studies" (rus. *naukovedenie*) became a subject of social scientists, which meant that more thought was put into how science and scientific education work (Allakhverdyan 2014: 32; Aronova & Turchetti 2016: 2).

However, the increased need for scientists neither meant that there were specific State campaigns to recruit women, nor did it mean that the role of parents as models or inspirations became less significant. Two examples will demonstrate this. I interviewed Elena Manaenkova (b. 1964) and Daria Gushchina (b. 1969) independently of each other, but the similarities between them are striking. They may point to a certain homogenization of the scientific system, but considering the small sample size no general conclusions should be drawn. Both of their parents were either academics working as professors in physics and chemistry, or, like Elena's mother, as a musician. Of all interviewees, only these two women, coming from the most privileged educational background, did their PhDs in climatology at the most prestigious place—the Moscow State University. Interestingly though, neither of the daughters pursued a career in physics—the profession of both of their fathers. In both cases, they were told by their friends and fellow academics to go for geography instead, since both women had expressed interest in traveling. The reasons for this guidance are not clear and neither of the women questioned it, but it is possible that their peers and elders tried to redirect the young women's enthusiasm away from the "male-dominated" science of physics toward less ambitious educational paths. Originally, both were supposed to become musicians according to their parents' wishes but finally decided against this in favor of science. Neither of them, however, placed great emphasis during interviews on the role of their parents on their decision to pursue an academic career. For both, it was another influential fact: they knew the building of the Moscow State University through their fathers, who were both professors there. As Elena Manaenkova explained: "I liked the place, it's like a cathedral. It's a place you want to be at, something special. And then I went to study geography".[8] Daria Gushchina told how she wanted to become a professor at that university since she was a child of 5 years, after she found out that professors get state apartments in one of the towers of the main building, she saw on her many childhood walks in the neighborhood. Even now, a professor herself, the main building with its columns at the entry, marble floors, and portraits inside gives her "energy, inspiration, strength and the wish to go to work, to give lectures".[9] Both women based their success on their scientific passion and vigorous work ethics, but they also noted a stable support network to manage parental duties and work—aspects older generations did not mention. Here again, it was not the State and its services they could count on as is often presumed about Soviet childcare services, but the family structure. Moreover, Elena Manaenkova's scientific career took off in the chaotic 1990s, when it was financially secured by her husband.

There were, however, also exceptions to the rule of family ties, at least in the post-war generation. Vera Pavlova (b. 1953), educated as an applied mathematician, for instance, had no parents working in academia. She attributed her career path in academia to coincidence. "I only became a researcher, because I found myself in such an environment… I did not think about it [her PhD]. I only went to work. But because I got into such a scientific collective, and the science [agrometeorology] developed so rapidly, I, willingly or unwillingly, joined this stream".[10] As applied mathematicians, students were sent off to different work places after graduation, which was perhaps what some meant by Soviet women having been forced into science ("Ladies of the Lab" 2019). At the same time, her career cannot be entirely compared with that of Elena Manaenkova, for example, who is now the Deputy General Secretary of the World Meteorological Organization; or with that of Daria Gushchina, who works as a full professor at the Moscow State University. Instead, she works at the Institute for Agrometeorology in Obninsk, outside of Moscow, where she provides applied research to the State Hydrometeorological Services of the Russian Federation. It can be argued that with the new science policy of the Soviet State beginning in the 1960s, the family and social backgrounds of women in the science system became more diversified as the state needed more scientific workers.

Soviet Gender Politics in Biographical Narratives

Although the State has been absent in these accounts, the reflections of the women show the power of discourse. The State's decision to render the question of gender equality as solved has determined the ways people spoke and thought about being a female scientist. Moreover, the discourse on women being like men not in need of special consideration has turned their invisibility into reality (Foucault 1981: 31–112). The women I interviewed speak, behave, and think in the way that discourses told them to. For instance, Daria Guschina's dream as a five-year-old demonstrated that Soviet women did not think per se that science was not for them. Often, they did not question it in their narratives; and recent studies confirm the same for contemporary Russian female scientists (Bullock 2017). But as the interviews here have shown, women, especially of the older generation, often still thought of themselves as less capable of doing science than men. This suggests a tension between an official view of women in science as normal and ordinary; and a view of a female scientist as being intellectually weaker than their male colleagues. It can be argued that the official state view of women being equal to men in the work force penetrated female scientist's self-awareness only partly. The gap between the professed state ideology and the ambient discourse in society was visible. As a consequence, women considered themselves as equal to men in having the same professional opportunities, but they did not think of themselves as equal to men in terms of their abilities and capacities. Instead, they presented a diminished sense of self.

As elsewhere in the world, the science system seemed gender-neutral to the Soviet women. None of the interviewees referred spontaneously to cases of discrimination, "glass ceilings" or inequalities. Russian academia today is known for its resistance to feminist studies, considering them as unscientific and politically biased. Therefore, broaching gender-related discrimination may be problematic for some (Shnyrova 2000: 232–233). However, this does not mean that interview subjects did not speak about it. They sometimes did not recognize discrimination as such. For instance, in her interview Nina Kobysheva remembered that "in Soviet academia, there was an equality of rights", but shortly after she stated, that "naturally, for leading positions, men were preferred".[11] She integrated gendered norms, transmitted through the state's discourse, without being aware of them. When speaking about the position of laboratory assistants, Kobysheva said that both men and women would do the same tasks in the beginning, but that men were promoted more often, and never stayed for long as laboratory assistants.[12] At no point did these examples strike her as a contradiction to gender equality in Soviet academia. As a consequence, women like Kobysheva also never thought about empowerment for other women; none of the women in leading positions had ever especially promoted women, nor did they support women specifically. As Anna Mesherskaya pointed out: "what counted for me, were the results. I did not care whether they came from a man or a woman".[13]

In her autobiography, Maria Gavrilova revealed an even more obvious blindness to gender. In her lengthy autobiography, she never referred to difficulties she encounters because of her gender. Except for once: when she entered her PhD program at the Main Geophysical Observatory in 1958, her supervisor Mikhail Budyko, a well-known Soviet climatologist, warned her that "[g]irls have greater difficulty getting their PhDs. Especially, once they are married" (Gavrilova 2007: 270). She did not comment on it but left it as an anecdote requiring in her eyes no further explanation.

These observations stand in strong contrast to the awareness of female researchers in the United States toward gender inequalities. Consider for example the autobiographical notes in the archive of Joanne Simpson (1923– 2010), the first American woman to have received a PhD in meteorology. Throughout her career, Simpson had become an "advocate for female scientist empowerment and an outspoken critic of the existing power hierarchy within the atmospheric sciences" (Fleming 2020: 121). She wrote articles on that matter and conducted surveys among female atmospheric scientists and meteorologists to prove her points. Her forms of protest were in line with the women's movement in the United States, when women's organizations were fighting against anti-nepotism rules and general discrimination against women in science (Rossiter 1995). The awareness of gender discrimination was evidently much higher in the United States, arguably because gender equality had not been officially declared by the State, thus it was not made a non-subject. This enabled and made it necessary for women in the West to demand improvements, which arrived soon after.

In the Soviet Union, however, women could not demand improvements, because officially there was nothing to improve. The State narrative had also presented the Soviet Union as a world leader in the matter. To demand more than equality was politically not possible. To fit into the discourse on women's role in science, they assume subject positions in order to be perceivable and able to act and to avoid sanctions and exclusion (Foucault 1981: 78–92). On different occasions, the interviewed scientists stressed the role of nature in the disadvantaged physical and mental position of women. Nina Kobysheva, for example, argued that "nature [was] built in a way, so that women have less interest in mathematics".[14] For her, this fact was a natural given just as the fact that men have easier lives and career paths, because women carry the weight of birth as she argued. This focus on the body can be similarly detected in Daria Gushchina's statements. When she explained that women were not allowed to become oceanographers, the only obvious discrimination against women in climate science, she justified it through the physical strength of men, which would allow them to lift heavy instruments out of the sea onto the research vessel. In doing so, she turned obvious discrimination into a mere physical problem, according to commonly held opinions about women's physical weakness. In fact, women often went with men on research expeditions on the ocean, but they went as laboratory assistants on the vessel, instead of as women oceanographers. In the same vein, the interviewees regarded their own gender often as deficient. For instance, explaining why women had less high-ranking positions in academia, Anna Mesherskaya argued that "women were less modern, less smart, less creative and less intelligent than men".[15] Similarly, Nina Kobysheva found female scientists in general to be "less serious, less mathematical" than men.[16] While gender is completely ignored in Soviet State policies and more or less rendered insignificant in biographical narration of scientists, it served as a means to express deficiencies.

Conclusion

Soviet Russia's example can show how early proclamations about the completed emancipation of women effected female scientists and their progression through the scientific hierarchy. In contrast to their fellow women from the late nineteenth century, Soviet scientists did not have the support of progressive statesmen or a movement like the nihilists that would have helped to improve their access to high-ranking academic positions. No concerted efforts were taken within scientific or technical institutions to improve the situation of female academics. The reason was simple: gender equality had been regarded as a given since 1919. Thus, while the statistics showed and still show high participation of women in science, especially in the STEM fields, something else became visible in this small study: female scientists were rendered invisible as a consequence of the state discourse on women's alleged equality that similarly claimed to have emancipated women and robbed them of any ability to fight for real gender equality to exercise their academic profession to fullest. While official claims and legal emancipation have led to a general higher female participation in science, women within the Soviet science system have been subject to disciplinary and hierarchical segregation that stuck female scientists into employment as lab technicians without a chance to climb the career ladder or sidelined them in "lesser" sciences within the hierarchy of disciplines. The Soviet system thus differed little in this regard from the one to be found in the West. More importantly, it seems that the discourse around female scientists lacks and the lack of state support in raising awareness about the particularity of women's situations has led to the continuation of inferior self-perceptions of female scientists and the ignorance of gender-related discrimination. It is an example of the power of discourse in the Foucauldian sense. It was indeed so powerful that the myth about equal female participation in Soviet science is as alive as ever.

Notes

1 Nina Kobysheva died in 2021 aged 97.
2 See for example articles in *Soviet Woman* of the 1940s and 1950s.
3 Considered were all monthly issues of the years 1974–1982. On average an issue contains 19 articles, of which between 2 and 3 were written by women as single authors or as second co-authors.
4 Nina Kobysheva (b. 1925), Anna Mescherskaya (b.1932), Vera Pavlova (b. 1953), Elena Manaenkova (b. 1964), and Daria Gushchina (b. 1969), interviewed between 2018 and 2021.
5 For more case studies see also Olga Valkova's article on Vera Varsanofieva in this volume.
6 Interview Nina Kobysheva (b.1925), 26 September 2018, Saint Petersburg.
7 Interview Nina Kobysheva (b.1925), 26 September 2018, Saint Petersburg .
8 Interview with Elena Manaenkova (b. 1964), 16 March 2021, Geneva.
9 Interview with Daria Gushchina (b. 1969), 2 March 2021, (Moscow) telephone.
10 Interview with Vera Pavlova (b. 1953), 29 April 2020, (Moscow) telephone.
11 Interview Kobysheva, 26 September 2018, Saint Petersburg.
12 Interview Kobysheva, 26 September 2018, Saint Petersburg .
13 Interview with Anna Mesherskaya (b. 1932), 24 February 2021, (Sankt Petersburg) telephone.
14 Interview Kobysheva, 26 September 2018, Saint Petersburg.
15 Interview Pavlova, 29 April 2020, (Moscow) telephone.
16 Interview Kobysheva, 26 September 2018, Saint Petersburg.

References

Aivazova, Svetlana. 2003. "Liberty and Equality for Women in the Socialist Countries of Eastern Europe, 1960–1980." In *Political and Historical Encyclopedia of Women*, ed. Christine Fauré, 624–648. London: Routledge.

Allakhverdyan, Aleksandr G. 2014. *Dinamika nauchnykh kadrov v sovetskoi i rossiskoi nauke*. Moscow: Kogito-Tsentr.

Aronova, Elena, and Simone Turchetti. 2016. "Introduction: Science Studies in East and West – Incommensurable Paradigms?" In *Science Studies during the Cold War and Beyond*, ed. Elena Aronova and Simone Turchetti, 1–20. New York: Palgrave Macmillan.

Bonatti, Enrico, and Kathleen Crane. 2012. "Oceanography and Women: Early Challenges." *Oceanography* 25 (4): 32–39.

Buckley, Mary. 1989. *Women and Ideology in the Soviet Union*. Ann Arbor: University of Michigan Press.

Budyko, Mikhail I., Nikolai N. Gribanov, Vladimir Ya. Nikandrov, Nikolai P. Rusin, Oleg G. Sorochan, Nikolai S. Shishkin, and Mikhail I. Yudin. 1967. *Glavnaya geofizicheskaya observatoriia imeni A.I. Voeikova za 50 let sovetskoi vlasti.* Leningrad: Gidrometeoizdat.

Bullock, Carolin. 2017. "Why is Russia so good at encouraging women into tech?" *BBC News* [on-line] https://www.bbc.com/news/business-39579321

Butenko, Raisa. 1980. "Vashe kreslo v akademii" *Sovetskaya Rossia,* 136, 23 April: 4.

Creese, Mary, and Thomas Creese. 2015. *Ladies in the Laboratory IV: Imperial Russia's Women in Science, 1800–1900: a Survey of Their Contributions to Research.* Lanham MD: Rowman and Littlefield.

Dasgupta, Nilanjana, and Jane G. Stout. 2014. "Girls and Women in Science, Technology, Engineering, and Mathematics: STEMing the Tide and Broadening Participation in STEM Careers." *Policy Insights from the Behavioral and Brain Sciences* 1 (1): 21–29.

Dorokhova, Galina. 1975. "Education: That Is How Equality Is Attained." *Soviet Woman* 11: 12–13.

Evans, Janet. 1981. "The Communist Party of the Soviet Union and the Women's Question: The Case of the 1936 Decree 'In Defense of Mother and Child'". *Journal of Contemporary History* 16: 757–75.

Eveleth, Rose. 2013. "Soviet Russia Had a Better Record of Training Women in STEM Than America Does Today." *Smithsonian* [online] https://www.smithsonianmag.com/smart-news/soviet-russia-had-a-better-record-of-training-women-in-stem-than-america-does-today-180948141/

Fleming, James. 2020. *First Woman. Joanne Simpson and the Tropical Atmosphere.* Oxford: Oxford University Press.

Foucault, Michel. 1981. *Archäologie des Wissens.* Frankfurt: Suhrkamp.

Graham, Loren. 1992. "Big Science in the Last Years of the Big Soviet Union." *Osiris* 7: 49–71.

Goldman, Wendy. 2002. *Women at the Gates Gender and Industry in Stalin's Russia.* Cambridge: Cambridge University Press.

Grin'ko, Ivan, and Anna Shevtsova. 2019. "Nikomu ne nuzhno: kak sovetskikh utshenykh videla ofitsialnaya karikatura." *Al'manakh „Etnodialogi"* 58: 82–110.

Grin'ko, Ivan, and Anna Shevtsova. 2018. "'Dyatel dokladodolbyashchiy' i drugue zhivotnie: nauka v zerkale sovetksoi karikatury." *Istoricheskaya ekspertiza* 15: 276–291.

Gavrilova, Maria. 2007. *Moe pokolenie s obozhzhennymi kryl'yami.* Yakutsk: Bichik.

Hargittai, Magdolna. 2015. *Women scientists: Reflections, Challenges, and Breaking boundaries.* Oxford: Oxford University Press.

Ilyina, Valentina. 1980. "Her road to Science." *Soviet Woman* 1: 36–37.

Kalemeneva, Ekaterina, and Julia Lajus. 2018. "Soviet Female Experts in the Polar Regions." In *The Palgrave Handbook of Women and Gender in Twentieth-Century Russia and the Soviet Union*, ed. Melanie Ilic, 267–283. New York: Palgrave Macmillan.

Karavaeva, Galina, and Valentin Karavaev. 1989. *Mastera sovetskoy karikatury. Al'bom.* Moscow: Sovetksiy khudozhnik.

Khanukaeva, Anna, and Lilli Di Puppo. 2021. "Entrepreneur or Group Member? Women in Science and Career Strategies in Russia and Germany." In *Gendering Post-Soviet space. Demography, Labor Market and Values in Empirical Research,* ed. Tatiana Karabchuk, Kazuhiro Kumo, Kseniia Gatskova, and Ekaterina Skoglund, 257–275. Singapore: Springer.

Kneen, Peter. 1984. *Soviet Scientists and the State.* London: Macmillan Press.

Kosheleva, Inna. 1983. *Women in Science.* Moscow: Progress Publisher.

Kotkin, Stephen. 1997. *Magnetic Mountain.* Berkeley: University of California Press.

Koblitz Ann Hibner. 2000. *Science, Women and Revolution in Russia.* London: Routledge.

"Ladies of the Lab". 2019. *The Economist.* 20 July.

Novikova, Anna. 2019. "Zhenshchina-uchenyy na ekrane: superzhenshchina ili supergeroy?" *TRV Science* [online] https://www.trv-science.ru/2019/03/novikova-gambschet/

Organization for Economic Cooperation and Development. 2008. *Encouraging Student Interest in Science and Technology Studies.* [online] https://doi.org/10.1787/9789264040892-en

Pakaln, Karl. 1970. "In Penguin Land." *Soviet Woman* 4: 28–29.

Randall, Amy. 2011. "'Abortion Will Deprive You of Happiness!': Soviet Reproductive Politics in the Post-Stalin Era.", *Journal of Women's History* 23: 13–38.

Rendle, Matthew. 2008. "The Problems of 'Becoming Soviet': Former Nobles in Soviet Society, 1917–41." *European History Quarterly* 38: 7–33.

Rossiter, Margaret. 1978. "Sexual Segregation in the Sciences: Some Data and a Model." *Signs* 4: 146–151.

Rossiter, Margaret. 1995. *Women Scientists in America. Before Affirmative Action 1940–1972.* Baltimore, MD: Johns Hopkins University Press.

Ruthchild, Rochelle. 1983. "Sisterhood and Socialism: The Soviet Feminist Movement." *Frontiers: A Journal of Women Studies* 7: 4–12.

Selezneva, Evgeniya S. 1989. *Pervyye zhenshchiny – geofiziki i meteorologi.* Leningrad: Gidrometeoizdat.

Shnyrova, Olga. 2000. „Problemy vospriyatiya gendernykh issledovaniy v akademicheskoy srede." In *Pol. Gender. Kul'tura. Nemetskiye i russkiye issledovaniya. vyp.* 2 ed. Karolin Haider and Elisabeth Chaeauré, 232–233. Moscow: RGGU.

Sylvester, Roshanna. 2011. "'Let's Find Out Where the Cosmonaut School Is'. Soviet Girls and Cosmic Visions in the Aftermath of Tereshkova." In *Soviet Space Culture. Cosmic Enthusiasm in Socialist Societies*, ed. Eva Maurer, Julia Richers, Monika Rüthers, and Carmen Scheide, 121–138. Basingstoke: Palgrave Macmillan.

Sylvester, Roshanna. 2013 "Russian space history: soviet women in stem fields", *Russian History Blog.* https://russianhistoryblog.org/2013/12/russian-space-history-soviet-women-in-stem-fields/

Valkova, Olga A. 2019a. "Gendernaya istoriya nauki v Rossii. Nachalo." *Nauchnie Vedemosti, Seriya: istoriya. Politologiya*, 46: 544–554.

Valkova, Olga A. 2019b. *Shturmuya tsitadel' nauki. Zhenshchiny-uchenyye Rossiyskoy imperii.* Moscow: Novoe literaturnoe obozrenie.

Valkova, Olga A. 2008. "The Conquest of Science: Women and Science in Russia, 1860–1940." *Osiris* 23: 136–165.

Yurchak, Alexei. 2005. *Everything Was Forever, Until It Was No More: The Last Soviet Generation.* Princeton University Press.

Zharkynbayeva, Roza, Ardak Abdiryaimova, and Gulnara Dadabaeva. 2020. "The Evacuation Policy of the Soviet Union during the Second World War." *The Journal of Slavic Military Studies*, 33: 403–419.

7 The Women Who Count

Gendering Calculations in the Soviet Atomic Project

Galina Orlova and Aleksandra Kasatkina

Telling the story of the Soviet hydrogen bomb, Vladimir Gubarev, the Soviet science journalist, and chief chronicler of the Atomic Project uses the framework of a fairy tale to recount the extraordinary daily life in a secret settlement in the Southern Urals:

> Physicists reflected on the mystery of matter, penetrated its depths, and discovered the essence of enormous energies and the highest temperatures. They tried to tame the fiery elements to put it in their service. And they succeeded! The girl-calculators, using their primitive computers, tried to help the physicists as best they could, and in the evenings, they kissed them, so that they could later get married and stay there forever. In general, there was no place more fabulous on this planet.[1]

The male world of physicists, which was created alongside the Bomb, was a space of cultural heroes of the nuclear age, whose scientific work was represented in terms of masculinity. The Soviet atomic agency, solving technological and biopolitical problems, mobilized young women as human calculators and life companions. Through fictionalization, Gubarev makes visible the gender pattern that characterizes the style of governing the Soviet Atomic Project in its first decades. Retrospective accounts of the displacement of women calculators, and their use as support staff in the development of nuclear weapons, contradictorily combine socialist mobilization with gender inequality and emancipation with objectification.

According to Elena Malinovskaya's recollections, in the arrangements of the Atomic Project, a *woman who counts* was subjected to mechanical objectification—her talents, agency, creativity, and personality were reduced to the body of a computer:

> When I studied at Moscow State University, the so-called practical training in maths was first introduced, where we were told to calculate something with a manual arithmometer, *Felix*. I remember a Komsomol meeting and a debate on the topic: 'What is happiness, and how

DOI: 10.4324/9781003562597-10

> not to become an appendix to the arithmometer?' For the first two years I was, indeed, mostly an appendix to the arithmometer. However, I acquired a sense of the problem calculation.[2]

Our chapter is about those women computers who, like Malinovskaya, performed auxiliary functions, calculating measurements for bombs and reactors with arithmometers in the 1940s and 1950s of the Atomic Project. Until now, their work in a male team, and life in closed cities, their experience and agency have not been subjected to academic scrutiny. We show that human calculators were engaged in the Soviet nuclear program within a service-oriented and gender-marked framework. Using a collection of institutional histories, nuclear memoirs, declassified files, and transcripts of interviews with Vladilena Dmitrieva, the first calculator at a research institute in the atomic city of Obninsk, we consider the feminization of calculations within the Atomic Project and their sexist devaluation as a gender bias in governing the large-scale Soviet scientific and technological programs.

Not a Woman's Business?

Observing the debate among American nuclear weapons experts, anthropologist Carol Cohn described how the anesthetic properties of the techno-strategic language allowed them to discuss the nuclear war in gendered terms of abstraction, alienation, and phallogocentrism (1987). The masculinity of the Soviet nuclear discourse was produced through swearing and greasy humor, i.e. genres of speech which were harmonious with the spirit of the command economy. For example, the nuclear Minister Slavsky, known as "an exceptional swearer", once failed to chair a meeting concerning a plutonium factory construction in the presence of a woman who stood in for her boss. Slavsky could return to his management style, bursting out with the choicest swearing, only after she had left.[3] The moral of this story is that governing the Atomic Project was not women's business.

If only because of Maria Skłodowska-Curie and Lise Meitner, whose contributions as pioneers of studying radioactivity and artificial fission were unconditionally recognized on both sides of the plutonium curtain, it seems the situation in nuclear science would have been different. However, Meitner faced Brockhaus's refusal to countenance an encyclopedia article on radioactivity from a female author and never won the Nobel Prize for discovering uranium fission.[4] As for Skłodowska-Curie, it was extremely frustrating that her female students, who provided "a better role model for the potential women physicists" than their supervisor, were totally neglected by biographers (Rayner-Canham and Rayner-Canham 1990).

Even less was said about women's impact on the applied use of *nuclear*, which was inseparable from the development of weapons. Howes and Herzenberg (2003) open the discussion on the invisibility of 1000 women[5]—physicists, chemists, biologists, officers, typists—in the Manhattan Project, and the gender asymmetry of American nuclear history. Criticizing the position from a feminist point of view for legitimizing militarism, Broadhead turned to another invisibility—the contribution of those scientists (women and men) who refused to participate in the development of the bomb (2009).

In the Soviet case, gender equality in higher education and science was proclaimed immediately after the October Revolution. By the end of the 1920s, women accounted for a third of all academics, and by the 1970s for half. However, they occupied predominantly low positions and faced a glass ceiling (Pushkareva 2011; Dolgova and Streltsova 2019).

Only a handful of women made academic careers in early Soviet physics (Stepnov 2019). The sharp increase in demand for physicists with the beginning of the Atomic Project, which occurred after the end of WWII, and an overall shortage of personnel contributed to the social mobilization of educated women into the new field. In 1940, there were 100 scientists specializing in nuclear physics in the USSR[6]; in 1947, 300 students comprised just one cohort attending the course offered by the Faculty of Physics at MSU. Among the best students transferred to the department of solid-state physics from all over the country for accelerated preparation in special work was Tamara Kazachkovskaya. A student of the Dnepropetrovsk Faculty of Physics, who continued her studies after returning from evacuation, was sent to Moscow in 1946.[7] Galina Lovchikova, who graduated from the secret faculty of the Leningrad Polytechnic Institute in 1951, was assigned to Laboratory "V".[8] Kazachkovskaya, nicknamed Tsarina Tamara by Kurchatov himself, headed the experimental group and defended her PhD thesis. Lovchikova was the first Soviet woman to obtain a Habilitation in Nuclear Physics. However, these success stories stand out from the boys' club that nuclear physics had remained.

Galina Nikolaeva's notebooks confirm this. In 1962, she was gathering material for a novel about a young and talented female physicist, the *Soviet Madame Curie*. In search of a prototype, Nikolaeva talked with female researchers from the Joint Institute for Nuclear Research in Dubna, seeking a woman's path to nuclear science. The records consist mainly of complaints about the deep-rooted lack of confidence in female physicists which manifested itself "from the very first test" taken at university, further discrimination in the lab, and above all, the secondary roles they were given.[9]

Kasperski and Josephson (2023) are cautiously optimistic about the emancipatory potential of the Soviet nuclear industry against the backdrop of post-Soviet *Miss Rosatom* pageants, in which the participants competed not in professionalism, but in attractiveness. As an example of a female with a nuclear career, the story of engineer-physicist Galina Petkevich is mentioned. She supervised the launch of the Kola NPP unit and participated in the top management of the Soviet nuclear power industry. A single woman's success elucidates the mandatory ritual of the Soviet politics of representativeness, which has no connection to full-fledged nuclear emancipation.

While Nikolaeva searched unsuccessfully for a *Madame Curie* among young nuclear physicists, the chemist Maia Vladimirova found one in the field of radiochemistry: Zinaida Ershova. She was identified as the *Russian Madame Curie,* who stood at the origins of nuclear material science and received the Stalin Prize for developing technology for plutonium recovery.[10] However, she never led the NII-9 Institute specializing in metallic plutonium and was praised, albeit only in comparison to men, by her male biographers as "the only woman among the leaders of the Soviet Atomic Project who was not inferior to men" in her expertise, contribution, and professionalism.[11]

At the beginning of the nuclear age, women made up about 60% of ordinary radiochemists engaged in painstaking and hazardous work with fissile materials (Melnikova 2017). After half a century of silence, the nuclear veterans admitted that plutonium was held "in young women's hands".[12] As we show, Soviet radiochemistry was not the only nuclear area where the main workload was performed by women. Another realm was mathematical computing.

Calculating the Atomic Project

There are many potential starting points for the Atomic Project (Holloway 2008; Kojevnikov 2004): first, the creation of the Commission on the Uranium Problem at the Soviet Academy of Sciences (AS USSR) headed by Vitaly Khlopin in June 1940; second, the appointment of Igor Kurchatov as the scientific supervisor of the *Problem* after the USSR had obtained its first intelligence data on the Manhattan Project in February 1943; third, the establishment of the Soviet nuclear agency, or the First Chief Directorate (PGU) after the US bombing of Japanese cities in August 1945. Whichever version we choose, it does not include any consideration of mathematical calculations.

The applied use of nuclear energy was an unexplored and expensive area. In the absence of technology, the strong demand for rapid product development turned theoretical physicists into the main experts on taming the atom. It was they who first modeled the behavior of radioactive matter in a nuclear explosion without any possibility of experimental verification. Since February 1946, when the guru of Soviet theoreticians, Lev Landau, started to calculate the output of an atomic bomb, his group consisted of two male physicists and 30 young women calculators.[13]

Sergei Sobolev, the first mathematician in the Atomic Project, constructed the models for isotope separation in the mid-1940s.[14] In the early 1950s, Mstislav Keldysh, unofficially called the "chief mathematician of Sredmash", carried out the general mathematical support of the problem in the Institute of Applied Mathematics AS USSR (IAM).[15]

Mathematical groups joined the nuclear labs with the development of thermonuclear weapons, which required huge and complex calculations.[16] A group headed by Modest Agrest was organized at the theoretical division at KB-11 in 1949, and he requested an increase in staff in 1950, to cope with the amount of work.[17] Lab "V", which also participated in bomb calculations, was reinforced with mathematicians in 1950 and in 1953. Advanced mathematicians came from other units and dozens of graduates from eight-month "courses of mathematicians-calculators" *to special topics* were hired.[18] Their arrival precipitated the shift from the method of approximate computations by physicists to more precise ones made by mathematicians (Vizgin 2011), which required not only heuristic mathematical models for accuracy but also a new scale of calculations. Men developed the models requiring collective bodies of calculators using female workers.

The word *calculator* had different meanings in the Soviet nuclear discourses. The most outstanding physicists at the top of the research pyramid were "virtuoso computers (*vychislitel', masculine*)", who "enthusiastically delivered lectures on quantum mechanics with very little use of mathematical apparatus".[19] The *calculator* (*rassshchetchik*, masculine) referred to engineers and designers with a tertiary degree who used mathematical methods to define the technical specifications. Sometimes it was applied to cyberneticians engaged in modeling, or mathematicians who built algorithms. In the *firma* specialized in reactor design, there was no clear division of responsibilities between designers, calculators, and testers.[20] Specialists, including young ones, were entrusted with the development of a unit from performing calculations to assembling the unit at a plant.[21]

The *calculators* (*rasschetchitsy*, feminine) occupied a special position with a job description that implied the technical execution of computational mathematics on semi-automatic devices. The advent of machine computers turned *calculators* into computer operators.

Counting Factories

Moving down the staff hierarchy, the number of women involved in computations increased, while the number of female names left in Soviet nuclear history decreased. Computer-technical positions for nuclear developments were occupied exclusively by young women. Similarly to typists or telephone operators (other "female professions" that served modern organizations with technologies and were perceived as machine attachments), *calculators* were united in bureaus. Bureaus serving different organizations were called *counting factories*. Before World War II, there was such a bureau at the Central Statistical Office (among other things, they calculated the census results). Another served military topographer.

There were several factories serving the needs of the Atomic Project. Thirty to fifty female proletarians of computational labor were employed as auxiliary personnel to perform standardized line calculations. Calculators were supervised by men with higher education and directed by an eminent male mathematician. After hydrogen bomb testing, calculators received awards together with physicists and mathematicians (Vizgin 2011: 11).

The Kantorovich division—the counting factory at the Leningrad branch of the Mathematical Institute—was the highest-ranked.[22] Among the participants of the Atomic Project, geologists and geophysicists searching for uranium were the first who organize a computing bureau. Staff from Lab No. 8, which calculated the explosions of plutonium and uranium balls for a year, was part of the Complex Geophysical Expedition (Sirotinin 2005). It was located in a building on Moscow's Kirov Street, under a sign reading "Retail Vegetable Warehouse", as was usual for secret subdivisions of the PGU.

The calculators occupied 4–5 small rooms on the ground floor. The mathematicians Samarsky, Goldin, Yanenko, and Rozhdestvensky coordinated the work of anonymous female workers employed "instead of computing machines".[23] Parallel computations—were introduced by Samarsky for the accelerated development of a "numerical model of the Bomb":

> I am proud of my idea of parallel computations. There were thirty girls working under me. And there were several hundred equations. Approximately ten equations per girl. They counted independently but passed their data to each other... [...] The parallelization of calculations enabled us to carry out the computations in two months, we made the work process about fifteen times faster. I consider this the biggest achievement in the first year of work on the Bomb.[24]

Comparing the American organization of computations with the Soviet one, Samarsky suggests that "in Los Alamos, it was the physicists who carried out the calculations. This is a fundamental difference".[25] Richard Feynman, a witness to the advent of computing at IBM in the Manhattan Project and one of its coordinators, describes the gender division of automated labor of skilled male mathematicians and unskilled female calculators.[26]

The idea, which was "a kind of new thing then", was to concentrate enough IBM machine in one room and "taking the cards and putting them through a cycle"[27] belonged to Stanley Frankel. This work was carried out by a Special Engineering Detachment recruited from "clever boys from high school who had engineering ability"[28] among military personnel. Feynman called them "my boys" and considered them as equals, linking the quality of their work with an understanding of the calculated problem and, consequently, with a deliberate weakening of the secrecy. Since automatic calculations were new and unreliable, they were checked manually by a "living computer"—a group of female calculators without special education working in the next room. Feynman was concerned with their efficiency, not awareness: unlike IBM, the girls could not continuously verify the results of calculations without getting tired.[29] Women were recruited from among employees' wives at Los Alamos (Light 1999). A double—sexist and capitalist—employment logic is revealed in the gender economy of recruiting: "We hire girls because they work better and they're cheaper" (Metropolis and Nelson 1982: 352).

Using the binary opposition in accordance with the spirit of the Cold War, Samarsky set the Soviet low-skilled labor of many against the highly-skilled

American labor of a few and distributed collective human action against an imperfect automatic device. In the USSR, young women included in the chains of computations made up for the lack of machine computers. In the United States, they corrected the quality of IBM's work. However, the generic connection between a woman and a machine underlined both national visions of computing development through the objectified narratives of the "living computing machines" performed by the male voices of mathematicians and mentors. Can "living computing machines" speak?

Agency through Calculation

We have no accounts from the women employed at the Moscow or Leningrad counting factories. The calculators assigned to the mathematical divisions at Arzamas-16 and Chelyabinsk-70 left memoirs about their bosses (mathematicians) or husbands (physicists), published in the *Sredmash* volumes celebrating the anniversaries of the defense industry's fathers [sic].

The mathematical bureau established at the theoretical division of KB-11 in 1949, was one of the first computing units within nuclear infrastructure. Before 1954, when "young women, who were experienced geodesists-calculators, mainly transferred from Leningrad", was known as "the domestic period".[30] By 1951, the bureau already resembled the counting factories in terms of size, organizational architecture, and masculine language, which objectified the intellectual labor of female calculators. Just like in Los Alamos, they were recruited from the wives of physicists (Vizgin 2011). The female calculators' ability to act autonomously was defined in terms of their limited agency, where the intuition of an ignorant subject "feels the solution without knowing its meaning".[31] A male mathematician might not have known the physics-related meaning of the task either, but his ignorance was explained with secrecy.

Writing the history of the mathematical division of KB-11, male mathematicians not only mention female calculators by name but also recognize the merits of their female colleagues Vladimirskaya, Kemper, Vesnina, and Zhernak in training staff and developing new calculating methods. The latter coordinated the work of calculators, connected them with theorists, developed their own careers, defended PhD theses, headed working groups, and produced female versions of the organization's history, offering the most detailed story of a counting bureau in the Atomic Project ever written.

Elena Malinovskaya was the only one from her classes who was assigned to KB-11 in 1949 after graduating from the Faculty of Mechanics and Mathematics of Moscow State University (MSU). She described her career trajectory at KB-11 as a movement from technical objectification to the acquisition of professionalism. Malinovskaya became the first engineer-mathematician admitted to the counting group. Recollecting, she did what her male colleagues did not—listed name by name all the computers and lab assistants and described how they were motivated through a course of lectures on the prolegomena of analytical geometry and mathematical analysis: "After examinations, they were promoted in their positions and salaries.[32]

Malinovskaya painstakingly described the computational routine as an uncomfortable workplace, a gender-marked gratitude of customers, and a timescape of emergency and mobilization:

> There were 5–7 people and the same number of counting machines in the room. Two breaks of half an hour. Eternal emergency like 'Come on, come on!'. But only in extreme cases we stayed after the eight-hour working day. An average calculation task took about a month. The end of a project was a celebration, when the customer presented the calculators with sweets.[33]

From this perspective, parallel computations of "living computers" look rather like a musical ensemble, than an assembly line: "The calculators Voroshilova and Bronnikova began four-hand calculation of a one-dimensional shock wave on the *Mercedes* as a duet".[34] Traditions became as important here as algorithms. The duration of a calculation indicated the imperfection of technology but also the personal courage of a calculator, while their completion was described in terms of both e(a)ffectivity.

Security restrictions are commonplace in the memoirs of the Atomic Project participants, and they recalled how the secrecy taught them not to trust paper. The story of Malinovskaya became special twice. First, when she described the classified notebook where physicists wrote project assignments as looking like a girl's autograph album.[35] Second, when the calculator asserted her own right to the agency under nuclear secrecy: "Adamskiy wondered if I knew what particles we were counting, asking me: 'What is the last letter?' 'The letter N'. 'Wow, she knows!' he said with surprise".[36]

The taleworld of Vera Avrorina, that of a computer from Chelyabinsk-70, "the youngest closed city", was different. The Center for Nuclear Weapon Development, established in the Urals in 1955 to keep Azamas-16 in a competitive tension, was staffed mainly with KB-11 youth. Fourteen 18-year-old women, who recently graduated from the one-year calculator course at MSU, were sent to be at the disposal of the physicists. They came to be named "14 × 18".

Parallel calculations—breaking into sectors, counting four-hands, comparing results, and passing the work to the next female worker—are mentioned by Avrorina. It was her way to tell how she completed the task faster, sparing time to play tennis with physicists. Avrorina treated her work as the interplay of figures. The physicist provided the figure obtained from the calculations on a slide rule with two decimal places, while the calculator, in the act of checking, "extended" this figure to 12, and sometimes 16 decimal digits.[37] She recalled a prominent physicist who "stood behind her and breathed into the back of her neck" waiting for the outcome and depending on it.

In keeping with the genre, Avrorina's story revolves around her meeting of her future husband, one of the USSR's leading nuclear weapons designers. In her perspective, the first meeting with physicists, which turned into a quick marriage for the two young calculators, was love at first sight: "And indeed, a week later the acquaintance continued. In a year there will be a wedding of the Vakhromeyevs, and in another month, there will be a wedding of the Avrorins".[38] In the version narrated by one of the husband's classmates, the former's choice was determined by more entangled logic, including the limited marriage options owing to the biopolitical predestination of living in the closed nuclear city.

Girls with Mercedes

In the nuclear narratives, male physicists calculated with slide rules, while female calculators used arithmometers. Slide-rule calculations were analog, rounded off, and approximate. The length of the rule influenced the calculation quality becoming a subject "of special envy": "Everyone had twenty-five-centimeter rules, and he [July Khariton] had a half-meter one".[39] The supervisor of Arzamas-16 appeared in the memoirs as "a short person" with "the largest slide rule ever seen", which represented the symbolic capital of both, man and mission.[40]

Everything was different for female calculators, who existed in male narratives only if an arithmometer was mentioned as in the *girls with Mercedes* case. Machine calculations were digital, accurate, and determined by the properties of the arithmometer (the availability of an electric drive and keyboard, the number of operations per second, and its automatic counting function). Paired up with the calculation device in a hybrid body of counting, nameless, and invisible female calculators made the specifications of the device visible.

If it was the mechanical arithmometer known as *Felix,* produced in the USSR since 1927, and named after the head of the Cheka secret police, Felix Dzerzhinsky,[41] the young women moved the levers and "turned the handle" to calculate the result. The *Triumphator* machine (from the German Democratic Republic) compared favorably to the *Felix* because of the former's convenient tens indicator on the revolving counter and its heavy stand that prevented the device from moving around.[42] *Rheinmetall*, an automatic counting machine with an electric drive, was jokingly called "the miracle of computing" because of its low speed, and "one floating point operation per second" without the time required for data entry.[43] Without any distinctions in the counting device specification, groups of young women constantly tapping out numbers all at the same time made a tremendous noise "like a machine gun, so the bureau was in a special room with padded walls".[44]

The keys and continuity of the operation helped to identify the *Mercedes Euklid* modèle, an electromechanical machine produced by *Mercedes Büro-Maschinen Werke A.G.* since 1913 and "obtained from Germany through reparations" as "the fastest machine employed for especially voluminous calculations".[45]. The keyboard, supporting 12 positions, facilitated operations with multi-digit numbers and precise calculations that extended the approximate computations of theoreticians to the tenth decimal place. The files of the Atomic Project mentioned the purchase of *Mercedes Euklid-37 MS* and *R-38* series, which were produced from 1937 to 1944, for PGU needs, under the item "various equipment, instruments and complex calculating machines" of the Trade Agreement between the USSR and the GDR in 1948,[46] 1951, and 1953.

The first samples of the domestic *Strela* computer were handed over to Keldysh's IAM in 1954 and to KB-11 three years later. While the norm of a qualified operator of an arithmometer was 800 operations per day, *Strela* performed 1,000–2,000 operations per second.[47] No wonder that arithmometers were rapidly becoming outdated together with the professional occupation of a calculator. Some people, like Avrorina, retained a generic connection with changing technology and moved to the programming division to become a machine operator in the late 1950s. However, in the memories of the nuclear veterans, they remained forever "girls with *Mercedes*".

Revising Nuclear Hierarchy

Both the physicists and mathematicians reinforced the gender positioning of young female calculators engaged in the Atomic Project in the mid-twentieth century through epistemic hierarchies, service status, invisibility, and technical reduction. The same patterns were evident in the attitude of physicists to mathematicians, regardless of sex or age. The mathematicians were outraged that the physicists valued them in a utilitarian way as "an appendix to arithmetic counting skill", stayed blind to the beauty of their science,[48] and reduced the original theoretical result to technician's craft.[49] The revision of disciplinary hierarchies, and contributions, was produced by the last President of the AS USSR, Guriy Marchuk, in vivo of his academic youth and by the Marchuks couple in vitro of their memoirs.

After her graduation from the Chemistry Faculty at Leningrad University, Olga Marchuk was assigned as a radiochemist to a PGU industrial unit near Moscow, working hard and raising a son. Her husband, after his PhD thesis defense, built mathematical models for meteorologists at the Institute of Geophysics in Moscow, visiting his family on Sundays. She dreamed of reuniting the family unable to leave her job because of an obligation of every Soviet student to work for three years after graduation in an assigned location. He was not ready to give up a scientific career in the capital, by moving to be with his wife and family in an industrial town.

In 1953, the Marchuks reunited at the secret nuclear Object "V" between Moscow and Kaluga. He had to leave a job he liked and to master the nuclear field. She remained a radiochemist and gave birth to two more sons. For ten years, Guriy Marchuk worked in Lab "V", creating a powerful mathematical department and developing a system of "numerical calculation" for the reactors that questioned the hegemony of physicists.[50] Before leaving, the Marchuks witnessed how a settlement of physicists transformed into the city of Obninsk, and the Lab "V" became the Institute of Physics and Power Engineering (IPPE). The spouses recall both together and separately, two episodes that redefined the relationships between gender, atom, calculation, and power.

The first episode concerns the genealogy of Marchuks' appearance at Object "V". Spouses agreed that they ended up there by the will of PGU. However, Olga treated it as a generous response of the nuclear authorities to the request of a young woman driven to despair by the long separation from her husband.[51] At the same time, Guriy told a spy story about mobilization without any mention of matrimonial issues. In broad daylight, a silent stranger took him from his office in Moscow and brought him to a secret lab in a picturesque forest. A man in uniform informed the startled mathematician about his new workplace. Hearing about Dmitriy Blokhintsev, a well-known theoretician, Marchuk became convinced that "there is science" here. "The case takes a completely different turn" when his wife was transferred to the Object.[52] Both versions omitted the context that mattered.

In 1953, the research staff at Lab "V" was strengthened with mathematicians for implementing the projects, which needed voluminous calculations, i.e. the so-called *First in the world* NPP development and a thermonuclear bomb counting (Vizgin 2011: 64). Marchuk was among those, who "for two years were engaged in thermonuclear matters".[53] However, he did not associate his transfer with these "matters". His wife also did not comment on the gender bias of radiochemistry, nor did she mention the need for qualified radiochemists at the budding nuclear lab. The same resolution that transferred Guriy to PGU, assigned 240 people to the staff of Lab "V". Olga was one of these anonymous units.[54] Her request might have made nuclear authorities notice a promising mathematician from a non-core institute. Nevertheless, the nuclear mobility of the Marchuks corresponded with the HR agenda of the Atomic Project. Private and public, gender and mobilization, applied and fundamental science, women's dissatisfaction with family separation, and the need for mathematicians to calculate a bomb are entangled in their case.

The second episode deals with the triumph of the mathematicians over the physicists. In half a year, Marchuk evolved from an "absolute layman" who did not understand the theoretical seminars and the physicist's jargon to a unique expert who was able to see the shortcomings of the "consolidated method" and correct poorly posed problems.[55] He started his self-education in nuclear physics with books borrowed from his wife's collection, who for the first time felt "her superiority over her husband at least in anything".[56] Then he fought against "physical illiteracy" in his laboratory: twice a week, the mathematicians read the handbook on nuclear reactors chapter by chapter. As a result, Lab "V", where the physicists reigned and "all the other departments were considered auxiliary",[57] recognized the right of a mathematician to produce critical knowledge on reactors, and Marchuk was nominated for the Lenin Prize.

Marchuk recalls as Lev Usachev, one of the top theoreticians of Lab "V", articulated the surrender of the physicists' class: "Before Marchuk, the mathematicians were the slaves of the physicists, while under Marchuk, the physicists became the slaves of the mathematicians".[58] Forty years later, he added ruthlessly that "there were no new interesting problems for him anymore in the mathematical theory of the reactor's calculations".[59] Thus, the mathematician stated the exhaustibility of the atom, contrary to Lenin's famous formula, featured in the Obninsk billboards.[60]

The Soul of a Research Institute

> An employee can be nominated for the medal "To the First Calculator of IPPE" if (s)he: a) has worked at IPPE for at least thirty years, b) is in fact the first calculator of IPPE, c) can perform calculations, incl. division and extraction of the square root, long division and multiplication on a piece of paper, with slide rules of all models, with a *Phillips* arithmometer and other instruments hardly suitable for counting, d) knows the blind method of working with Barlow's and Semendyaev's tables, e) has at least 1 year work experience at the Mathematical department of IPPE, f) is personally acquainted with G. Marchuk, g) is quite active and sociable in the team (knows at least 75% of the institute), h) skilfully uses the qualities listed above for the benefit of production.[61]

The paper with this comic algorithm kept in the home archive was a 50th birthday present to Vladilena (Gudkova) Dmitrieva (1929–2019) from her colleagues. The algorithm identifies Dmitrieva not only as the first calculator at Lab "V" but also as an oligoptikon in the female world of calculations within the Atomic Project too (Latour 2007: 181).

Vladya Gudkova moved with her family to the future Object "V"/Obninsk in 1944, two years before the arrival of the physicists. Her oral narratives, saturated over 70 years, tie together positions, practices, artifacts, and feelings scattered around the other stories of female calculators within the Atomic Project we have touched upon: a school graduate counts for physicists; ignorance of secrecy; hunting the digits after the decimal point; tight schedule and enthusiasm of mobilization.[62]

In 1947, Vladya Gudkova, who dreamt of becoming a doctor after finishing school, worked as a barmaid at Lab "V" to help her family. Alexander Leipunsky, the scientific leader of Lab "V", noticed how skillfully the young woman manipulated the abacus and invited her "to calculate something". Her father, once a maths teacher, gave her his slide rule.[63] The calculator fit perfectly into the team of young theoreticians, drawing funny posters for seminars, hiking, and skiing with them. Before the Marchuks arrived, the "first computing bureau" at Lab V was a homemade assemblage of school maths skills, canteen abacus, and dad's slide rule.

People and relationships were much more important for Dmitrieva than her connection with the Atomic Project. She did not know what she was counting when she performed calculations for the *First in the World* NPP. The point was not the exoticization of nuclear secrecy, but a friend Lev Usachev who told her the secret out loud in the middle of the street right after the newspapers reported about the birth of *the peaceful atome,* in June 1954.[64] She brought the atmosphere of warm and equal relationships from her youth at the secret Lab "V" to a much more subordinate era of IPPE.

Like the other "girls with *Mercedes*", Gudkova worked on the counting machines, translating the sequence of devices in a professional biography:

> There was a *Mercedes*, after that a *Rheinmetall*. It is like 'te-te-te-te', *(imitates a machine sound)*. And I was able to count also on the *Triumphator* before the machine finished. On the two machines simultaneously! Like a machine operator (*stanochnitsa, feminine*)! *(smiling)*.[65]

There was not an irony in her smile, but a working-class pride of a virtuoso, which marked the industrial logic of calculations in nuclear science.

The "domestic period" of calculations without special training, staff, scale, and infrastructure was best elaborated in both corporative nuclear memory and Dmitrieva's own collection of tales. She told about her "life afterwards" in a less detailed and more formulaic manner. Like many pioneers of calculations, Dmitrieva mastered changing technologies by herself and supervised younger IPPE computers. Through the sequence of devices, recognized as a sequence of possibilities, she came to feel her work as creative and intellectual:

> This is the M-220 already, yes. Three-address. It's... even better than that one, *Ural*. So, the programming is very interesting! You do a cycle in a cycle, a cycle after a cycle – and the program is reduced to several punched cards! *(eagerly)* So great! What is interesting... if you haven't written the program in detail, you don't understand what you programmed. It means, the program, which you wrote, no one else can use.[66]

In the 1960s–1970s, when programming provided personal creative solutions, the calculator felt herself truly expert, able, and irreplaceable, but still in a serving position. Her assistance was essential to those physicists, who could not count. In the late 1960s, Dmitrieva worked for a year in "dissertation slavery", making complex calculations for an unsociable physicist in the next-door department. "Slavery" was a joke, but positioning calculation as a service was quite real. The acknowledgment of the mathematics that Guriy

Marchuk enjoyed did not change the attitude of the nuclear hegemon toward day-to-day calculations.

Decades later, the unsociable physicist kept greeting Dmitrieva, presenting small gifts, while his female colleagues revealed how they call her "our Little Sun". Within Vladilena's narrative her nickname and motto ("I like to feed and medicate people") performatively turned a slave into a helper.[67] For Dmitrieva, who devoted herself to her family and never defended her dissertation after graduating as an engineer-physicist, the horizontal interconnections of friendship, saturated narrative memory, and the moral strength of caring were the way to overcome the calculator's low position both in the hierarchy of the IPPE and in the historical atomscape of Obninsk.

Conclusion. Theoreticians Start and... Lose?

In 1966, Nikolai Rabotnov, a 30-year-old physicist from Obninsk, underwent six-month training at the Niels Bohr Institute, University of Copenhagen. An intellectual, an expert in New Wave Cinema, and a co-author of the "thaw" bestseller *Physicists Are Joking*[68] brought with him the sexism of a Soviet theoretician to Denmark being convinced that calculating was women's work:

> I worked even harder, today I was at the Institute until seven o'clock. It's too bad there are no assistants here. They do not know what a calculator [*rasschetchitsa*] is, so I will have to master the local machine, otherwise these numbers here and there make me mad. Having a computer, they consider it madness to keep girls employed as well. As far as I know, there is only one mechanical device [*arithmometer* – G.O., A.K.] in the whole Institute, and I am the only one who uses it.[69]

The ability to use computers and write programs had already become a common skill among Rabotnov's foreign colleagues. The diaries that he kept in Copenhagen provide insights into how a Soviet theoretician under the pressure of circumstances discovered, recognized, and partially eliminated his technological lag, while simultaneously abandoning a gender-labeled discourse to describe calculations. He wrote and debugged programs, discussed the electronics quality, and learnt to operate a machine—first with a student who worked part-time as an operator, and then on his own. Returning home, Rabotnov continued to program and used the computer jargon to compare his soul with software.[70] One physicist had changed, but Soviet atomic science did not, revealing its archetypal gender order through Gubarev's fairy tale.

Three points are significant for the story on the gendering of computations within the Atomic Project.

First, in the 1940s–1950s, the groups performing voluminous joint and distributed calculations for the Soviet physicists were homosocial. They consisted mainly of young women with secondary, vocational, or (less often) tertiary education. Thus, the history of calculations in the Atomic Project turns into a gendered history.

Second, the invention of "living computers" became the response to the Soviet thermonuclear weapons development accompanied by the decreased role of intelligence and the increased impact of mathematical calculations. With "living computers", the new form of gender-marked objectification emerged, i.e. the reduction of women's labor, participation, and personality to a mechanical device. At the same time, physicists literally stood behind young women who made calculations, waiting for them to provide both figures and solutions. It means that the gender asymmetry, where a female calculator was both invisible and significant, requires revision in terms of joint work and collaboration.

Finally, the calculations within the Atomic Project acquired a service and gender-marked character regardless of the status and sex of those who counted. At the start of the digital age, when the computational paradigm arose, the epistemic rigidity and conservatism of Soviet physicists, including gender labeling, strongly restrained knowledge production. By increasing the visibility of female calculators and their contribution to the large-scale technoscientific program, we tackle an issue that is both ethical and epistemological.

Acknowledgments

This article is the output of a research project implemented as part of the Basic Research Program at the National Research University Higher School of Economics (HSE University).

Notes

1 Gubarev, V. 2012. *Superbomba dlya superderzhavi*, 163. Moscow: Algorithm. (Author's translation).
2 Malinovskaya, Y.V. 1995. "O stanovlenii matematicheskogo otdeleniya". *Khochesh mira —bud silnym*, 75. Arzamas-16.
3 Zhuravlyev, P 2003. *Moi atomnyi vek*, 93–95. Moscow: Chronos-Press.
4 Meitner, L. 1960. "The Status of Women in the Professions". *Physics Today*, 13(8): 16–21.
5 In the 2010s, women made up "less than a quarter of the workforce in the nuclear sector worldwide" according to the NGO *Women in Nuclear*, which has 35,000 members and fights to achieve gender parity in the industry (Gaspar and Dubertrand 2018).
6 Postanovlenie GKO № 7572ss/ov. 2002. "O podgotovke spetsialistov po fizike atomnogo yadra". 21 fevralya 1945. *Atomnyy proekt SSSR: dokumenty i materialy*, v1(2): 223–225. Moscow: Nauka.
7 Kazachkovsky, O. 2010. *Zapiski fizika o voine i mire*, 256. Obninsk.
8 Galina Lovchikova, interview with A. Kasatkina, Obninsk, 17 July 2013: GNZ - RF FEI. From the Obninsk Project Archive (OPA).
9 RGALI. F. 2292. O.3, Ed.36. Galina Nikolaeva. *Tetrad s zapisjami besed s fizikami*. 1962.

10 Vladimirova, M. 2004. *Pervaya ledy sovetskoi atomnoi nauki*. Moscow: VNIINM.
11 Kiselyev, G.V. 2005. "Veterany pishut istoriyu atomnoi otrasli" *Bulleten' po Atomnoi Energii*, 3: 75.
12 Sokhina, L.P., Y.N. Kolotinskiy and G.V. Khalturin. 2003. *Plutoniy v devichikh rukakh*. Yekaterinburg: Litur.
13 Landau's calculation group was organized based on the decision of the Technical Council of the Special Committee under the Council of People's Commissars of the USSR on February 11, 1946. "Protocol No. 18". *Atomnyy project SSSR USSR: documenty i materialy*, v2(4): 74.
14 Smorodinskiy, Y. 1998. "Eto bylo neobychayno interesnoe i udivitelnoe vremya". *Istoriya sovetskogo atomnogo proekta*, v1: 202. Moscow: Yanus-K.
15 Bogunenko, N.N., A. Pelipenko and G.A. Sosnin. 2005. *Geroi atomnogo proekta*, 183. Sarov: VNIIEF.
16 Greshilov, A.A., N.D. Yegupov and A.M. Matushchenko. 2008. *Yadernyy shchit*, 400–419. Moscow: Logos.
17 Vodopshin, A.I. 2012. *31 god, 2 mesyatsa i 3 dnya raboty s akademikom Yu. B. Kharitonom*, 193. Sarov: Interkontakt. KB –*konstruktorskoe byuro* or design bureau. KB-11 (or Arzamas-16) was the codename for the core engineering unit engaged in the development of the Soviet (thermo)nuclear weapons.
18 In 1953, the courses were organized at the IAM with 1 million roubles allocated to train 60 calculators. "Rasporyazhenie SM SSSR № 756-rs, 12 yanvarya 1953". *Atomnyy proekt SSSR: dokumenty i materialy*, v2 (5): 504. Moscow: Nauka, 2005.
19 Golovin, I.N. 1995. "O nastavnike v nauke i v zhizni". *Vospominaniya o I.Ye. Tamme*, 149–168. Moscow: IzdAT.
20 *Firma* (from English *firm*) is, in the jargon of Soviet engineers, a design bureau operating under the leadership of chief designers (e.g., Korolev) at the top of technopolitical hierarchy.
21 Sinyavskiy, V.V. 2009. "Vospominaniya". *K istorii sozdaniya i ekspluatatsii BR-5 (BR-10), 1959–2009*, 134–138. Obninsk: FEI.
22 Khariton, Y., V.I. Alferov. 2007. "Pismo L. Beriya. 19 dekabrya 1950". *Atomnyy proekt SSSR: dokumenty i materialy*, v2(7): 201. Moscow: Nauka.
23 Samarskiy, A.A. 1997. "Pryamoy raschet moshchnosti vzryva". *Istoriya sovetskogo atomnogo proekta (40-e–50-e)*, v1: 229. Dubna.
24 Gubarev *Belyy arkhipelag Stalina*.
25 Gubarev *Belyy arkhipelag Stalina*.
26 Feynman, Richard. 1985. *"Surely You're Joking, Mr. Feynman!": Adventures of a Curious Character*. New York: W.W. Norton.
27 Feynman 1985: 126.
28 Feynman 1985: 127.
29 Feynman 1985: 126.
30 Vladimirov, V.S. 1995. "Vospominaniya (1948–1956)". *Khochesh mira —bud silnym*, 88–96. Arzamas-16.
31 Samarskiy 1997: 227.
32 Malinovskaya 1995: 79.
33 Malinovskaya 1995: 79.
34 Malinovskaya 1995: 79.
35 Malinovskaya 1995: 79.
36 Malinovskaya 1995: 79.
37 Avrorina, V.A. 2003. "Dushoy ostalsya na Urale". *Lev i Atom*, 188. Moscow: Voskresenye.
38 Ibid.

39 Toshinskiy, G.I. 2014. "My ne boyalis zadavat glupykh voprosov", Interview with E. Ger. Retrieved 10 September 2021 from http://memory.biblioatom.ru/persona/toshinskiy_g_i/toshinsky/.
40 Veretennikov, A.I. 2005. "Iz vospominaniy". *Yuliy Borisovich Khariton. Put' dlinoyu v vek*, 312. Moscow: Nauka.
41 Loparev, S.Y., Smirnov G.A. and Y.N. Barmakov, eds. 2014. *V nachale bolshogo puti*, 128. Moscow: Kodeks.
42 Dmitriev, N.A. 2002. "Nachalo. Vospominaniya ob atomnoy bombe". *Nikolai Alexandrovich Dmitriev: vospominaniya, esse, statyi*, 32–34. Sarov: VNIIEF.
43 Nikitin, V.A. 2003. "Pervye oboroty puchka uskoritelya". *Vladimir Iosifovich Veksler*, 96. Dubna: OIYaI.
44 Kulikov, V.V. 2007. "Vspominaya byloe". *Vospominaniya veteranov OKBM*, v3: 56. Nizhniy Novgorod.
45 Greshilov et al. 2008: 400–419.
46 Postanovlenie SM SSSR no. 1990-774ss/op. 10 iyunya 1948". *Atomnyy proekt SSSR: dokumenty i materialy*, v2(1): 445–448. Moscow: Nauka, 1999.
47 Sofronov, I. D. 2002. "Yarkiy talant". *Nikolai Alexandrovich Dmitriev: vospominaniya, esse, statyi*, 99. Sarov: VNIIEF.
48 Malinovskaya 1995: 83.
49 Arnold, V.I. 2008. "YaB i matematika". *Yakov Borisovich Zeldovich: vospominaniya, pisma, dokumenty*, 225–230. Moscow: Fizmatlit.
50 Marchuk, Guriy. 1961. *Metody rasscheta yadernykh reaktorov*. Moscow: Atomizdat.
51 Marchuk, O.N. 1998. *Obninskaya istoriya*, 5–11. Moscow: TsISN.
52 Marchuk, G.I. 1995. *Vstrechi i razmyshleniya*, 12–13. Moscow: Mir.
53 Marchuk, G.I. 2000. *Zhizn v nauke*, 26. Moscow: Nauka.
54 "Postanovlenie SM SSSR no. 1429–574ss/op. 8 iyunya 1953. *Atomnyy proekt SSSR: dokumenty i materialy*, v2 (5): 553–555. Moscow: Nauka.
55 Marchuk 1998: 27–29.
56 Marchuk 1998: 27–29.
57 Marchuk 1998: 19.
58 Marchuk 2000.
59 Marchuk 2000.
60 "The electron is as inexhaustible as the atom". Lenin, V.I. 1970, 269. *Materialism and Empirio-Criticism: Critical Comments on a Reactionary Philosophy*. N.Y.: International Publishers.
61 Polozhenie o medali "Pervomu vychislitelyu FEI" [The Statute on the Medal for the First Calculator of the IPPE]. The Archive of the Obninsk project.
62 Her best stories entered the canon of nuclear memory, when the journalist and writer Ergaly Ger recorded and published them at the Biblioatom.ru. Dmitrieva, V.S. 2014. "Roza dlya Glazanova" [A rose for Glazanov]. Retrieved 10 September 2021 from http://memory.biblioatom.ru/persona/dmitrieva_v_s/roza_dly_glaznova/.
63 V.S. Dmitrieva, interview with A. Kasatkina, Obninsk, 30 May 2013. From the Obninsk Project Archive.
64 V.S. Dmitrieva, interview with A. Kasatkina, Obninsk, 30 May 2013. From the Obninsk Project Archive.
65 V.S. Dmitrieva, interview with A. Kasatkina, Obninsk, 30 May 2013. From the Obninsk Project Archive.
66 V.S. Dmitrieva, interview with A. Kasatkina, Obninsk, 26 August 2013. From the Obninsk Project Archive.
67 V.S. Dmitrieva, interview with A. Kasatkina, Obninsk, 26 August 2013. From the Obninsk Project Archive.

68 Konobeev, Y.V., V.A. Pavlinchuk, N.S. Rabotnov and V.F. Turchin. 1966. *Fiziki shutyat*. Moscow: Mir.
69 Rabotnov, N. n.d. 21 November 1966. *Dnevniki*. Retrieved 15 September 2021 from https://prozhito.org/person/57.
70 Rabotnov, N. n.d. 21 November 1966. *Dnevniki*. Retrieved 15 September 2021 from https://prozhito.org/person/57, 21 October 1976.

References

Broadhead, Lee-Anne. 2009. "Our Day in Their Shadow: Critical Remembrance, Feminist Science and the Women of the Manhattan Project". *Peace and Conflict Studies* 15(2): 38–61.

Cohn, Carol. 1987. "Sex and Death in the Rational World of Defense Intellectuals". *Signs: Journal of Women in Culture and Society* 12 (4): 687–718.

Dolgova, Eugenia, and Elena Streltsova; 2019. "Welcome to the Club: Position of Women in Soviet Science in the 1920s". *Sotsiologicheskie issledovaniya* 2: 97–107.

Gaspar, Miklos, and Margot Dubertrand. 2018. "Toward Closing the Gender Gap in Nuclear Science". *IAEA Bulletin* 11: 21–22.

Holloway, David. 2008. *Stalin and the Bomb: The Soviet Union and the Atomic Energy, 1939–1956*. New Haven: Yale University Press.

Howes, Ruth, and Caroline Herzenberg. 2003. *Their Day in the Sun: Women of the Manhattan Project*. Philadelphia: Temple University Press.

Kasperski, Tatiana, and Paul Josephson. 2023. "Women, Reactors, and Nuclear Weapons: From Revolutionary Liberation to the 'Miss Atom' Pageant in (Post-) Soviet Russia". *Technology and Culture*, 64(3): 791–822.

Kojevnikov, Alexey. 2004. *Stalin's Great Science: The Times and Adventures of Soviet Physicists*. Singapore: World Scientific.

Latour, Bruno. 2007. *Reassembling the Social. An Introduction to Actor-Network Theory*. Oxford: Oxford University Press.

Light, Jennifer. 1999. "When Computers Were Women". *Technology and Culture* 40(3): 455–483.

Melnikova, Natalia. 2017. "Zhenshchiny v realizatsii sovetskogo atomnogo proekta". *Demograficheskiy potentsial stran EAES* 1(8): 118–124.

Metropolis, Nicholas, and Eldred Nelson. 1982. "Early Computing at Los Alamos". *Annals of the History of Computing* 4(4): 348–357.

Pushkareva, Natalia. 2011. "Zhenschiny v sovetskoi nauke, 1917–1980e". *Voprosy intorii* 11: 92–102.

Rayner-Canham, Marlene, and Geoff Rayner-Canham. 1990. "Pioneer Women in Nuclear Science". *American Journal of Physics* 58(11): 1036–1043.

Sirotinin, Eugene. 2005. *Moskovsky universitet I sovetsky atomnyi proect*. Moscow: MGU.

Stepnov, Alexey. 2019. "Pervye uchenye-zhenschiny v sovetskoi fisike: gendernye aspekty professionalno-adaptatsionnykh praktik". *Istoria* 57: 183–192.

Vizgin, Vladimir. 2011. "Vzaimodeystvie fizikov i matematikov v sovetskom atomnom proekte". *Istoriko-matematicheskie issledovaniya* 14(49): 53–76.

8 "One Woman Started It All". Gendered Approaches to the Governance of Knowledge in Post-War Greece

Loukas Freris and Maria Rentetzi

September 1960. Leon Carnovsky, Professor of library science at the University of Chicago Graduate Library School, and his wife, Marian Satterthwaite-Carnovsky, travel on a ship that has set sail from Italy to Greece. Carnovsky, "a forceful and persuasive lecturer, helpful adviser, and consummate critic", with unprecedented experience for his time of the structure and government of public libraries abroad, was often in demand by UNESCO to advise foreign governments on library programs and education. In 1955, he had counseled the Israeli government about the feasibility of opening a library school in Jerusalem, and pleased with the success of the Israeli experiment, UNESCO invited Carnovsky to undertake a similar mission to Greece five years later. At the Greek port of Piraeus, the Carnovskys were welcomed by Stella Peppa-Xeflouda, head of libraries of the Greek Ministry of Education.[1] That day Carnovsky "dutifully recorded that, for the first time in the peninsula's history, guests were delivered by bookmobile to an Athens hostelry". Indeed, Peppa-Xeflouda gave the couple a ride to Athens on a bookmobile, part of the country's mobile library program (Haygood 1968).

Peppa-Xeflouda had first met Carnovsky in 1948, when, as a guest of the British Council, she took a course at University College London's School of Librarianship (in a program now known as Library and Information Studies) and visited several specialized libraries in Britain, France, and Scandinavia (Collingham 1952: 13). Through the UNESCO scholarship she also participated in the first international summer school for librarians in Manchester and London, organized by UNESCO in collaboration with the International Federation of Library Associations and the government of the United Kingdom. One of the five instructors of the summer school was Carnovsky, while Peppa-Xeflouda was among the 50 librarians from 19 countries who took part in the meeting, which lasted four weeks.[2] These seminars helped her become the first person in Greece with a modern library education, and it seems that she played a prominent role at the time.

In a lengthy article published in 1952, the *UNESCO Courier* introduced Peppa-Xeflouda as the person largely responsible for the founding of the first modern public lending library in Greece in 1948. "Two years ago the Greek island of Aegina had no public library. Today it has. This is largely due to the

DOI: 10.4324/9781003562597-11

initiative of one woman" (Collingham 1952: 13). The year of Carnovsky's visit to Athens, the *UNESCO Courier* celebrated Peppa-Xeflouda as the woman in charge of the UNESCO Mobile Library program in Greece: "The cultural warmth that radiates from the Mobile Library of Mrs. Peppa-Xephlouda is a conquest over the arid intellectual soil of some of the country districts" ("Books on Wheels" 1960). Peppa-Xeflouda, then, was among those who implemented UNESCO's plan in Greece and governed knowledge in a deep and significant way. Hired by the Ministry of Education, she dedicated her entire life to the creation and promotion of the institution of the public library in the country, accepting high levels of responsibility and shaping an entirely novel and successful educational program.

After World War II, UNESCO, presenting itself as a modern Prometheus of sorts, championed the idea of public libraries as a weapon that would liberate people from the bonds of their ignorance by teaching them their rights. Promoting a universal culture of reading, UNESCO offered technical aid to its member states through literacy programs and the development of public libraries. Public libraries could be perceived as regulatory places constitutive of concepts of globalization and standardization of knowledge. At the same time, however, the public library (mobile or not) was a vehicle through which governments created national identities. In other words, as André Maurois, the famous French author closely associated with UNESCO, argued, "a library is not only a valuable tool for the nation's use – it helps shape the nation itself" (1961: 18).

While the Greek government supported Peppa-Xeflouda's effort to establish public libraries throughout the country, and, with particular emphasis promoted the institution of mobile libraries, it did not show the same willingness when it came to educating the people who would staff these libraries. It is not by chance that reports from UNESCO envoys about the establishment of a library school in Greece were locked away in ministries' desk drawers not long after they saw the light of day. Given that the vast majority of professional librarians were women, librarianship was considered to be a "soft" female occupation, probably temporary, involving bookkeeping, and not requiring an academic background.

Feminization was indeed a critical aspect of the public service sector in the country. Since their late massive entry into the Greek public sector in the 1930s, women occupied lower positions in the hierarchy, performed monotonous tasks, were given fewer opportunities than their male colleagues, and were paid much less (Avdela 1990). In this gendered context, the resulting gap in the education of librarians was again filled by women's initiatives.

From 1961 to the late 1970s the Young Women's Christian Association (YWCA) of Athens provided vocational training in librarianship, helping young women to enter the job market. In these seminars, instructions were provided on a variety of key library topics, such as cataloging, classification, bibliography, and library management. From Peppa-Xeflouda and the YWCA to the young women who staffed public libraries; in every way women held the institution of public libraries in their hands.

Based on Peppa-Xeflouda's achievements in implementing the mobile library in Greece, one of the most ambitious UNESCO educational programs, this chapter explores the gendered governance of knowledge after World War II through institutional archives. We argue that international organizations such as UNESCO played an important role by creating opportunities not only for women's education but also for promoting women to leadership positions in their country of origin. By doing so they at least attempted to alter the gendered governance of knowledge in fields that were already considered as female disciplines. In Greece, the government promoted the establishment of public libraries because it considered them an ideal space in which the "enlightenment" of the people could be achieved; a space that would prove that Greekness was (both historically and culturally) directly linked to the pro-American policy of Greek post-war governments. Mobile libraries and libraries in general were seen as sites of Cold War propaganda.

The Dissemination of Knowledge in Female Hands

Stella Peppa's first post-secondary studies were a two-year seminar in library science at the National Library of Greece from 1930 to 1932. These studies likely contributed to her appointment as head of the Department of Libraries and Archives at the Ministry of Education in 1933. As William Carter, head of UNESCO's Bureau for the Exchange of Persons, described her role in the late 1940s, Peppa-Xeflouda became from early on a *fonctionnaire*, a civil servant of higher status, which later raised some eyebrows in UNESCO when her case was discussed.[3] In 1933, she was also appointed secretary of the General Council of Libraries of Greece, an institution founded in 1931 with the aim of "organizing, specializing and arranging libraries in Greece in a spirit of harmony in accordance with the modern principles of librarianship".[4] In addition, in 1937, she enrolled in the Panteios School of Political Science, from where she received her Bachelor's degree in 1940.

Peppa-Xeflouda's first important contact with new approaches to librarianship was through attending UNESCO's first international seminar for librarians. The seminar was held at the Manchester Public Library and headed by Arne Kildal of Norway in 1948. It brought together participants from some 20 countries and included lectures from the staff, university professors, and guest speakers; providing discussions and demonstrations of public library services. As Everett N. Petersen, UNESCO's first Head of Public Library Development, reported, the seminar was "an eye-opener for librarians from countries where public library services are little developed" (Petersen 1953:533). Subjects discussed, included: "the philosophy of public librarianship... the organization and administration of public libraries, technical processes, personnel, and finance" (Ibid.). Immediately after the completion of these summer courses, Peppa-Xeflouda was awarded a UNESCO reconstruction fellowship. That brand-new program was established by UNESCO to "assist and stimulate educational, scientific and cultural reconstruction" (Drzewieski 1948: 5) especially in war-devastated countries planning to rebuild their facilities. The fellowship supported Peppa-Xeflouda's tour in various European countries from October 1948 to March 1949. She had the opportunity to "visit a large number of institutions: public libraries as well as National – General and Special Libraries and to study librarianship" in "Great Britain, Denmark, Norway, Sweden, Holland, Belgium and France".[5]

On her return to Greece, Peppa-Xeflouda immediately began to promote and transmit the knowledge she acquired. In 1949, she wrote an article in the magazine O *Bibliofilos* in which she suggested the urgent need to establish a school for librarians in Greece:

> The education of library staff requires special training and presupposes the existence of Library Schools... if one examines the programs of the library schools in America and in European countries, one will see that they pay the most attention to the practical part of the librarians' education. Because they all agree that while the librarian is of course a scientist, he is also a technician and an expert (Vakalopoulou 1994: 1).[6]

At the same time, Peppa-Xeflouda organized Greece's first public lending library in Aegina, her home town. The fact that this library was installed in Aegina was symbolic in several ways. In 1829, Governor Ioannis Kapodistrias founded the first National Library of the newly formed Greek state also in Aegina.[7] Thus intended or not, Peppa-Xeflouda's action demonstrated the country's strong library tradition. Previously, in Ottoman Greece, there had been libraries in Greek schools, established under the names *scholio*, *gymnasion*, or *academia* and usually operated by an Orthodox monastery. Many of these libraries were maintained after the establishment of the new Greek state in 1832, while new libraries, such as the National Library and the Library of the University of Athens, were added (Tsirimokos 1934). However, these libraries were aimed at a specialized and educated audience as they were used mainly by academics, researchers, or students. In contrast, the public library that Peppa-Xeflouda organized was—by definition—aimed at a larger population; a group also targeted by state propaganda.

Thus, Aegina, a small island in the Aegean Sea just South of Athens, became a point of transition from a classic library to the first public lending library in Greece. "And the emphasis is on the words modern and lending", noted the *UNESCO Courier* when it welcomed the establishment of the library. Although public libraries already existed in Greece, they were "only public in the sense that they are property of the municipality… Readers may not borrow, nor are they free to browse amongst the shelves" (Collingham 1952:13). The lending system and the principle of free access to shelves were some of the innovative features of this library which would later be adopted by older state libraries (Peppa-Xeflouda 1958). In addition, since its establishment, it housed a collection for children, making it the first and only library in Greece to provide such services for many years (Thanopoulou 1995).

Although the Aegina library started informally and on a voluntary basis, its establishment, along with Peppa-Xeflouda's influence at the Ministry of Education, was instrumental for the 1949 legislation that concerned the nation's library system. The governance of knowledge through the new public system for the distribution of books, and the establishment of library studies in the country were both strongly shaped by Peppa-Xeflouda and women's organizations. Yet, the politics of knowledge remained in the hands of the government.

The Politics of UNESCO's Technical Assistance Programs

In 1949, Konstantinos Tsatsos, the Minister of Education, noted that

> The Greek state has so far witnessed a long delay in the establishment and organization of libraries in the capital and in the provinces. However, any further delay in this matter, which makes us lag behind other Balkan countries, many of which have maintained a well-organized library for many years, we do not think has a valid justification (Tsouderou-Papadatou 1962a).

As Alfred Reisch (2013) has argued with respect to the activities of the U.S. Information Service (USIS), the distribution of books to key individuals, libraries, and research institutions during the Cold War could be considered a type of political warfare against the Soviet bloc. John Matthews (2003) has described this program as the "secret Marshal Plan for the mind". On a smaller scale, the establishment of libraries and the distribution of books through mobile units in Greece served a similar aim: to affect the political behavior of the public and reduce the power of the country's Communist Party.

In Greece, the end of World War II coincided with the beginning of the Greek Civil War between the right-wing Greek government army (backed by the United Kingdom and the United States) and the left-wing Democratic Army of Greece (discreetly backed by the Soviets). The involvement of the great powers in this civil conflict has led many historians to consider the Greek Civil War as one of the first "hot" moments of the Cold War. As the historian André Gerolymatos put it, the Greek Civil War was "an international civil war, at least by proxy" (2016: xiii). The conflict ended in 1949 with the victory of the Greek government and its Western allies. Although at the time, Greece's political system was a "royal parliamentary democracy" power was essentially in the hands of non-parliamentary powers stronger than the official government: The king and his entourage, the intelligence services directly controlled by the CIA, and the military directly dependent on NATO and the

United States were the key players during this period (Mazower 2000; Liakos 2019). According to Peter Grose, biographer of Allen Dulles, the first civilian director of the CIA, "the most important center of power in Athens was the German-born Queen Frederika, manipulative queen mother of the future King Constantine" (1994: 450). But the defeat of the left in no way signaled the elimination of leftist views from the whole of Greek society. Just before the start of the Civil War, in the elections held on 31 March 1946, the power of the left had been measured by its absence. The decision of the leadership of the Communist Party of Greece (KKE) to abstain from the elections resulted in a general electoral abstention estimated at 20–25% of the electorate—a dynamic that did not disappear during the Civil War (Mazower 2000: 7; Klapsis 2011).

In this context, the government's support for libraries served not just the reorganization of the nation; it entailed an attempt to reproduce the official discourse of the state. Enacted in November, Emergency Law 1362 of 1949 proposed the establishment (and reconstruction) of state libraries across the country, so as to create (or restore) and to uniformly organize publicly supported libraries.[8] The idea was to establish a standardized national library system, and the law described in detail the management, operation, and supervision of local libraries. A major concern was the recruitment of educated librarians who could classify and register new books, offer borrowing services, and preserve the new, precious holdings. Thus, one of the provisions referred to the "special training of executives for the acquisition of a specialized staff of librarians and classifiers".[9] According to the law, the training courses had to be conducted by the National Library, the country's major library, founded in 1829, after Greece's War of Independence.

Despite the legal provisions for the standardized management of libraries, the training of librarians had not yet become a national task. Instead, in the early 1950s, the USIS established four libraries in the country, in cities where the KKE was strong: Athens, Patras, Salonica, and Kavala. The "American Libraries" in Greece function as information nodes in an extended transnational network of libraries across the globe. By the end of the 1950s, USIS had established 161 libraries in 65 countries (Bournazos 2019: 283). Besides books, these libraries offered a window into American culture, music, and civilization. As Wilson P. Dizard—who served in the U.S. State Department and USIS at the time—put it, "Our worldwide system of libraries and cultural centers presents more than an image; they communicate directly and personally by mind and heart" (Dizard 1961: 191).

In collaboration with the Greek Ministry of Education, USIS soon offered training workshops for young librarians throughout the country: a two-day seminar in Athens in 1955, followed by five more seminars in other Greek cities in 1956, and finally, in 1957, a two-week training course at Athens College, a prestigious private school co-established by the Greek merchant Emmanouil Benakis, and American philhellenes in 1925. USIS's activities were running in parallel with UNESCO's initiatives in Greece, which had offered the first international training opportunity to Peppa-Xeflouda in 1948 and which donated the first mobile library unit to the country in 1957. In the early 1960s, UNESCO sent two internationally renowned librarian educators to Greece to assist in assessing the country's library system as part of its technical assistance program (Krikelas 1982).

In the first post-war decades, providing technical assistance to developing countries by sending experts from United Nations family organizations was common. Technical assistance was crucial in the co-construction of the late-developing State. According to David Webster, "Technical assistance as a new form of diplomacy was just as much about spreading ideas of modernity from North to South as it was about delivering economic development" (2011: 250). As Webster further notes, "in selling expertise, technical assistance advisors were also selling a specific model of development to receptive, like-minded elites in developing countries" (2011: 271). In the context of the development and modernization policies advocated by the then-Prime Minister of Greece, Konstantinos Karamanlis, the Greek government considered UNESCO's consulting as another bridge between Greece and the West.

The first UNESCO expert to arrive in Greece was Leon Carnovsky, Professor at the University of Chicago's Graduate Library School. Almost immediately after receiving Carnovsky's report, the Greek Ministry of Education asked UNESCO for another expert to reassess the situation. The second study was conducted shortly afterwards, in 1962, by Preben Kirkegaard, Director of the Royal Danish Library School in Copenhagen (Krikelas 1982). Although both experts were sent by UNESCO, their approaches differed significantly, and each came to their own conclusions about where to establish the new library school in Greece and whether it should be autonomous or part of a renowned institution.

Carnovsky experienced difficulties in identifying the right location for the school of librarianship. However, as he noted, "if the perfect site were demanded before establishing a library school, no school would be established at all" (1961: 12). In the end, he proposed a location contrary to the centralization tendencies of the time, which favored the capital, Athens. Carnovsky considered that the University of Salonica in Northern Greece was the best choice for the new library school since this institution was "young and adventurous in spirit" (12). He believed that the University's "large and well-organized library"—together with Salonica's Municipal Library, which he also praised—would provide the necessary background for enhancing the teaching of library courses (12).

Carnovsky suggested that the new school "be incorporated with the School of Philosophy" at the University of Salonica (24). According to the UNESCO expert's recommendations, prospective students planning to enroll in the library school would first have to attend the courses of the School of Philosophy for two years, while the actual training in library science would take place in the third and fourth year (14–15). He also stressed the need to organize seminars for the training of employees already working in libraries(20). The main reason Carnovsky suggested that the library school be part of an existing school was his conviction that a new endeavor like this should be sponsored by a prestigious educational institution (Gerolimos 2011). On leaving Greece, Carnovsky submitted his report to the Ministry of Education in the hope that this would form the backbone of the new school. However, the Greek government did not find his proposal satisfactory.

Two years later another representative of UNESCO, Preben Kirkegaard, came to Greece with the same purpose: to study the conditions in the lending library system in Greece, and to submit his ideas for the establishment of a formal program of librarian education. In his report, Kirkegaard proposed, unlike Carnovsky, the creation of an independent school in Athens, a solution he arrived at after consulting with the Ministry of Education as well as some librarians. In the Ministry, he "found very little support for the plan of affiliating the library school with the University of Salonica" (Kirkegaard 1964: 4). Taking into account this reluctance concerning a library school in Northern Greece, and seeing that it would be rather difficult to find enough professional librarians in Salonica to support the school's curriculum, Kirkegaard came up with the Athens solution. Athens was home to a significant percentage of the Greek population and also had more libraries, both functioned as additional reinforcements (Gerolimos 2011).

Finally, one of the most compelling points of the Kirkegaard report was that UNESCO would provide technical assistance to help organize the new school of librarianship.[10] Starting in the academic year 1963–1964 and for the next three years, UNESCO would allocate an annual fund for sending experts to Greece, providing fellowships for the training of young Greek librarians abroad, and for the provision of equipment. The total cost in Kirkegaard's plan amounted to $44,500 (Kirkegaard 1964: 10). However, despite any hopes that might have been fueled by these reports, neither proposal was implemented by the Greek Ministry of Education. It would take many more years before the country's first Department of Library Science was established in 1977 (Krikelas 1982).

The vacuum created by the state's failure to establish a library school was filled partly by the newly founded Library School of the YWCA in Athens. YWCA Athens had started operating in the early 1920s and acquired a high-rise building in the city center in the 1950s, which enabled them to better organize programs aimed at the development and support of women. Under Chairperson Athena Tsouderou-Athanassiou, daughter of a former Greek Prime Minister, Emmanouil Tsouderos, the YWCA Athens housed social services, a boarding school for female students, a hostel for women travelers, a restaurant, and a lending library. One of the most important aspects of the YWCA's work consisted in providing educational and training programs that aimed to educate young Greek women and thus enable them to "practice some 'female' job". Thus, foreign language courses were taught within the YWCA, and a school for secretaries as well as the library school were established.[11]

From September 1961 until 1977, when the Library Science Department at the Technological Educational Institute of Athens was established, the YWCA was the only way to be trained in librarianship. Their annual study program was the first professionally organized training in the field of libraries in Greece. In its first year of operation, 12 people took part in the courses at the YWCA—the same number of Greek graduate students who had studied librarianship abroad that year. The following year the interest increased with student numbers doubling (Tsouderou-Papadatou 1962b). In total, 288 people graduated from the school from 1961 to 1977 (Gerolimos 2011).

Because the YWCA school was the only provider of librarian education in Greece, the Greek government was forced to officially recognize it. The school received state certification in 1964.[12] While granting permission to operate, the government did not mention the graduates' professional rights. Since the school was not an institution of higher education; the government considered it to be a vocational high school for young women who had a basic education and worked (or wanted to work) in libraries. Even after attending this school, young librarian graduates were considered unskilled by the Greek State. Prerequisites for admission were a high school diploma, an interview, an entrance essay, and working knowledge of English. Courses were taught by local librarians and lasted three hours per day. Attendance costs 250 drachmas per month (Tsouderou-Papadatou 1962b). At the time, the minimum monthly wage for 18-year-old female office workers was set at 1.250 drachmas, as reported in the centrist newspaper *Eleftheria* (1961).

Since the YWCA School of Librarianship did not provide participants with a state-recognized certificate, this was both a serious handicap for the YWCA in securing funding and also for bestowing upon librarians a recognized professional status in the eyes of academics and the government (Krikelas 1982). The journalist Virginia Tsouderou-Papadatou, sister of the then president of the YWCA, commented strongly on this problem at the time. In an article (1962b) in *Eleftheria*, she accused that the project of establishing a state-supervised library school was still stagnant despite the government having passed the relevant law in 1949. However, the downplaying of the "female profession" of librarianship was not just a Greek phenomenon (Brand 1983). In 1972, Dee Garrison argued that the feminization of public librarians in the United States, which had begun in the mid-nineteenth century, resulted in a lower professional status than it might have had if left in the hands of men (1972). Thus, librarianship was perceived as feminine, reproductive work, in contrast to the productive work of research and book writing done by men (Mars 2018; Higgins 2017; Simon 1994). Barbara Brand also reminded us in 1983 that women had played an important role as institution builders, and she drew attention to their networking efforts providing mutual support and advancement (Brand 1983). In twentieth-century terms, governing knowledge included not only institution building but also gender strategies for social advancement and recognition.

This state of affairs concerned all the young women who staffed the public libraries of the country. An exception to this rule was the staff of the library of the Greek Atomic Energy Commission (GAEC), which had been established at about the same time. The foundations of the Commission were laid down slowly in the early 1950s when the Greek government aimed at the country's participation in the creation of CERN, the European nuclear research center in Switzerland. At the same time, the influence of American foreign policy as expressed by the "Atoms for Peace" program gave a great impetus to the Greek Atomic Energy Commission (Rentetzi 2009, 2010). One of the first recruitments made by GAEC was Noria Christoforidou, hired in March 1959—"Employee number 36!", she recalled humorously—as a self-taught librarian "with foreign language qualifications".[13] The Greek AEC took care of her education by sending Christoforidou to CERN; in 1960, her training lasting one and a half years, and in 1963, for another year "to widen her experience of library and documentation services" (see Figure 8.1).[14] The recognition of the importance of a systematic and scientific education

Figure 8.1 Noria Christoforidou provides information to Kurt Gottfried, a visiting scientist from Harvard, CERN 1963 (Courtesy of CERN, https://cdsweb.cern.ch/record/40366).

for the librarian made it plain that other librarians' lack of state-organized education was not a matter of ignorance, but a clear political choice that implicated the Greek Crown and government, as well as US interests (Rentetzi 2009).

Thus, during the first decades after World War II, the Greek state did not treat librarians as a profession that employed scientific methods and, therefore, was in need of systematic education. Librarianship was considered an easy and unskilled job which consisted merely in acting as a caretaker or guardian of a book collection. Accordingly, public libraries were not considered institutions capable of educating people. The Greek state saw libraries instead as a means of anti-communist propaganda, "penetrating and influencing the rural population" (Lialiouti 2019: 180). This political importance of libraries became even more evident in the case of mobile libraries.

Mobile Libraries and Cold War Propaganda

In late 1957, UNESCO, in collaboration with the Greek Ministry of Education, launched a program to expand the free library service throughout the country and invited Stella Peppa-Xeflouda to govern its implementation. Having obtained the only Fulbright Scholarship intended for Greek specialists and youth leaders, she spent 1955 at the prestigious library school of Simmons College in Boston—acquiring the appropriate qualifications to lead a lending library program.[15]

As part of the new program, UNESCO donated a four-ton mobile library (or "bookmobile") to Greece in 1957 which was also equipped with a projector and screen to show films. With some 15,000 books at her disposal, Peppa-Xeflouda undertook the establishment of book collections in towns and villages around the country, starting from the extended area of Athens and gradually moving to other inland cities and islands. In each town that the mobile library visited for the first time, the librarians left a collection of 50 to 150 volumes, "usually in the office of the town clerk", which were exchanged for other books on later visits. The bookmobile visited the suburbs around Athens monthly, and more distant parts of the country once every two or three months (Carnovsky 1961: 6–7) (see Figure 8.2).

The Greek press warmly welcomed the initiative. Ioannis M. Panagiotopoulos, one of the most famous writers and educators of the time, wrote a laudatory article about mobile libraries. "An enlightened lady, Stella Peppa-Xeflouda", he related, "has devoted valuable forces and knowledge to the organization and promotion of these mobile libraries in the farthest corners of the country". He further noted:

> Mobile libraries lend books, pass them on, return the borrowed ones, share others, pass them again, collect what they have scattered, scatter new ones. The villages welcome them at first with curiosity and cautiously, then they look forward to it. Mobile libraries are a cultural project the importance of which we may not have yet realized.
>
> (Panagiotopoulos 1959: 1)

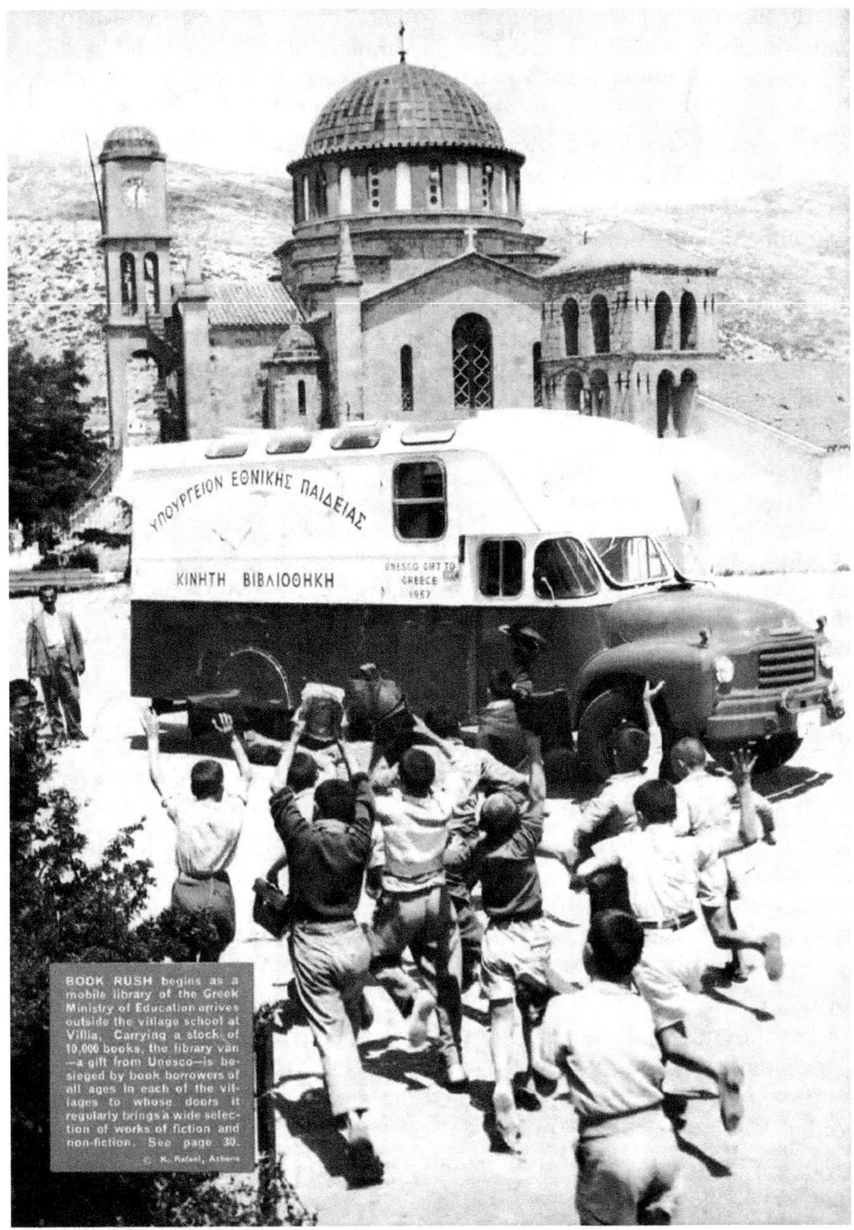

Figure 8.2 Arrival of the Greek bookmobile at the school of Vilia, Greece, 1958 (Courtesy of UNESCO, *UNESCO Courier* 13(1), 1960 36).

While Panagiotopoulos's readers might not have realized the importance of mobile libraries, the Greek government understood how they served its own interests. While Greek authorities seemed reluctant to organize a state-supervised school of librarianship, the situation was completely different regarding the mobile libraries. Inspired by UNECSO's idea, the government organized four more vehicles that toured the country from September 1957 and throughout the next decade. And just like the original UNESCO bookmobile, these mobile libraries doubled as mobile cinemas, an institution with a long tradition in Greece.[16] During the Marshall Plan, mobile units of the armed forces, properly equipped by the United States, toured the Greek countryside, showing propaganda films and videos of news and current affairs (Loukopoulou 2018)—"celluloid weapons", as Thomas Doherty aptly described them, in the "Cold War struggle for hearts, minds and, perhaps most of all, stomachs" (2005: 151). Based on this tradition, Greek authorities perceived the new cine-library units as an opportunity for widely distributing government, and especially anti-Communist propaganda.

Despite the constant persecution of the Communists from the end of the Civil War onwards, in the late 1950s the Greek left was gaining momentum. In the elections of 1958, the Communist Party of Greece had been declared illegal, so the United Democratic Left (EDA) became the leading opposition party with 24.4% of the vote (Nikolakopoulos 2013). Despite the pro-American and anti-Communist orientation of post-war Greek governments, "the predictable political presence of the left and the repercussions of its ideology" posed a challenge to both the Greek government and their American allies (Lialiouti 2019: 40)—a challenge to which the establishment of public libraries could potentially be a powerful response.

The central figure in this strategy was Konstantinos Tsatsos, a former professor of law who would later serve as President of Greece from 1975 to 1980. In addition to being an experienced politician, Tsatsos was also one of the foremost Greek intellectuals. During the interwar and post-war periods, he had been instrumental in forming a discourse on Greekness which, according to the historian Antonis Liakos (2019: 379), combined Greek nationalism with an ideology of the free world.[17] The common denominator was anti-Communism. According to Tsatsos and his allies, "Communism was not only a political but also a racial enemy" (Liakos 2019: 389; Papadimitriou 2006). Thus Communism was presented as an ideology that might suit the Slavs but certainly not the Greeks, seeing that all of Western culture had been built on Greek antiquity.

In the first government formed by Prime Minister Konstantinos Karamanlis in 1955, Tsatsos took over the Ministry of the Presidency, a top ministry in post-war Greece, overseeing subordinate departments and acting as the Prime Minister's staff. The Greek elections of 1958 again brought victory to the conservative faction. However, in the same elections, the left-wing EDA managed to considerably extend its share of the vote. The government's response was "to create invisible mechanisms for controlling political views that involved the Secret Services" (Liakos 2019: 389). Thus, the Ministry of the Presidency, under Tsatsos's leadership, drafted a proposal "for the organization of the Greek Enlightenment" which aimed, on the one hand, at "keeping the population in the national coalition" and, on the other hand, at "deploying a percentage of those affected by Communism and returning them to this faction". According to this proposal, one of the key tools for achieving these goals was mobile libraries, which were overseen by the Ministry of the Presidency and the Ministry of Education.[18]

Using vehicles and drivers from the Greek army and projection machines provided by the USIS, the government thus achieved an unprecedented diffusion of selected knowledge to the people in rural Greece. In this mountainous terrain, where many areas are geographically isolated, the contribution of the Greek army was decisive. Their military vehicles, built to withstand even the most difficult conditions, could reach the most inaccessible mountain villages. This capability was particularly important for the government propaganda apparatus if we consider that EDA had a strong performance both in urban centers and in the countryside (Rizas 2008). In the second half of 1957 alone, mobile cine-libraries visited a total of 264 villages. They lent about 10,000 books, donated another 7,500 books and magazines, and held 201 movie screenings. Much of the material distributed by these libraries was indeed propagandistic and anti-Soviet, as a look at the titles of some of the books makes clear: "Meet Communism", "Have at you Greece!", "Blame the Communist Party of Greece", "The Red Jackals", "Red Agents and Greek Workers", "In the Thieves of Siberia", etc.[19]

It would be naive to think that the Greek government based its propaganda machine entirely on some books and anti-Communist brochures, nor should we underestimate the impact of the mobile library program, and dismiss it as an expression of Cold War paranoia. These books were an important link in a chain. The goal of the Greek government was to dissuade those that "had been infected by Marxist materialism during the Occupation" (Liakos 2019: 380–381). For the sake of such political intervention, many institutions were used, first and foremost the schools, in which children are taught the values that the government chooses to transmit. As the then Minister of Education Konstantinos Tsatsos put it in 1949, "pedagogy" was nothing but "the continuation of the same war [that is, the civil war] in a new form" (Katsoudas 2018: 195). But state-sponsored nationalism was also expressed in churches, on the radio, in courtrooms, in national anniversary ceremonies, and wherever there was social life (Liakos 2019; Karakasidou 2000; Arkadas 2017). In this context, the books, magazines, and brochures that the Greek government distributed via the mobile libraries functioned as propaganda weapons.

Conclusion

On 3 January 2019, UNESCO tweeted about its recent video on the history of its educational programs and promised to take us on a visual journey around the world. In this 40-second-short clip, an image of a mobile library stands out. Taken in Colombia in 1955, the image features an "eager queue of girls" waiting to visit the UNESCO mobile library in Medellin.[20] This bus-like library attracted our attention. Our recent involvement with the history of the International Atomic Energy Agency's mobile radioisotope laboratory made us want to learn more about this type of mobile unit (Rentetzi 2021). We soon discovered that such a mobile library had been donated by UNESCO to Greece in 1957 to bring books directly to people in areas where there were no local infrastructures. One of the photographs taken at that time shows a group of young children running happily toward the mobile library (see Figure 8.2). Our further involvement in the issue made us realize that this was not a simple photograph but a palimpsest with multiple historical and social layers.

At first glance at the photo, we can see the inscription on the door of the bookmobile: "UNESCO gift to Greece 1957". Promoting a universal culture of reading and public libraries—and thus governing the dissemination of knowledge on a global level—UNESCO offered technical assistance to member states through literacy programs and the development of public libraries. Looking closer at the palimpsest, the figure of a woman appears, that of Stella Peppa-Xeflouda. This woman, a librarian at the Ministry of Education, was the driving force behind the development of public libraries in Greece. Taking advantage of the opportunities for studies abroad, i.e., through the provision of scholarships by UNESCO and other international actors, Peppa-Xeflouda developed leadership skills, and then she headed the mobile library program in Greece. Given the lack of archival sources, it has been impossible to access how many women (and men) were able to benefit from this program. What became obvious however, is that going deeper into the layers of the palimpsest, the figures of many more women are revealed; women who participated in the librarianship courses organized by the YWCA. The absence of any state concern about training librarians made the YWCA the sole choice of prospective librarians for over a decade. Thus, education in librarianship was left in the hands of Greek women.

Next, other figures emerge, such as Leon Carnovsky or Preben Kirkegaard, who visited Greece—separately—in the early 1960s to advise the government on setting up a library school. These experts reveal the role played by international organizations such as UNESCO after World War II. By sending experts to developing countries or by training young scientists through the institution of fellowships, international organizations tried to transfer a specific model of development to these countries. However, the indifference of the Greek government regarding the proposals of the UNESCO experts reveals that development is a deeply gendered issue.

The last layer of the palimpsest invites us to study the history of post-Civil War Greece through the operation of public libraries (mobile or not). The Greek government orchestrated an anti-Communist symphony around the libraries. The composer and conductor of this symphony was Konstantinos Tsatsos, one of the most important politicians and intellectuals of the time. Greece, indifferent to the education of the women who staffed its libraries, set up with great care a well-tuned mechanism for distributing thousands of propagandistic "hot books" (Reisch 2013) in every city and village in the country.

Acknowledgment

This publication is part of the "Living with Radiation: The Role of the International Atomic Energy Agency in the History of Radiation Protection" (HRP-IAEA) project that has received funding from the European Research Council (ERC) under the European Union's Horizon 2020 research and innovation program (Grant Agreement No. 770548).

Notes

1 Stella Peppa-Xeflouda, born Peppa in 1912, married Stelios Xefloudas (1901–1984), a Greek writer of the Generation of the Thirties.
2 The Work of the Libraries Division of UNESCO, 1948 Conference: UNESCO/IFLA International Summer School for Librarians, Manchester and London, LBA/CONF.2/6, UNESCO Digital Archives. Retrieved from https://unesdoc.unesvco.org/ark:/48223/pf0000147429? posInSet=8&queryId=5fa863b1-a66d-44a6-a9a8-a2964212fe16.
3 UNESCO Archives, Box Reg 331, 376 (495) "-56" Peppa.
4 F.E.K. (Government Gazette) [in Greek]. 210/1931 A, Regarding the establishment of the General Council of Libraries of Greece.
5 UNESCO Archives, Box Reg 331, 376 (495) "-57" Peppa.
6 All translations from the Greek are our own unless otherwise noted.
7 The official independence of Greece took place in 1832 when, with the Treaty of London, the country became a monarchy named the Kingdom of Greece. The Greek revolution had commenced in 1821, but until the arrival of Kapodistrias in 1828, the Hellenic Republic was not recognized as a state; it was called the "Provisional Administration of Greece" as long as it fought for independence from the Ottoman Empire. In 1826, Aegina was designated the provisional capital of the newly formed Greek state, a status it maintained until 1829, when the capital was moved to Nafplio. For an overview of the history of Modern Greece see Koliopoulos and Veremis (2010).
8 Emergency Law 1362/1949. Retrieved from http://www.et.gr/idocs-nph/search/pdfViewerForm.html?args=5C7QrtC22wH6SCddhZcqgXdtvSoClrL88bwy3eyVpoJ5MXD0LzQTLWPU9yLzB8V68knBzLCmTXKaO6fpVZ6Lx3UnKl-3nP8NxdnJ5r9cmWyJWelDvWS_18kAEhATUkJb0x1LIdQ163nV9K--td6SIuUu1fBzJwL7HKBaEEb9xk3oSQy3OD0Y1jX0-GLI6LR8C.
9 Emergency Law 1362/1949. Retrieved from http://www.et.gr/idocs-nph/search/pdfViewerForm.html?args=5C7QrtC22wH6SCddhZcqgXdtvSoClrL88bwy3eyVpoJ5MXD0LzQTLWPU9yLzB8V68knBzLCmTXKaO6fpVZ6Lx3UnKl3nP8

NxdnJ5r9cmWyJWelDvWS_18kAEhATUkJb0x1LIdQ163nV9K--td6SIuUu1fBzJwL7HKBaEEb9xk3oSQy3OD0Y1jX0-GLI6LR8C.

10 Through the early United Nations technical assistance program, UNESCO sought to lift developing states—like Greece—out of poverty and into Western-derived development trajectories (Webster 2011).

11 "Athena Tsouderou-Athanassiou (1920–2008) President of YWCA Greece & President of the World YWCA". Retrieved 10 November 2021 from https://xen-athinon.gr/xen-athinon/athina-tsoyderoy-athanasioy-1920–2008-lt-br-gt-proedros-chen-ellados-amp-proedros-pagkosmias-chen/; "YWCA Athens. History". Retrieved 10 November 2021 from https://xen-athinon.gr/history/.

12 F.E.K. 409/1964 B, Regarding the granting of a license for the establishment and operation of the vocational library school of the Christian Union of Young Women in Athens

13 Noria Christoforidou, interview with Maria Rentetzi, Athens, Greece, 17 February 2005.

14 https://cdsweb.cern.ch/record/40366.

15 For a list of Greek Fulbright Scholarships 1954–1955, see http://www.fulbright.gr/images/alumni/pdf/GR_Grantees_1954–1955.pdf

16 Gennadios Library, K. Tsatsos Archive, Subsection III, Folder 56, Subfolder 4, "Mobile Lending Libraries", 1957. [in Greek].

17 Other important figures of this intellectual elite were Ioannis Theodorakopoulos and Panagiotis Kanellopoulos, both collaborators and old acquaintances of Katsos from their days as students at the University of Heidelberg. For an analysis of "the correlation between [Greek] nationalism and Neo-Kantian philosophy of history in the interwar period" see Papari (2015).

18 Gennadios Library, K. Tsatsos Archive, Subsection III, File 56, Sub-file 5, "Proposals for the organization of the National Enlightenment submitted to the Prime Minister", 23/2/1959.

19 Gennadios Library, K. Tsatsos Archive, Subsection III, File 56, Sub-file 5, "Proposals for the organization of the National Enlightenment", 23/2/1959.

20 "Gallery: Education transforms lives". Retrieved 10 November 2021 from https://en.unesco.org/galleries/education-transforms-lives. For the full publication see Müller and Stanca-Mustea (2018).

References

Arkadas, Dimitris. 2017. "Constructing Patriots: Sunday Schools [ekklisiastika katichitika] as a Field of Religious Anti-Communism," *Thesis* 138: 69–91. [in Greek].

Avdela, Efi. 1990. *Female Public Servants: Labour Division by Gender in the Public Sector, 1908–1955*. Athens: Emporiki Bank of Greece [in Greek].

"Books on Wheels: A Library Comes to a Greek Village." 1960. *UNESCO Courier* 13(1): 30–31.

Bournazos, Stratis. 2019. "The Congress for Cultural Freedom (CCF) and Its Activity in Greece, 1950–1967: Cultural Cold War and Anti-Communism." PhD dissertation, University of Crete, Greece [in Greek].

Brand, Barbara. 1983. "Librarianship and Other Female-Intensive Professions." *The Journal of Library History (1974–1987)*, 18(4): 391–406.

Carnovsky, Leon. 1961. *A Library School for Greece: A Prospectus*. Paris: UNESCO.

Collingham, Beatrix. 1952. "One Woman Started It All." *UNESCO Courier* 5(7): 13.

Drzewieski, Bernard. 1948. "UNESCO to Promote Reconstruction Funds." *UNESCO Courier* 1(1): 5.

Dizard, Wilson P. 1961. *The Strategy of Truth: The Story of the U.S. Information Service.* Washington: Public Affairs Press.
Doherty, Thomas. 2005. "A Symposium on the Marshall Plan Films in New York City." *Historical Journal of Film, Radio and Television* 25(1): 151–154.
Eleftheria. 1961. "New salary thresholds for private employees." 28 March: 8 [in Greek].
Garrison, Dee. 1972. "The Tender Technicians: The Feminization of Public Librarianship, 1876–1905." *Journal of Social History* 6(2): 131–159.
Gerolimos, Michalis. 2011. "Greek Higher Education for Librarianship." *Techmirion* 10: 11–60 [in Greek].
Gerolymatos, André. 2016. *An International Civil War: Greece, 1943–1949.* New Haven: Yale University Press.
Grose, Peter. 1994. *Gentleman Spy: The Life of Allen Dulles.* London: André Deutsch.
Haygood, William Converse. 1968. "Leon Carnovsky: A Sketch." *The Library Quarterly* 38(4): 422–428, doi:10.1086/619698.
Higgins, Shana. 2017. "Embracing the Feminization of Librarianship." *LIS Scholarship Archive.* November 10: 67–89.
Karakasidou, Anastasia. 2000. "Protocol and Spectacles: National Ceremonies in North Greece." In *After the War Was Over: Reconstructing the Family, Nation and State in Greece, 1943–1960,* ed. Mark Mazower, 243–267. Princeton University Press.
Katsoudas, Kostas. 2018. "Moral Panic and Anti-communist Discourse in the first Post-War Years." In *The Unruly: Texts on the History of Youth Arrogance in the Postwar Period*, ed. Kostas Katsapis, 183–205. Athens: Okto.
Kirkegaard, Preben. 1964. *Greece: Library Development.* Paris: UNESCO Digital Archives. https://unesdoc.unesco.org/archives-fr
Klapsis, Antonis. 2011. "The elections of March 31, 1946." *Kathimerini* 18 September: 31 [in Greek].
Koliopoulos, John S., and Thanos M. Veremis. 2010. *Modern Greece: A History Since 1821.* Chichester: Wiley-Blackwell.
Krikelas, James. 1982. "Education for Librarianship in Greece." *The Library Quarterly: Information, Community, Policy* 52(3): 227–239.
Liakos, Antonis. 2019. *The Greek 20th Century.* Athens: Polis [in Greek].
Lialiouti, Zinovia. 2019. *The 'Other' Cold War: American Cultural Diplomacy in Greece 1953–1973.* Heraklion: University Press of Crete [in Greek].
Loukopoulou, Katerina. 2018. "'A Campaign of Truth': Marshall Plan Films in Greece." In *Cinema's Military Industrial Complex*, ed. Haidee Wasson and Lee Grieveson, 321–338. Berkeley: University of California Press.
Mars, Patricia. 2018. "Gender Demographics and Perception in Librarianship." *School of Information Student Research Journal* 7(2): 1–14, doi:10.31979/2575-2499.070203.
Matthews, John. 2003. "The West's Secret Marshall Plan for the Mind." *International Journal of Intelligence and CounterIntelligence* 16(3): 409–427.
Maurois, André. 1961. *Public Libraries and Their Mission.* Paris: UNESCO.
Mazower, Mark., ed. 2000. *After the War Was Over: Reconstructing the Family, Nation and State in Greece, 1943–1960.* Princeton: Princeton University Press.
Nikolakopoulos, Elias. 2013. *The Cachectic Democracy: Parties and Elections, 1946–1967.* Athens: Patakis [in Greek].

Panagiotopoulos, Ioannis M. 1959. "The Reader from the Provinces." *Eleftheria*, 26 July: 1 [in Greek].

Papadimitriou, Despina. 2006. *From Law-Abiding People to a Nation of the Nationally Minded: Conservative Thought in Greece, 1922–1967*. Athens: Savvals [in Greek].

Papari, Kate. 2015. "The Crisis of Historicism and the Correlation Between Nationalism and Neo-Kantian Philosophy of History in the Interwar Period." *Journal of History Research* 5(3): 141–155.

Peppa-Xeflouda, Stella. 1958. "Libraries in Greece." *UNESCO Bulletin for Libraries* 12(2-3): 29–31.

Petersen, Everett N. 1953. "UNESCO and Public Libraries." *Library Trends* 1(4): 531–542.

Reisch, Alfred A. 2013. *Hot Books in the Cold War: The CIA-Funded Secret Western Book Distribution Program behind the Iron Curtain*. Budapest: Central European University Press.

Rentetzi, Maria. 2009. "Gender, Science, and Politics: Queen Frederika and Nuclear Science in Postwar Greece." *Centaurus* 51: 63–87.

Rentetzi, Maria. 2010. "'Reactor Is Critical': Introducing Nuclear Research in Postwar Greece." *Archives Internationales d' Histoire des Sciences* 60(164): 137–154.

Rentetzi, Maria. 2021 "With Strings Attached: Gift-Giving to the International Atomic Energy Agency and US Foreign Policy." *Endeavor* 45(1–2): 100754.

Rizas, Sotiris. 2008. *Greek Politics after the Civil War: Parliamentarism and Dictatorship*. Athens: Kastaniotis [in Greek].

Simon, Jane. 1994. "The Construction of Femininity in Discourses of the Woman Librarian: 1890s to 1940s." *The Australian Library Journal* 43(4): 257–271.

Thanopoulou, Katerina. 1995. "Proposals for the Relocation of the Public Kapodistrian Library of Aegina." *Vivliothikes kai Pliroforisi*: 12–13 [in Greek].

Tsirimokos, Markos. 1934. "Greek Libraries" in the *Calendar of Great Greece* [Imerologion tis Megalis Ellados], ed. Georgios Drosinis, 161–177. Athens: Sideris [in Greek].

Tsouderou-Papadatou, Virginia. 1962a. "Libraries for the People." *Eleftheria* 16 December: 9 [in Greek].

Tsouderou-Papadatou, Virginia. 1962b. "Libraries for the People." *Eleftheria* 18 December: 5 [in Greek].

Vakalopoulou, Maria. 1994. "Vocational Education of Librarians." *Vivliothikes kai Pliroforisi* 10–11 [in Greek].

Webster, David. 2011. "Development Advisors in a Time of Cold War and Decolonization: The United Nations Technical Assistance Administration, 1950–59." *Journal of Global History* 6(2): 249–272.

Part III

Individual Paths in Governing the Technosciences

9 "She Was Only a Post-Doc"

Governing Science by Lab Directors and the Vanishing Credit of Women in the Discovery of RNA Splicing

Pnina Geraldine Abir-Am

The discovery of RNA splicing offers a challenge to our understanding of science governance by rules and principles, especially when it comes to allocation of credit in textbooks, recognition with awards of institutional, monetary, and prestige value such as the Nobel Prize, and the sedimentation of a shared public memory (e.g. at anniversaries such as its 40th which was marked by the scientific community in 2017) which is seen as matching the historical record.[1]

The paradox at the heart of the discovery of RNA splicing stems from the fact that though women were the lead co-authors of two parallel discovery papers, (lead co-author being the scientist most responsible for the reported content of such papers) the 1993 Nobel Prize awarded for this discovery included no women, being further limited to two male lab directors, even though a Nobel can be shared by up to three scientists.[2]

The question persists as to not only why the two women were not recognized in ways commensurate with their decisive contributions, (e.g. as part of the Nobel or other comparable prizes; and/or job offers in prestigious institutions, as befits those mainly responsible for a major discovery) but why they were further marginalized. While the two laureates spent their careers in some of the most coveted regions for basic research in the United States; (Greater Boston and Greater New York) the two women spent the greater part of their careers in peripheral regions such as the "Deep South" and the "Gulf of Mexico".[3]

Since credit in science is the currency of career building, its allocation or misallocation further becomes a key aspect of the persistent problem of women's under-representation in science.[4] Given science's major role in society as an engine and symbol of economic progress, credit allocation, or misallocation, thus underlies persisting gender inequality more widely. Indeed, in the context of the prolonged public debate on women's under-representation in science which unfolded in 2005–2006 and has continued since that time, attention was specifically drawn to the disproportionate role of men of science in power, such as chairmen of science departments, hiring and promotion committees, and lab directors, in maintaining the status quo of women's

DOI: 10.4324/9781003562597-13

under-representation, "glass ceiling", and related phenomena of covert gender discrimination, which came to prevail once overt discrimination had been deemed illegal.[5]

A great deal of research has focused on structural barriers and gender bias, but less attention was given to the predicament of women scientists who managed to overcome structural barriers, but who could not advance to leadership in science, because the credit due to them for making discoveries was often taken away precisely by those who were supposed to protect them as a newcomer, gender-based, minority in science. Long accustomed to cultivating male protegees, lab directors rarely encountered potential female protegees prior to the 1970s, when the affirmative action legislation of 1972 greatly increased the number of women aspiring to careers in science.

The case study examined in this chapter is therefore historically significant not only because it deals with a major discovery in the central science of molecular biology—the discovery of RNA splicing,[6] but also because it illuminates the historicity of gender in the mid- and late 1970s, when the impact of affirmative action legislation of 1972 just began to be felt. Unlike the preceding major discoveries of DNA structure in 1953, and messenger RNA in 1961, which featured very few women, six women served as co-authors of several publications claiming the discovery of RNA splicing in 1977.[7] Three of these women acted as first or lead co-authors, i.e. the co-authors most responsible for the reported contents of a given publication.[8]

This chapter focuses on the perspective of a woman scientist, Susan (Sue) Berget, who served as the first or lead co-author of a publication emanating from a lab in the Cancer Research Center (hereafter CRC) at MIT, (Massachusetts Institute of Technology) in Greater Boston, USA, and published in the *Proceedings of the National Academy of Sciences* (hereafter *PNAS*) in August 1977 (Berget et al. 1977). Based on interviews with her and her colleagues by the author and several research assistants, it documents Susan (Sue) Berget's perspective on her road to the discovery, before seeking to understand how the process of science governance enabled Berget's lab director to distance her so completely from the credit for this discovery; as well as why she, or others who had an institutional obligation to protect post-docs or dependent scientists, could not better protect her rights to her discovery from such a total appropriation. It thus aims to explain how issues of science governance ranging from the epistemic paternalism associated with the position of lab director, as a key cog in science governance, to a pervasive lack of moral accountability, further enabled this epistemic injustice not only to be created in the first place, but to persist for more than four decades.

Susan (Sue) Berget's Long Silenced Voice

A paper by a team at MIT-CRC composed of post-doc fellow Sue Berget, technician Claire Moore, and lab director Philip Sharp was one of several discovery claims published in 1977, mostly in *PNAS*, as contenders for the discovery of RNA splicing. Other papers came from teams based at the Cold Spring Harbor Laboratory on Long Island, NY; (hereafter CSHL) a cross-institutional collaboration between teams at the National Cancer Institute (hereafter NCI) at the National Institutes of Health (hereafter NIH) in Bethesda, Maryland and the Dept. of Genetics at the Weizmann Institute for Science (hereafter WIS) in Rehovot, Israel; the Rockefeller University in New York City; and a team from the Dept. of Virology at the Weizmann Institute of Science. These teams totaled two dozen co-authors, six of them women. All of these teams published in 1977; the 1993 Nobel was given to two team members, one from MIT-CRC and one from CSHL.

Berget arrived as a post-doc at MIT-CRC in 1975 under particularly auspicious circumstances, which may have blinded her from seeing the difficult prospects of women in a university which, as a prestigious school of engineering, had very few women students, and fewer women faculty.[9] As a major defense contractor, MIT was obligated after 1972 to provide equal opportunities to women and minorities, which it did not do until the 2000s.[10] After Sue Berget did her undergraduate studies at Duke University, in Durham, North Carolina, she continued for Ph.D. at the University of Minnesota in Minneapolis, arriving there as a trailing spouse married to another graduate student who also pursued graduate studies there in molecular biology. She worked on gene expression using bacterial viruses but developed an interest in studying the more complex gene expression in eukaryotes, a rising theme at the time among both students and mature scientists looking for new challenges. In mid-1974, as she was completing her Ph.D., Victor Bloomfield, a member of her Ph.D. committee heard of her interest in studying eukaryotes, and offered to inquire whether a former student of him at the University of Illinois in Urbana, Philip A. Sharp, who was about to relocate to MIT, might have space for her as a new post-doc.

Berget was able to get departmental support to attend the large annual conference on DNA Tumor Viruses held at the Cold Spring Harbor Laboratory (CSHL) on Long Island, NY, in the summer of 1974. There, she met Sharp and his collaborator S. Jane Flint, a two year more senior post-doc who planned to relocate with Sharp to MIT-CRC from CSHL where these two collaborated since 1973. Berget interviewed with them and arrived at MIT-CRC in January 1975, as soon as she completed her Ph.D. and also secured a three-year NIH post-doctoral fellowship.

In addition to MIT's general reputation, which was widely flaunted in science magazines and noticed by science hopefuls, its Department of Biology had been reorganized in 1959 with the arrival of Salvador Luria, a founder of molecular biology. In 1969, Luria shared the Nobel Prize for his contributions to microbial and phage genetics and, in 1971, he became the founding Director of the new Cancer Research Center at MIT, established by him as an institutional initiative responding to new opportunities in funding provided by the "War on Cancer" (Luria 1984; Selya 2022; Abir-Am 2020a).

In the early 1970s, following the affirmative action legislation as well as a demand by students and post-docs that the Department of Biology provide a permanent slot for Annamaria Torriani,[11] a well-known scientist employed there as a research associate without a formal appointment for the preceding 15 years, new women faculty were hired, such as Mary-Lou Pardue in 1971 and Nancy Hopkins[12] in 1973. As a result of her exposure to the Torriani case, Alice Huang left her position as a spousal research assistant at MIT-CRC for an assistant professorship at the Harvard Medical School, despite being repeatedly advised that it would be easier for her to remain a research associate [of her husband] at MIT-CRC.[13]

Though MIT-CRC included quite a few women, at the time only Nancy Hopkins was (junior) faculty, (and lab director) with Sue Berget being one of very few women post-docs.[14] The majority of women were graduate students and technicians, forming about one-third of the staff. Sue Berget was again very fortunate to encounter another woman post-doc, S. Jane Flint, a Ph.D. from University College, London who had already moved with Sharp to MIT-CRC in 1974, when Sharp was hired there and became a lab director.

Flint taught Berget the techniques used in their lab, how to write well, and she even became Berget's mentor and friend, co-authoring two papers with her, both in 1976 (Flint et al. 1976, Berget 1976). The work Berget did with Flint was a direct continuation of the work Flint and Sharp did prior to their arrival at MIT in mid-1974. Flint began her post-doc at CSHL in 1973 after a lab director there, Joseph Sambrook, recruited her in London. At CSHL she began collaborating with Sharp who preceded her in Sambrook's lab by two years.[15] Sharp began a second post-doc at CSHL in 1971 and became a staff member in 1972. Flint continued her post-doc with Sharp at MIT-CRC when he moved there in mid-1974. The work of Flint and Sharp, much as other work in Sambrook's lab, focused on mapping viral-specific RNA in virus-infected and virus-transformed cells. The mapping of tumor genes was a special interest of Sambrook, as well as being the rationale for the large NIH grant to CSHL for which Sambrook served as Principal Investigator. That grant supported the bulk of research at CSHL, making Sambrook a key figure there, some sort of "second-in-command",[16] while also prioritizing comparative research on normal and cancerous cells.

Flint and Sharp's collaboration with Sambrook and other members of Sambrook's lab was very productive, co-authoring 10 papers, five at each institution, (CSHL and MIT) in the period 1974–1976. They continued to collaborate with both Sambrook and a member of his lab, well into 1976, or two years after they moved to MIT, until Flint left MIT-CRC late in 1976 for a position as assistant professor at Princeton.[17]

Flint's imminent departure persuaded Berget that she would have to do something different in order to get a comparable job, because she felt she could no longer capitalize on the techniques which she thought landed Flint her job at Princeton. Therefore, Berget began hunting for something new and interesting, "running 2–3 projects on her own".[18] Post-docs were expected to work on research projects set by their host lab director, as Berget did indeed in the first year and a half when she collaborated with Flint, a more senior post-doc.

However, Flint's departure, as well as Sharp's increasing responsibilities stemming from the arrival in 1976 of two post-docs and a graduate student, all males, compelled Berget to pursue independent research of a scouting nature, "on her own".

The Annual DNA Tumor Virus Meeting

A turning point came at the annual DNA Tumor Virus Meeting, held in August 1976 at CSHL where Berget encountered three new findings presented by other scientists, findings which would lead her to the discovery of RNA splicing within roughly eight months.[19] The first finding which caught Sue Berget's eye, in part because she was exploring the same direction, pertained to biochemical evidence that all the mRNAs had the same sequence at the 5′ terminus.[20]

Berget's recollections are interesting for several reasons: first, she immediately understood the importance of this finding, which was contested at the time among most other researchers, including those at CSHL. Second, as a trained biochemist, she also understood the importance of working with purified or enriched RNA. Third, her reference to this finding as having come from CSHL as a whole is strange because this finding was produced by a post-doc in the Nucleic Acids Lab at CSHL, Richard Gelinas, who was Berget's perfect contemporary, having also started his post-doc in 1975 (Gelinas and Roberts 1977). Possibly, because Gelinas later became a co-author of a parallel or competing discovery paper from CSHL (Chow et al. 1977), Berget was keen to avoid the impression that she was influenced by a contender post-doc from another institution.

Perhaps this is also why she did not mention her interaction at that Meeting with yet another CSHL scientist, Thomas Broker, chief of the Electron Microscopy (hereafter EM) Lab there, who discussed with her in detail yet another new finding that attracted her attention at that Meeting, namely the then recently developed technique of R looping (Thomas et al. 1976). Broker, who began using this new technique already in the spring of 1976, together with his collaborator spouse, Louise Chow, also an electron microscopist, and their two additional co-authors, explained its intricacies to Berget who was totally new to it.[21]

Like many biochemists, Berget thought of EM as boring. Nevertheless, she realized that the then-new R looping technique was both doable and promising. Once Berget and Moore used the EM to good effect, she still noted that its use in RNA splicing was "probably the first time that EM was used to prove something really important".[22] Berget's reservations about EM were so widely shared that leading scientists in RNA processing missed the discovery of RNA splicing because they didn't bother to use the EM, which was best suited to visually capture the splicing process. Hence, the R loops as seen and measured by EM came to be seen as the "discovery proof" or an obligatory component for any team seeking to claim the discovery of RNA splicing. Berget's team was one of only two teams to survive this process; the other being a team from CSHL which included Broker and Chow, EM experts who teamed up with nucleic acid biochemists, the above-mentioned post-doc Gelinas, a contemporary of Berget, and his post-doc adviser, Richard Roberts, director of the Nucleic Acids Lab at CSHL (Berget et al. 1977; Chow et al. 1977; Abir-Am 2020b).

Despite her initial reservations about EM, Berget pursued her plan of hybridizing DNA and RNA and following the outcome by EM, while enlisting the help of their lab's EM technician Claire Moore, with whom she was good friends. This course of action shows Berget to be enterprising, as well as boldly branching out into new areas she had no experience with, in firm pursuit of her own ideas and opportunities. Berget's decision was part of "her own two-three projects, hunting for something interesting", and not an idea given to her by anyone else. As she put it: "But what I can tell you is, nobody told me to do the experiment. I was not instructed by Phil [Sharp, lab director] to use the electron microscope".[23]

Interestingly enough, Berget's collaboration with Claire Moore is the second instance, (after her collaboration with S. Jane Flint) in which women helped each other in a research setting dominated by male scientists. At MIT-CRC, all but one of the lab directors were male, and so was a majority of post-docs and graduate students. Berget was thus very fortunate to find the assistance she needed from other women, especially since, as she was to find out soon, lab directors favored male post-docs.

Shortly after Berget began her experiment with Moore's assistance, these two made an observation that can be said to have been the first phase of the discovery, namely that some of the RNA was not pairing to DNA in the RNA-DNA hybridization experiment, but was dangling as a single-stranded "tail" at the 5′ [RNA] terminus; moreover, they noticed that those "tails" always had the same length. Initially, their lab director Sharp considered those "tails" to have been artifacts, but Sue Berget and Claire Moore confirmed the validity of their findings in numerous control experiments. As Moore put it: "Biggest contribution was finding the tails and repeating experiments many times to confirm that they were not a mistake".[24] Moore also confirmed that Berget, who was in charge of mapping the late genes of adenovirus (a post-doc who joined the lab a year later than Berget was in charge of mapping the early genes), had an integrative grasp of the project, being further able to see the "big picture".[25]

Moore viewed herself and Berget as a good team because Moore considered herself to be a "plotter" but together they "were able to understand the discovery", combining the different types of skills and approaches needed, namely to see the big biochemical picture and confirm beyond doubt that the EM pictures were not artifacts. At the time, Moore participated in the RNA splicing experiments as a technician with M.Sc. level education only.

Though neither Berget nor the lab director, had any idea as to the origins of the mRNA tails on the 5′ terminus, not even that it would be critical to find out the precise location of their coding sequences on the genome, they agreed that the observation and confirmation of the "tails" merited publication. Berget suspected that the "tails" hinted that part of the mRNA might be coded elsewhere on the genome, but she considered this possibility as something to be clarified in the future.

Since it ran against the accepted wisdom of DNA-RNA co-linearity, this assumption had already stirred fierce opposition when it was floated in the literature in 1975 (Lavi and Shatkin 1975; Abir-Am 2020b). Others who also made such an observation tended to delay its pursuit, because the prevailing view was that such results were an artifact.[26]

Much as Berget was in a hurry to get a good job (her three-year post-doc fellowship was to end with calendar 1977), her lab director Sharp was in a hurry to get tenure. Hence, he contemplated a fast-track publication process, opting for *PNAS* a journal that did not use the time-consuming peer review process but required that papers be submitted by a member of the National Academy of Science. Sharp thus approached his former post-doc adviser at Caltech, Norman Davidson an expert in various biophysical methods including EM and a NAS member, to submit to *PNAS* a paper he wrote on the basis of Berget and Moore's results.[27]

To her surprise, Berget saw a visibly shaken Sharp hearing from Davidson that he had already received a similar request from CSHL, and hence, he could not submit the paper from MIT-CRC. Davidson further advised that they publish back-to-back with the CSHL team. Davidson's advice did not suit Sharp who apparently had no confidence in the CSHL team. During his time at CSHL (1971–1974), Sharp belonged to a lab directed by Joseph Sambrook, who had strong opinions about other lab directors there. Sharp may have been exposed there to negative views of competing labs, including Roberts' lab, which was part of the CSHL team which approached Davidson and which would publish a parallel paper on the discovery of RNA splicing. CSHL had a reputation of being "balkanized" with competition and distrust prevailing among its various labs.[28]

Sharp still believes that the CSHL team of Chow, Gelinas, Broker, and Roberts somehow "stumbled" on the discovery. In Sharp's view, the CSHL team was not interested in the type of RNA processing questions which Darnell pioneered, questions which in Sharp's view led his lab to the discovery, precisely because they followed Darnell's lead, of solving how a very long high nuclear (hn) RNA became many short messenger RNA.[29] Though Sharp maintained contact with his former colleagues at CSHL, he seemed to have missed the progress made at CSHL due to the synergetic collaboration between two labs, those of Gelinas and Roberts (the Nucleic Acids lab) and Broker and Chow[30] (the EM lab).

Given Davidson's revelation of the existence of parallel results elsewhere, and his refusal to submit to *PNAS* a paper from Sharp's team without a back-to-back joint publication with the CSHL team (which also included Davidson's former post-docs Chow and Broker), Sharp approached instead the MIT-CRC Associate Director who hired him in 1974, David Baltimore, also a NAS member, and asked him to submit to *PNAS* the paper from his lab.[31] Baltimore consulted James E. Darnell of Rockefeller University, the leading RNA processing scientist at the time, who as a NAS member could have provided the required second opinion. Darnell requested that the MIT-CRC co-authors explore the origins of the RNA "tails" in the genome. By his own account, Darnell was reluctant to give up the principle of co-linearity, this being a reason, though not the only one, for his missing the discovery of RNA splicing despite being best positioned to make it. Perhaps Darnell hoped that looking for the tails' genomic origins would ultimately confirm, rather than refute, co-linearity.[32]

Baltimore passed Darnell's request to Berget, Moore, Sharp, and their lab mates during a lab meeting, though Sharp recalls being so informed alone. Both versions are plausible; in the context of dispensing wisdom to members of MIT-CRC's labs, Baltimore, by 1977 the MIT-CRC star scientist with a recent Nobel (1975) for a major discovery was accustomed to pontificate to groups of scientists. However, in the context of revising a manuscript, Baltimore would have talked to Sharp only since Sharp took sole charge of writing the paper.

Despite the inventiveness, resourcefulness, and hard work that Berget invested in her project, and despite the intriguing results she obtained together with Moore, Sharp wanted to take the project away from her and give it to a male post-doc, Arnold Berk. Berk developed a new technique of S1 mapping as a diagnostic of where RNA hybridizes to DNA, a type of work that Sharp was well positioned to value, since Sharp himself was best known at that time for his contributions to the development of techniques for the separation of DNA strands. Sharp's interest in developing techniques was influenced by his close rapport at Caltech with Jerome R. Vinograd (Davidson 1976; Sinsheimer 2007), who served as his mentor. Vinograd perfected the cesium ultracentrifuge that enabled Meselson and Stahl to prove the semi-conservative duplication of DNA, a landmark experiment in molecular biology (Abir-Am 2014b).

In the event, Berget was spared the indignity of having her project taken away, when Baltimore suggested that she perform her experiments on a bigger piece of DNA, so that the coding regions for the "tails" could be detected. Berget vividly recalled this critical moment many years later:

> I was not happy about having my project passed of to somebody else… And David finally turned to Phil and said 'well why don't you just give Sue a bigger piece of DNA.' And that's what I did and the rest is history.[33]

Berget expected to find one loop near the promoter (the DNA site where RNA transcription starts), but she experienced a great thrill when she saw several loops which meant that mRNA was pieced together from RNA transcribed from non-contiguous genome segments. It wasn't an indirect result. It was very direct picture of single molecule.[34] The piecing together or splicing, as it came to be known, is a different mechanism for the production of mRNA than the mechanism known from bacteria, where the co-linearity between mRNA and DNA is preserved after transcription. Hence, the discovery of RNA splicing came as a big surprise, because it demolished the universality of co-linearity, until then a pillar of molecular biology.[35]

Berget's Assessment of Her Lab Members

Sue Berget remains particularly grateful to Claire Moore, the technician, who was crucial in materializing Berget's opportunity to make an important discovery via their joint EM experiments. Since Berget was aware that technicians were not getting credit for their work, she insisted that Moore's name be added as a co-author. Berget remains very clear about Moore's essential role: "I would have been a footnote to history without her".[36] Ironically, her gratitude to Claire Moore would contribute to blurring her own role as the primary planner and interpreter of the RNA splicing experiments at MIT-CRC. Berget's decision to include Moore as a co-author contributed to the confusion of both women's key role in the discovery, some sort of "guilt by association"; if one woman was a "mere" technician, and the other was "only a post-doc", then both could be, and were reduced, as it was common at the time, to the lowest common denominator.[37] If one of them was a technician, then the other one must have also been a support personnel rather than a discoverer. Perhaps a single woman first co-author might have been seen as comparable in contribution to her male last co-author, but two women as a majority of co-authors were viewed against a background of most women participating in science as "supportive cast". At the time, most women occupied positions of dependent scientists, with few only advancing to positions of independent staff scientists.[38] This meant that those unfamiliar with the details of the co-authors' involvement in the discovery, were likely to assume that the lab director was the actual discoverer.

Even more so since the scientific community was further exposed to the lab director's imposed monopoly on presentations of the discovery by himself, as work coming out of his lab, without naming the actual discoverers or elaborating on their respective roles.[39] The two women co-authors remained voiceless and faceless, as if the lab director was embarrassed by their great success. At a key international conference on Chromatin in June 1977, where the presentation of the EM discovery proof in papers from MIT-CRC and CSHL determined those two institutions' priority over all other contender teams; the lab director did not come with his two woman co-authors but with two male post-docs who had no connection to the discovery but served as proof that the director's lab had a wider research program.[40] Possibly, the lab director felt compelled to highlight the size of his lab, since the competing paper from CSHL included an even larger number of co-authors and labs.[41]

It is also plausible, as Berget observed, that at the time the lab director believed the work of another post-doc to be of greater value; this explains why the lab director did not focus on Berget's efforts, just to be surprised that they led to the discovery while his favorite post-doc's innovation remained just that, a technical advance not a conceptual one, like Berget's. As Berget observed, the lab director was primarily interested in another post-doc's invention of a technology, as a type of contribution he best related to.[42] Plausibly, the lab director and his entire lab's preference for a new technology, also explain why neither Berget nor her lab director Sharp thought about that crucial next step in her project. Such a preference was acquired during the lab director's time as a post-doc closely mentored by the scientist who perfected the cesium ultracentrifuge which played a key role in proving the semi-conservative duplication of DNA.[43] Indeed, at this moment, Sharp was known for developing a technique for separating DNA strands.[44]

The next crucial step in Berget's project required a more conceptual orientation, which was evidently lacking in that lab. In the event, that next crucial step was supplied by the MIT-CRC associate director, David Baltimore, whose crucial help was acknowledged in the discovery paper.[45] Under normal circumstances, such crucial help should have entitled Baltimore to co-authorship, but being at the time (1977) a recent Nobel co-laureate for a major discovery which modified the Central Dogma of molecular biology, Baltimore did not seek such a co-authorship.[46] Evidently, he preferred to gift his insights, either out of a general spirit of generosity, or perhaps for other reasons.

Berget believes that Baltimore was able to provide that key next step because Baltimore was more experienced with thinking about gene and genome organization, as well as the relationship between RNA and DNA. Baltimore was then a star scientist with special expertise in unconventional relationships between RNA and DNA. He discovered, in parallel to his Nobel co-laureate Howard Temin, that RNA can serve as a template for making DNA; prior to that discovery, in 1970, it was believed that only DNA serves as a template for RNA. It is thus possible that Berget could not help but repeat the common scientific "wisdom" at the time, and later, which attributes to Baltimore a supreme ability to solve other people's problems. But she also contradicted herself when she stated that Baltimore's contribution was an obvious one, one which every post-doc might have thought about. While it is plausible that Berget, and/or her lab director might have eventually zeroed in on that key step, the fact remains that in the context of competition with one or more other labs, Baltimore's guidance, whether obvious or not, was certainly most timely. Without it, Berget, Moore, and Sharp's team might not have been able to publish before other teams, including the CSHL team which informed Davidson of their discovery before the MIT-CRC team of Berget, Moore, and Sharp did.

The issue of whether Baltimore's ability to provide the lead to the discovery stemmed from his greater knowledge as a leading scientist, or from other factors, is discussed elsewhere. Here, in the context of establishing Berget's rights to credit for the discovery, it is sufficient to clarify that the lead came from someone other than the lab director, and therefore the latter cannot be said to have had more rights to the discovery than Berget.

"He Tried to Kill Me Getting a Job"

Berget saw the lab director's role in the discovery as managerial, or someone who "paid the bills", wrote the paper, and negotiated publication in a fast-track outlet, thus narrowly beating the competition from other research institutes. She is not upset at being excluded from the Nobel Prize, not even at her lab director taking credit for her work while preventing her from presenting her results, since she believes that the thrill of making such a discovery was worthwhile everything:

> I have never felt short changed in the Nobel department. I would not have enjoyed the lifestyle of a Nobel laureate... Well it was Phil's lab. I would have never done the experiment if I hadn't been in the lab studying that problem. He paid the bills, he set the lab on this course, he gave me an environment where my mind could do what I did. So that has never bothered me. The fact that he tried to kill me getting a job that bothers me a lot.[47]

However, she remains very upset at her lab director as a result of her experience with him in looking for a job upon completion of her post-doc, which included several papers on the discovery of RNA splicing with herself as the first co-author in all of them (e.g. Berget et al. 1977, and Berget et al. 1978, among others).[48] Though Berget got interviews in many prestigious universities, speaking to standing room only audiences, still she received no single offer. One of her friends made some inquiries and discovered that the letters of recommendation from her lab director failed to mention her key role in the discovery of RNA splicing.

Only after confronting Sharp and demanding that he spell out her contribution to the discovery of RNA splicing, was Berget able to receive a couple of job offers. She accepted a position at Rice University in Houston, Texas, where her biologist spouse also found a position. A few years later, Berget relocated to Baylor Medical School, also in Houston, where she remained for the rest of her career, reaching the position of Chairperson in the Department of Pharmacology.

When Berget was promoted to full professor, she needed letters from all the people she worked for; her lab director, who remained at MIT for the rest of his career, was asked again to write one. A copy of his letter reached her and lo and behold, again, he said nothing about her key role in the discovery of RNA splicing.

Though Berget is understandably upset at her lab director's incomprehensible efforts to destroy her career both at its start and at the peak of full professor, she apparently internalized the ethos of science governance without accountability to such an extent that she even made excuses for the behavior of her lab director. She suggested that his harmful and unprofessional behavior does not reflect "who he was" but was the result of a "confluence of a lot of factors", especially his desire to get tenure at that time.[49]

Aside from the fact that one's desire for tenure does not justify denying credit that is due to the lead co-author of a joint paper, Berget seems to treat Sharp's denial of her role not only in letters of recommendation for jobs and promotions but also in presentations at conferences and allocation of prizes, as a lapse that owes its existence to mitigating circumstances. However, if we follow the methodical way in which her lab director systematically distanced her from her own discovery, then the outcome of the enormous disparity between his career at MIT and hers in Texas seems to be quite a calculated one. The question persists as to how such a calculated disparity was predicated upon the governing of science by lab directors who are not held accountable for their abuses of power.

As soon as the important results she obtained were validated, Berget, a post-doc, or a dependent scientist, was forced to relinquish any further agency; the lab director monopolized the conversion of her discovery into an asset for his own career, as well as a source of glory for his institution. As Berget noted, she was not allowed to present the discovery either inside or outside MIT-CRC. Aside from two small conferences, all presentations to the scientific community were made by the lab director, who not only "refined" the discovery as the product of the lab but further compiled and circulated a discovery narrative to this effect to many key people in the field.

Conclusion

Berget should have never agreed to be excluded from public presentations of the discovery. There was enough glory in this discovery for all three of them, as well as for contender scientists from other institutions. This should have been obvious to lab directors, as well as to their institutional Directors and Associate Directors. Robbing junior women scientists of their credit, thus depriving generations of girls and women scientists of suitable role models, and setting back gender equality in science and society should have never been an option for promoting one's career or the glory of one's institution.

Such a conduct may sound as something that could have only happened in the distant 1970s; however, the culture of denying opportunities to women by powerful male scientists seems to persist: in 2012, MIT rescinded a job offer from a young woman scientist, who won it in an international competition, at the request of a director of a research institute who said he did not welcome competition.[50] Is it a mere coincidence that the perpetrator of the 2012 episode, who was forced by public opinion to resign his directorship, is the same one who nominated Berget's lab director for the Nobel Prize without even contemplating to include her as the discovery's lead co-author?

Along these lines, a more recent attempt to rob the two women discoverers of Crispr by yet another director of a research institute at MIT, was barely thwarted (Lander 2016).

The question persists as to why the scientific community at large, and the Nobel Committee in particular, did not investigate the role of women first co-authors such as Berget and Chow when it began contemplating the prize for this discovery, in the period 1989–1993. Restoring epistemic justice should be on the Nobel agenda, not ratification of cannibalism of the scientific credit of women innovators, which deprived generations of girls of role models and precipitated the under-representation of women in science as a lasting legacy of systemic gender inequality.

"Thou hath murdered and also taken possession!" is the moral lesson with which the prophet Gad confronted King Ahab, who took over Navoth's vineyard by disposing of its owner who refused to sell his ancestors' land. This lesson should be inscribed on the marble façades of 800-pound gorilla research institutions in which the credit of woman scientists has been repeatedly appropriated by lab directors and institute directors who were supposed to protect them, or at the very least not hinder their precarious opportunities.

Acknowledgment

Research for this essay was sponsored by NSF-STS (National Science Foundation, USA, Program in Science, Technology, and Society) Award 1755024. Research assistance with conducting and/or transcribing oral history by Brandeis students (in chronological order) A.J. Mazella, J. Koerner, D. Robbins, A.B. Levison, A.Z. Unger, M. Hayford, and V.E. Richardson is gratefully acknowledged. Work-in-progress benefited from comments received at various conference presentations, including AAAS-2017; HSS-2018, 2021 & 2022, ISHPSSB-2019, 2021; ESHS-2020, 2022; as well as from the Editors and session commentator at a conference organized by the editors in 2021. Last but not least, the cooperation with the oral history of the scientists mentioned in the text and notes, and most notably of Professors Sue Berget and Claire Moore, is gratefully acknowledged. The author alone is responsible for this perspective which differs greatly from the secondary literature on the discovery of RNA splicing, as well as from the justifications offered for awarding the 1993 Nobel Prize to two male lab directors.

Notes

1 See Abir-Am and Elliott (2000) on public memory in many fields of science.
2 The Nobel Prizes, Presentations, Biographies, and Lectures, Stockholm-Sweden: Almqvist & Wiksell International Publishers, 1993; www.nobelprize.org/1993/Physiology or Medicine. On the reasons for many Nobel prizes being contested see Friedman (2002). See also Norrby (2019).
3 The laureates were associated with MIT/Cambridge, MA and CSHL/Greater NY; the women lead co-authors with the Universities of Alabama/Birmingham and Texas/Houston. See also Abir-Am (2020b).

4 NSF (2006); Baltimore et al. (2007).
5 See Witze (2020). The ADVANCE Program https://www.nsf.gov/crssprgm/advance/ at NSF (National Science Foundation) in the United States was designed to promote the careers of women scientists by involving high administrators in its grants. At a meeting held at the height of the public debate of 2005, a theatrical troupe based at the University of Michigan demonstrated, with the advisory help of faculty members, how such chairpersons manipulate hiring and promotion committees so as to avoid the hiring or promotion of women candidates. On the impact of NSF's Programs, see Rosser (2004). On the persistence of covert discrimination, see also Valian (1999), Martinez et al. (2007), Rossiter (2012), and Murgia and Poggio (2018).
6 RNA splicing is a mechanism by which a transcript of the genome is converted into a functional messenger-RNA, (hereafter m-RNA) which codes for life sustaining proteins. The discovery of RNA splicing is important because it replaced the principle of co-linearity between DNA and m-RNA, (a major tenet of molecular biology which was discovered in simple organisms such as bacteria) with a relationship of alternative splicing, in which non-contiguous segments of the genome are transcribed, cut and linked, to form functional m-RNAs in a variety of ways. This feature of modularity in RNA splicing explains why a smaller number of genes (~20,000) are sufficient to code for a larger number of proteins (~150,000) needed to maintain life in higher organisms, including humans. The main ramification of this discovery was that a major portion of genetic regulation was no longer confined to genes and DNA but had to be understood at the RNA level. The discovery also has ramifications for our understanding of the origins of life and of diseases stemming from splicing errors. On the discovery of RNA splicing, see Darnell (2011, chapters 4 & 5); Fry (2016); Abir-Am (2020b) https://www.americanscientist.org/article/the-women-who-discovered-rna-splicing.
7 There were no women co-authors in the nine scientist two teams claiming the discovery of messenger RNA in 1961; the discovery of DNA structure included a key woman protagonist—Rosalind Franklin, but it took the scientific community half a century to accept her as one of the co-discoverers, see Abir-Am (2004). The women co-authors of the RNA splicing discovery were, in alphabetical order: Sue Berget, Louise Chow, Mia Horowitz, Sara Lavi, Claire Moore, Sayeeda Zain. See also Abir-Am (2020b). For the 1977 publications on RNA splicing, see Berget et al. (1977); Chow et al. (1977); Aloni et al. (1977); Hsu and Ford (1977); Lavi and Groner (1977). The 1993 Nobel was given to R. Roberts, one of the four co-authors of the paper by Chow et al. (1977), and P. Sharp, one of the three co-authors of the paper by Berget et al. (1977).
8 Berget was the first author of Berget et al. (1977); Louise Chow was the first co-author of Chow et al. (1977); and Sara Lavi was the first co-author of Lavi and Groner (1977). On co-authorship in collaborative research see Beaver (2004); Biagioli (1999); Wray (2006).
9 On the education of women engineers in America, see Bix (2014).
10 https://www.nytimes.com/1999/03/23/us/mit-admits-discrimination-against-female-professors.html, http://web.mit.edu/fnl/women/women.html, https://news.mit.edu/1999/women-0331, https://news.mit.edu/2002/genderequity.
11 Torriani collaborated with Jacques Monod at the Pasteur Institute in the 1950s prior to arriving in the US; see "Longtime biology professor Annamaria Torriani-Gorini dies at 94, Microbiologist and lifelong social activist worked at MIT from 1960 to 1989", *MIT News*, June 5, 2013, https://news.mit.edu/2013/obit-torriani-gorini-biology.
12 Hopkins was hired in 1973. She played a key role in co-writing a Report on the discrimination of women scientists at MIT, a report which led MIT to give women pay back for wage discrimination and monitor diversity more widely, see The *MIT Faculty Newsletter*, Special Edition, A Study on the Status of Women

Faculty in Science at MIT, March 1999, vol. XI, No.4, http://web.mit.edu.fnl, For a comprehensive account of her role in combatting gender inequality at MIT and beyond see Zernike (2023).

13 She served as AAAS President in 2011–2012. Oral History with Huang by the author and R/As in January 2019; transcription by research assistant A. Levison.

14 See Sambrook's obituary in Dunn et al. (2019).

15 See Sambrook's obituary in Dunn et al. (2019).

16 Dunn et al. (2019). See also references to Sambrook as a scientist and mentor in the oral history conducted with his long time colleague Ashley Dunn by an oral history project based at CSHL's Library, https://library.cshl.edu/dunn.

17 Oral history with Professor S. Jane Flint, by P.G. Abir-Am, May 2019.

18 Berget, Oral history with P.G. Abir-Am and research assistants, 23 August 2019; 6–7 of transcript prepared by V.E. Richardson, or #9 in Updated Table of Summary of Main Themes, prepared by P.G. Abir-Am, 25 May 2021.

19 The discovery paper of Berget, Moore, and Sharp, published in August 1977 was submitted for publication in *PNAS* on 9 May 1977 or about eight months after the August 1976 Meeting.

20 Berget, Oral history with P.G. Abir-Am and two research assistants, 23 August 2019; 6–7 of transcript prepared by V.E. Richardson, or #9 in Updated Table of Summary of Main Themes, prepared by P.G. Abir-Am, 25 May 2021.

21 Oral history with T. Broker & L. Chow by P.G. Abir-Am and R/As, on 27 September 2019; transcripts by R/As M. Hayford and V. Richardson.

22 Oral history with Berget by P.G. Abir-Am and R/As, 23 August 2019: 29 of transcript.

23 Oral history with S. Berget by P.G. Abir-Am & R/As, 23 August 2019 : 43 of transcript.

24 Oral history with Claire Moore in March 2016 at Tufts University and transcription by research assistant A.J. Mazella.

25 Berk and Sharp (1977); see note 21.

26 Other teams who made such an observation were among the contender teams who published on RNA splicing in 1977, see note 9.

27 On N. Davidson see his obituary published by Caltech, https://www.caltech.edu/about/news/caltech-molecular-biologist-norman-davidson-dies-547.

28 Oral history with scientists from CSHL including (in alphabetical order) Botchan, Broker, Mathews, and Roberts by the author and R/As; oral history with Ashley Dunn by CSHL Library.

29 Oral history with Philip A. Sharp, 23 January 2019, P.G. Abir-Am and R/As; transcribed by A. Levison. On Darnell's pioneering role in RNA processing research, see Darnell (2011).

30 On this collaboration see Abir-Am (2020b).

31 Baltimore shared the 1975 Nobel Prize for his co-discovery of reverse transcriptase (in 1970) which was shown to transcribe RNA into DNA thus redefining the Central Dogma of molecular biology which postulated unidirectional transfer of information, from DNA to RNA only. On Baltimore as a key figure in American science see Crotty (2003).

32 Oral history with Darnell at Rockefeller University, 22 March 2019 by P.G. Abir-Am and R/As; transcription by R/A Victoria E. Richardson, Brandeis 2020. See also Darnell (2011).

33 Oral history with S. Berget by P.G. Abir-Am and R/As, 23 August 2019: 12 of transcript.

34 Oral history with S. Berget by P.G. Abir-Am and R/As, 23 August 2019: 29, #31 in Table of Summary of Main Themes.

35 Darnell (2011); Fry (2016).

36 Berget to Abir-Am, 24 (my emphasis).

37 The use of post-doc status as an excuse to deny credit to a scientist persisted even after several post-docs (and a graduate student) were included in Nobel Prizes in

1984, 2004, 2008, and 2009. For example, at the reception for the Rosenstiel award in 2016, https://www.brandeis.edu/rosenstiel/rosenstiel-award/past.html.

38 See the early experience of Elizabeth Blackburn, who shared the 2009 Nobel Prize in Physiology and Medicine; she was offered a position of research assistant despite having had two years experience as a post-doc, around the same time of 1977. For details see Brady (2007); Abir-Am (2014a).

39 Author's correspondence with James Darnell; also oral history with him, March 2019.

40 The two post-docs and a graduate student were also included in the Acknowledgment of Berget et al. (1977: 3715) for reading the paper before publication. While the expansion of the presentation to include other lab work, by other two post-docs may provide a wider context, at the same time such a strategy also shifts the balance of power from the discoverer to the lab director.

41 For a parallel attempt to attribute the discovery of RNA splicing by a woman scientist not just to a lab or lab director, but to an entire research institute, which took place at CSHL, see Abir-Am 2025.

42 Oral history with S.M. Berget, 36.

43 J.R. Vinograd, see Davidson (1976) and Sinsheimer (2007).

44 Sharp et al. (1973).

45 "We gratefully acknowledge the suggestion by David Baltimore that we map the RNA sequences in the 5′ tail by electron microscopy of RNA-DNA hybrids", Berget et al. (1977, Acknowledgement, 3175).

46 On Baltimore's discovery and his career more widely see Kevles (1998); Crotty (2003).

47 Oral history with S.M. Berget, 36.

48 Berget et al. (1977, 1978).

49 Oral history with S.M. Berget, 37–38.

50 S. Tonegawa, a 1987 Nobel laureate and Director of the McGovern Institute for Brain Research at MIT prevented a junior woman faculty from accepting a job at MIT on the ground that she would compete with him, see "Professor Allegedly Bullied MIT Prospect", *The Boston Globe*, by Marcella Bombardieri, Globe Staff|July 28, 2006; see also the link to this article in News and Views for 2012 at the author's website on the under-representation of women in science, http://people.brandeis.edu/~pninaga/sger/, and most recently, Zernike 2023, which covers the "fight for women in science" at MIT in the last two decades or so.

References

Abir-Am, Pnina G. 2025. "Visualizing the invisible women of the discovery of RNA splicing: A collective portrait to help restore epistemic justice?" in *(In)Visible Women and Science in the Twentieth Century*, ed. Amelia Bonea and Irina Nastasă-Matei. Manchester: University of Manchester Press.

Abir-Am, Pnina G. 2020a. "[Review of] Robin Scheffler *A Contagious Cause*", *Medical History* 64 (2): 295–297. doi: 10.1017/mdh.2020.11

Abir-Am, Pnina G. 2020b. "The Women Who Discovered RNA Splicing", *American Scientist* 108 (5): 298–306.

Abir-Am, Pnina G. 2014a. "Women Scientists in the 1970s." in *Writing about lives in science*, ed. Paola Govoni and Zelda A. Franceschi, 223–261. Gottingen: Vandenhoeck & Ruprecht.

Abir-Am, Pnina G. 2014b. "The Meselson-Stahl Experiment on the Semi-Conservative Replication of DNA." in *Encyclopedia of Life Sciences*. Hoboken: John Wiley & Sons. https://doi.org/10.1002/9780470015902.a0025093

Abir-Am, Pnina G. 2004. "DNA at 50: Institutional and Biographical Perspectives", *Minerva* 42: 191–213. Abir-Am, Pnina G., and C.A. Elliott. 2000. *Commemorative Practices in Science: Historical Perspectives on the Politics of Collective Memory.* University of Chicago Press.

Aloni, Yosef, Ravi Dhar, Orgad Laub, M. Horowitz, and G. Khoury. 1977. "Novel Mechanism for RNA Maturation: The Leader Sequences of Simian Virus 40 mRNA Are Not Transcribed Adjacent to the Coding Sequences", *Proceedings of the National Academy of Sciences*, 74 (9): 3686–3690.

Baltimore, David, Lawrence H. Summers, Susan Hockfield, Sirley M. Tilghman, John Hennessy, Robert Birgeneau, Mary Sue Coleman, Amy Gutman, RichardC. Levin. 2005. "Joint Statement by the Nine Presidents on Gender Equity in Higher Education." Cambridge: Massachusetts Institute of Technology. https://news.mit.edu/2005/nine-pres-letter

Beaver, Donald., 2004. "Does Collaborative Research Have Greater Epistemic Authority?" *Scientometrics*, 60 (3): 399–408.

Berget, Susan M., S. Jane Flint, J.F. Williams, and Philip A. Sharp; 1976. "Adenovirus Transcription. IV. Synthesis of Viral-Specific RNA in Human Cells Infected with Temperature-Sensitive Mutants of Adenovirus 5", *Journal of Virology*, 19 (3): 879.

Berget, Susan M., Claire Moore, and Philip A. Sharp. 1977. "Spliced Segments at the 5′ Terminus of Adenovirus 2 Late mRNA", *Proceedings of the National Academy of Sciences*, 74 (8): 3171–3175.

Berget, Susan M., Arnold J. Berk, T. Harrison, Philip A. Sharp. 1978. "Spliced Segments at the 5' Termini of Adenovirus-2 Late mRNA: A Role for Heterogeneous Nuclear RNA in Mammalian Cells", *Cold Spring Harbor Symposium on Quantum Biology*, 42 (1): 523–529. doi: 10.1101/sqb.1978.042.01.055. PMID: 277361.

Berk, Arnold J., and Philip A. Sharp. 1977. "Sizing and Mapping of Early Adenovirus mRNA by Gel Electrophoresis of S1`Enconuclease-Digested Hybrids", *Cell*, 12 (3): 721–732.

Biagioli, Mario. 1999. "The Instability of Authorship: Credit and Responsibility in Contemporary Biomedicine" in *The Science Studies Reader*. ed. Mario Biagioli, 253–279. New York: Routledge.

Bix, Amy S. 2014. *Girls Coming to Tech! A History of American Engineering Education for Women.* Cambridge: The MIT Press.

Brady, Catherine. 2007. *Elizabeth Blackburn and the Story of Telomeres: Deciphering the Ends of DNA*. Cambridge: The MIT Press.

Chow, LouiseT., Richard E. Gelinas, Thomas R. Broker, and Richard J. Roberts. 1977. "An Amazing Sequence Arrangement at the 5′ Ends of Adenovirus 2 Messenger RNA", *Cell*, 12 (1): 1–8.

Crotty, Shane. 2003. *Ahead of the Curve: David Baltimore's Life in Science*. Berkeley: University of California Press.

Darnell, James. 2011. *RNA: Life's Indispensable Molecule.* Cold Spring: Cold Spring Harbor Laboratory Press.

Davidson, Norman. 1976. "Obituary of J.R. Vinograd, 1913–1976", *Nature*, 263: 178.

Dunn, Ashley, Ricky W. Johnstone, and Bruce Stillman. 2019. "Joseph F. Sambrook (1939–2019)", *Nature Structural and Molecular Biology*, 26; 846–847.

Flint, S. Jane, Susan M. Berget, and Philip A. Sharp. 1976. "Characterization of Single-Stranded Viral DNA Sequences Present During Replication of Adenovirus Types 2 and 5", *Cell*, 9 (4): 559–571.

Friedman, Robert M. 2002. *The Politics of Excellence: Behind the Nobel Prize in science*. New York: Times Books.

Fry, Michael. 2016. *Landmark Experiments in Molecular Biology,* Amsterdam: Elsevier.

Gelinas, Richard E. and Richard J. Roberts. 1977. "One Predominant 5′-Undecanucleotide in Adenovirus 2 Late Messenger RNAs", *Cell*, 11 (3): 533–544.

Hsu, Ming-Ta., and John Ford. 1977. "Sequence Arrangement of the 5' Ends of Simian Virus 40 16S and 19S mRNAs", *Proceedings of the National Academy of Sciences*, 74 (11): 4982–4985.

Kevles, Daniel J. 1998. Scientific fraud and misconduct in American political culture: Reflections on the Baltimore case. *Engineering and Science*, 61(3): 10–19.

Lander, Eric S. 2016. "The heroes of CRISPR". *Cell*, 164 (1): 18–28.

Lavi, Sara, and Aaron J. Shatkin. 1975. "Methylated simian virus 40-specific RNA from nuclei and cytoplasm of infected BSC-1 cells", *Proceedings of the National Academy of Sciences*, 72 (6): 2012–2016.

Lavi, Sara, and Yoram Groner. 1977. 5'-Terminal Sequences and Coding Region of Late Simian Virus 40 mRNAs Are Derived from Noncontiguous Segments of the Viral Genome", *Proceedings of the National Academy of Sciences*, 74 (12): 5323–5327.

Luria, Salvador E. 1984. *A Slot Machine, A Broken Test Tube*. New York: Basic Books.

Martinez, Elisabeth D., Jeannine Botos, Kathleen M. Dohoney, Theresa M. Geiman, Sarah S. Kolla, Ana Olivera, Yi Qiu, Geetha V. Rayasam, Diana A. Stavrena, and Orna Cohen-Fix. 2007. "Falling Off the Academic Bandwagon. Women Are More Likely to Quit at the Postdoc to Principal Investigator Transition". *EMBO Reports*, 8 (11): 977–981. doi: 10.1038/sj.embor.7401110

MIT News Office. 2013, June. "Longtime Biology Professor Annamaria Torriani-Gorini Dies at 94, Microbiologist and Lifelong Social Activist Worked at MIT from 1960 to 1989", *MIT News* (online). https://news.mit.edu/2013/obit-torriani-gorini-biology

Murgia, Annalisa, and Barbara Poggio, eds. 2018. *Gender and Precarious Research Careers: A Comparative Analysis*. New York: Routledge.

Norrby, Erling. 2019. *Nobel Prizes: Cancer, Vision, and the Genetic Code*. Singapore: World Scientific.

NSF. 2006. *Beyond Bias and Barriers: Fulfilling the Potential of Women in Academic Science and Engineering*. Arlington: National Science Foundation.

Rosser, Sue V. 2004. "Using POWRE to ADVANCE: Institutional Barriers Identified by Women Scientists and Engineers", *NWSA Journal*, 16 (1): 50–78.

Rossiter, Margaret W. 2012. *Women Scientists in America: Forging a New World Since 1972*. Baltimore: Johns Hopkins University Press.

Selya, Rena. 2022. *Salvador Luriu, An Immigrant Biologist in Cold War America*. Cambridge: The MIT Press.

Sharp, Philip A., Bill Sugden, and Joe Sambrook. 1973. "Detection of Two Restriction Endonuclease Activities in *Haemophilus parainfluenzae* Using Analytical Agarose—Ethidium Bromide Electrophoresis", *Biochemistry*, 12 (16): 3055–3063. doi: 10.1021/bi00740a018

Sinsheimer, Robert L. 2007. "Jerome Vinograd, 1913–1976", *Biographical Memoirs of Members of the National Academy of Science*, 89: 1–13. https://www.nasonline.org/publications/biographical-memoirs/memoir-pdfs/vinograd-jerome.pdf

Thomas, Marjorie, Raymond L. White, and RonaldW. Davis. 1976. "Hybridization of RNA to Double-Stranded DNA: Formation of R-Loops", *Proceedings of the National Academy of Sciences*, 73 (7): 2294–2298.

Valian, Virginia. 1999. *Why So Slow? The Advancement of Women*. Cambridge: The MIT Press.

Witze, Alexandra. 2020, 2 July. "Three Extraordinary Women Run the Gauntlet of Science – A Documentary", *Nature*, 583: 25–26. (review of *Picture a Scientist, Film by Sharon Shattuck and Ian Cheney, Uprising Production*). https://www.pictureascientist.com

Wray, K.Brad. 2006. "Scientific Authorship in the Age of Collaborative Research", *Studies in History and Philosophy of Science Part A*, 37 (3): 505–514.

Zernike, Kate. 2023. *The Exceptions: Nancy Hopkins, MIT, and the Fight for Women in Science*, New York: Scribner.

10 When the "Lady of Washington" Explored Men's Work

Industrial Medicine, Labor, and Gender in Early Twentieth-Century America

Judith Rainhorn

> "Not only do you know how to see, Ladies, but you also know how to guess, you understand by half a word: your wonderful heart feels the reasons that are not said!"[1]

From the end of the nineteenth century onwards, scientific surveys conducted by women researchers gradually became a prerequisite for women's actions in the social sphere. Although women's research aimed at building up knowledge without a reformist objective was still considered transgressive, particularly in European Catholic circles, the mere fact of having observed social realities and bearing witness to them became part of a process of legitimization of research carried out by women: the feminine survey increasingly became an empirical tool for social action in reformist circles, relying on a supposed "female intuition" that would make women's contributions particularly useful to the development of knowledge.

This was especially true because stakeholders were quick to attribute to budding female investigators those qualities, conceived as specifically feminine, which would facilitate the practice of investigation: empathy, emotion, the ability to listen, and finally, that "heart [which] feels the reasons that are not said" (Bulmer, Bales, and Kish Sklar 1991). In the models that circulated on both sides of the Atlantic, whether it be the Leplaysian surveys (Kalaora and Savoye 1986) or those carried out as of 1899 by North American consumers' leagues founded by Jane Addams, Josephine Lowell and Florence Kelley (Chessel 2012; Haydu 2014), the investigations carried out by women very widely concerned the women's condition and female and child labor, aiming to convince the public authorities of the need to protect the more vulnerable part of the workforce: physiologically weaker, paid lower wages and less unionized than their male counterparts, women were considered as needing special protection from the violence of industrial society due to their role as mothers-to-be. In this well-known context, Dr. Alice Hamilton (1869–1970), although she was a close friend of the above-mentioned prominent American women and activists, adopted a very specific approach in her research, which focused on men's work in industrial settings: apart from the sweatshops of

DOI: 10.4324/9781003562597-14

the garment industry or the domestic workshops producing artificial flower in North American metropolises, Hamilton mostly investigated the large car factories in Detroit, the stone quarries in Ohio, the copper mines in Arizona, or the sanitary-ware enameling works in Chicago, all of which employed an almost exclusive male workforce. While young women working in can-making plants and suffering from benzol poisoning also received Hamilton's attention, her focus was not gender-oriented; she mainly investigated whatever harmful industrial process was in the spotlight.[2] She clearly stood apart from other female social reformists in America and Europe, by transgressing gender boundaries and nevertheless imposing a legitimate scientific discourse.

Hamilton's career was obviously not exempt from contradictions regarding the preferred fields of her medical practice. Her early responsibilities as an intern at the Minneapolis Hospital (1893) brought her into contact with obstetrics and diseases affecting mothers and children. However, she clearly preferred toxicological research to these medical fields based on the importance of the relationship with the female body. Instead of human contact with patients, she then favored a dialogue with test tubes in the silent laboratory.[3] Nevertheless, two decades later, on the eve of World War I, Dr. Hamilton had become an investigator of occupational diseases for the Federal Department of Labor in Washington, touring the factories, building sites, and workshops of the United States to track down the deplorable sanitary conditions in industry, coming as close as possible to workers and their activities, investigating their working environment.[4] Between these two moments, Alice Hamilton, a young woman from an upper-middle-class Presbyterian family in Indiana, had encountered the American industrial expansion of the Progressive Era and made it the focus of her medical practice. During a period of phenomenal growth in the US economy, the young Dr. Hamilton quickly became a pioneer in the emerging industrial medicine field, going against the prevailing discourse and ideology that saw industrialization as the source of human progress, and putting forward a critical analysis of the American industrial machine as a cause of disease and death.[5]

The turn of the twentieth century is considered the golden age of social reform in action in the United States, which a prolific historiography has extensively addressed (Leiby 1978; Rosenthal et al 1985; Fitzpatrick 1990; Muncy 1991; Sklar 1995; Skinner 2014). While she has usually been regarded as a secondary actor, Hamilton seems perfectly emblematic of the British-American social reform circles where middle-class and wealthy women played a prominent role, to a greater extent than in continental Europe. Scientifically trained in both the United States and Europe, Hamilton illustrates what Daniel T. Rodgers called the "Atlantic crossings" emphasized in his eponymous book, in which he cast light on an "Atlantic community" of social reformers from the Progressive Era until World War II (Rodgers 1998). Mobilizing the progressive and radical circles that were then emerging within the American bourgeoisie, the question of industrial labor became crucial, through the issues of European immigration, child labor, and occupational and public health. Although a minor public figure compared to Jane Addams,

Josephine Lowell, Florence Kelley, or Frances Perkins, Alice Hamilton was deeply involved in these reformist networks, of which Chicago was one of the major centers. However, apart from a few references to her work and career in books for the general public celebrating scientists or eminent women who marked the American twentieth century (Rossiter 1982; Morantz-Sanchez 1985; Zach 2002; Tichi 2011), despite the publication of her autobiography and part of her correspondence,[6] Hamilton is only mentioned briefly in scholarly works on industrial pollution or occupational health risks: Christopher Sellers gave the most comprehensive study of her life and work in a chapter of his *Hazards of the Job* (Sellers 1997: III). Recently, a short biographical sketch looked at Alice Hamilton's education and training from second-hand and published sources (Ringenberg et al 2019).

Through Hamilton's personal and professional itinerary, this chapter proposes to discuss what contribution gender makes to the scientific career of a woman who barely ever mentioned the gender issue as an obstacle in her public life, whereas being a woman obviously placed her in the position of having to resist attempts at intimidation or inferiorization and to attest to the scientificity of her discourse. This chapter is thus an attempt to question the role of gender in the construction of the scientific and academic career of a woman who, for half a century, shook up the American industrial order in the name of medical science. It sheds light on the stages in the construction of her often-questioned scientific legitimacy as an expert on the health conditions of industrial work. This was built from a position that was socially marginal in relation to scientific institutions—a community of women involved in social reform—within a group that conformed to gender norms (care and social support). In the second stage of her career, Hamilton moved into more masculine and technical—labor—and more legitimate—academic and political—spheres. In her career, she clearly used self-staging to assert the power of her discourse and actions and, in so doing, showed how a woman could occupy a prominent—albeit limited—position in the sphere of scientific expertise and governance.

This analysis is based on a wide range of largely unpublished and previously unanalyzed archives from Hamilton's papers, the bulk of it is held at the Schlesinger Library at the Radcliffe Institute for Advanced Studies (Harvard University) and the University of Illinois at Chicago, which includes numerous diaries and personal notes, handwritten and typed accounts of her investigations in industrial sites, private and professional correspondence with industrialists and researchers, drafts and manuscripts of speeches and public addresses, documentation on various subjects of industrial toxicology, and some of the publications resulting from Hamilton's work.[7]

Standing Where Expected: An Educated Woman in "Womanly" Spheres

Hamilton's early career placed her on the margins of the institutions of social reform, in a predominantly female environment reflecting in a wide extent the path of social and professional insertion that American society drew for women who engaged in social and political action at the end of the nineteenth century. Her family environment had obviously been favorable to women's individual emancipation. Born in 1869 in a Scottish Presbyterian family that had settled in the United States in the 1840s, she was raised in Indiana. Illinois, to the west, was a growing industrialized state whose main city Chicago experienced massive urban growth due to immigration in the final decades of the nineteenth century. Of the four Hamilton sisters and their two female cousins, all remained single and built successful careers in science, art, and education. Alice did her medical training in both the United States and Germany, which was one of the most important European centers for the growing microbiological field. These experiences placed her at the heart of the transatlantic scientific, technical, and intellectual crossings, where social welfare was definitely a major topic of interest. As a result, Hamilton spoke fluent German and showed great interest in German affairs, at a time when, under the impetus of Chancellor Bismarck from 1883 onwards, Germany was building a particularly protective social protection system that appeared to be a model to other industrialized nations.

At the time, Alice Hamilton began her medical training in 1892, being a female medical student was no longer an exception in the United States: in her class at the University of Michigan (Ann Arbor), more than a quarter of the students were women.[8] After her initial medical training, she did her advanced internship at Northwestern Hospital in Minneapolis, Minnesota, and at the New England Hospital in Boston, Massachusetts, before eventually joining the prestigious Johns Hopkins University School of Medicine in Baltimore.[9] She completed her training with two years of specialized education in bacteriology and microbiology at the German universities of Leipzig and Munich (1895–1896), while at the same time enduring the ostracism reserved for young women in the European academic world. Hamilton reports in her autobiography that at both the Universities of Leipzig and Munich—closed to German women at the time—her professors consistently seated her in a corner away from the male students. She was escorted by the professor into and out of the lecture hall and became the object of unpleasant curiosity within the local academic community (Hamilton 1943: 45–46). Studying at prestigious universities was not, however, a passport to a brilliant scientific career. Margaret Rossiter, in her landmark book *Women Scientists in America* (1982), has emphasized this paradox: the unprecedented rise of female education in the United States made the country the world leader in women's education in the second half of the nineteenth century, but this effort was driven by the underlying belief that higher education for women "would produce better wives and mothers for the American republic"[10] rather than skilled professionals. Women could and should be educated if they were careful to use all this learning in "womanly" spheres, such as in raising morally upright and patriotic sons or taking care of the vulnerable and needy.

Alice Hamilton's initial career path is undoubtedly part of this pattern. Soon after returning to the United States, Alice Hamilton became a resident member of Hull House in Chicago, which was the first community founded after the British Settlement movement.[11] Since 1889, in a suburban Chicago neighborhood inhabited mainly by immigrants from Italy, Bohemia, and Russia, and by African Americans migrating from the Southern states, the radical reformers Jane Addams and Ellen Gates Star had been bringing together a few dozen women deeply committed to social reform in action. At Hull House—as in Toynbee Hall in East London, founded in 1884—the reformers, most of whom were from the educated bourgeoisie, were very much concerned with the growing gap between social classes, the rising poverty of the working classes, and the brutal routinization of industrial work. The Hull House community was based on the commitment and dedication of dozens of people who supported themselves to live in the community house and devoted themselves to social work among the local community. For more than two decades from 1897 onwards, Hamilton's stay and work at Hull House both made her a full-fledged member of this "reformist nebula"[12] rooted in the American-educated middle class and profoundly modified her relationship to medical practice. From this long and rich community experience, Alice Hamilton wrote that she drew a quite exceptional human depth that allowed her to give "at last, a practical aspect to [her] knowledge which [had] always seemed abstract and academic".[13] During this lengthy community experience where she lived and shared the daily work of Jane Addams, Florence Kelley, Julia Lathrop,[14] and other famous women reformists, Hamilton joined the American social reform circles through the practice of local care and field investigation. The practice of medicine in a working-class environment was not so common at the time, and Hamilton's skills were part of a range of care and leisure activities offered to the people of the South Chicago neighborhood. In the community house of Hull House, according to their respective skills, members set up: a legal aid center for migrants, English and literacy courses, and various cultural activities offered free of charge to all those whom American society, on its path to prosperity, had left by the wayside. According to what was considered womanly fields, Alice Hamilton first instituted a maternal and infant medical consultation, care for toddlers and young children, and support for single mothers and underprivileged families, relevant to her early training. From 1897 to 1919, she thus took an active practical part in the reforming crucible of Hull House, met the ordinary women of urban America, visited hundreds of workers' homes in the underprivileged districts of Chicago, trying to teach domestic hygiene to the women and daily preventive measures to all, especially after the great typhoid epidemic of summer 1902 in the city. Through this field practice, working in close contact with individuals and families, Hamilton discovered the sanitary disasters linked to unhindered industrial development, listened, cared for, and provided comfort to those harmed by industrial work. Through her commitment to women and children, she slowly made her way to male industrial work: her visits and conversations with immigrant families made her encounter a huge issue she had previously been totally unaware of, that is harmful work conditions and occupational health and safety in the growing industrial sector. The latter was a completely invisible field at the national level at that time, which then became the core of her investigative and scholarly achievements in the next decade. Hull House experience thus appears to be the practical tool Hamilton used to achieve the physical and epistemological translation from private practice to the public sphere, allowing her to establish a reliable scientific knowledge and both a prominent and influential position in this field of expertise.

Can a Woman Tell the Truth About Men? Staging Herself as an Expert

Chicago then constituted one of the heartlands of the labor movement in the United States. The Haymarket protest (4 May 1886), which turned into a bloodbath, became one of the icons of the international social and political unrest for the improvement of working-class conditions. In 1884, the Bureau of Labor had been established in the Department of Interior on a national level and soon became an independent department before being incorporated into the Department of Commerce and Labor in 1903. The Industrial Workers of the World union was founded in Chicago in 1905 and the State of Illinois Department of Factory Inspection was founded two years later, as evidence of the political will to build a legal framework of social protection in the workplace, based on field research within the industrial sector.

Throughout her long stay at Hull House, Hamilton had experienced social welfare at a local level. The expertise she gained from this community-based experience in Chicago made her the perfect choice to run the Illinois state policy on occupational health and safety she had actually largely displayed on a local basis. In 1909–1910, the Democratic governor of Illinois, Charles Deneen, following the advice of Charles Henderson, professor in Sociology at the University of Chicago and author of a study on the German sickness insurance scheme, founded the Occupational Disease Commission. It was the first investigative and political body of that kind in the country, composed of two members of the State Bureau of Labor, one industrial employer, and five medical doctors, including Alice Hamilton (1943: 118–119). Deneen appointed Hamilton as Special Investigator on industrial poisons to lead the first large investigation—known as the *Illinois Survey*—conducted by a large team of doctors, medical students, and social workers, under her leadership. The *Illinois Survey* aimed at giving public policy-making an empirical basis with knowledge-based networks, mostly female in this case. Hamilton herself, in her autobiography, wrote to what extent this was a completely new way of working, building public policy at the state level from local experience and door-to-door field work, disclosing unsuspected ways and grounds of intoxication in the workplace: "It was pioneering, exploration of an unknown field. No young doctor nowadays can hope for work as exciting and rewarding. Everything I discovered was new and most of it was really valuable", she wrote three decades later (Hamilton 1943: 121).

Beyond the personal experience, the *Illinois Survey* was a tipping point in the consideration of workers' health in the United States; it was also clearly set up to implement public policy on work conditions at the state level, and it directly gave way to the passing of the first workers' compensation acts in the history of the United States (Wisconsin in 1911, Illinois in 1912, soon followed by other industrial states). The *Illinois Survey* was also a founding moment in Hamilton's career: it confirmed her vocation for field epidemiology and gave her institutional legitimacy, thus establishing her in the landscape of health reform to such an extent that the following year (1911) the Federal Bureau of Labor in Washington appointed Hamilton as Special Investigator on Occupational Diseases. Armed with these official mandates at the state and then federal levels, which were so many incentives for research and health action, Alice Hamilton developed and refined her epidemiological practice over the course of the next decade in the form of investigations in factories, mine shafts, and quarries across the country.

From light bulb and battery factories to stone quarries, from tile factories to building sites, all over the country, Hamilton's investigations and surveys focused mainly on male workspaces.[15] In any case, she did not adopt a gender perspective to study a variety of workplaces but envisaged the industrial context as a whole, with its local and sectoral variations. One of these was Arizona's "Copper Belt", a work setting renowned to be particularly harsh, with an exclusively male workforce and a significant share of Mexican immigrants, situated in a remote and land-locked region of the country. Arizona became the 48th state of the United States in 1912, and the Miami-Globe mining area remained, when Hamilton visited the copper mines, a *Far South*. Called in by a copper miners' union to investigate the pathologies affecting the workers who used compressed air hammers in the mine shafts, this frail, austere 50-year-old woman wearing a high collar dress crossed the entire United States from Chicago by train and automobile, passing through desert territories, skirting Hopi and Apache reserves and stopping overnight in boom towns that had recently sprouted up along the railway tracks.[16] In these remote places, so far from the metropolises of Chicago or New York she was familiar with, she encountered the daily occupational hazards: dressed in a miner's outfit, wearing a helmet and carrying a safety lamp, she descended into the bowels of the earth in the miners' cage, visited the galleries by walking or even crawling, climbed the rails and platforms above the gaping open pits, and carefully observed the workers. Above all, she insisted on trying out the tools herself, transgressing gender boundaries for what was by no means a female occupation, in order to reproduce the workers' movement on their instructions—hence the frequent statement "I tried it" in her written reports.[17] She took precise written notes of everything, recorded every evening, which are invaluable witnesses to an epidemiological practice in action:

> I saw two Americans or Spaniards [Mexicans] using an air hammer equipped with water. The water comes through a small pipe that fits

> on the hammer and sends a stream of water to the head of the tool. It is almost impossible to prevent leaks... *I tried* an unequipped machine, held it against my legs, my stomach, my chest. It's fine against the legs, although the vibration is very clear, but against the stomach and chest, the effect is very strange. *It was too heavy for me...*[18]

The account given of the brutal physical confrontation with the reality she intended to understand and describe was a fundamental aspect of Hamilton's investigation. Her writings, whether in the form of notes taken on the spot or in her later reformulated autobiography, were full of anecdotes and accounts of her physical involvement in the field, despite the genuine fear she felt and despite gender stereotypes that she adopted as her own, as during this visit to the Ajo copper mine:

> I am not a very courageous person and that visit was something of an ordeal. It began with a climb in the dark up to the third story of the crushing mill on a sort of glorified ladder running up the outside wall, with open treads and a high hand rail on one side only. Coming down was much worse than going up. The crushed ore was leached with dilute sulfuric acid and then by electrolysis the pure metal was recovered. The acid was in enormous tanks and I was escorted around them on a narrow path which ran along the edges, and here again there was only a hand rail between me and that evil-looking, dark, bubbling acid; and I had the feeling all the time that if I stumbled I could so easily slip under the rail and plunge into the acid. (Hamilton 1943: 212–213)

Telling her own fear, in the moment and in retrospect, was a way of assuming her feminine side while seeking to overcome it through action. Regarding her confrontation with the underground mine and workplace, she wrote many years later: "Indeed I was pleased and flattered by the way [the man leading me into the mine] assumed that I was as good as a man, and could be trusted to look after myself" (Hamilton 1943: 216). This experience is undoubtedly a performance that places Hamilton beyond the social and professional threshold assigned to her gender in the context of the American Progressive era and she attributed it to a cultural trait of European civilization that has remained deeply embedded in the culture of the eastern and southern states of the United States, while the wild West had evidently freed itself from it.

In her tireless investigations in many hazardous occupational settings, she found herself in an uncomfortable position and had to fight to be taken seriously despite being a woman in mainly—if not only—male trades and occupations. This meant being taken seriously by the workers with whom she talked informally and compiled a history to understand the origin of their ailments. In this respect, trade unions sometimes played a facilitating role, by relying on her reputation and asking for her help in unearthing work-related illnesses. This was the case in Arizona's Copper Belt, where Hamilton intervened at the request of the miners' union. She also sought to be regarded as an authority by industrial leaders, who were often reluctant to allow her to visit shop floors or construction sites, which were considered closed spaces. As a government representative but having no clear legal right to enter private property, she often had to fight to visit plants and was frequently confronted with managers who refused to open the door to her inquisitive eyes or who had intentionally cleaned up the place before she came (Hamilton 1943: 146–147). The "Lady from Washington" nonetheless continued her obstinate work, visiting facilities, surveying workshops, interviewing and examining workers inside and outside plants, and meeting with their union leaders, their wives—or their widows. "The Lady from Washington" is an expression Hamilton used several times in her autobiography because she knew that it was how she had been referred to on many occasions by stakeholders in the places she had investigated (Hamilton 1943:146). It showed an obvious ambiguity with regard to a woman who deserved the respect due to her social position as an educated woman ("Lady") and her institutional legitimacy (she was working for the Federal Labor Office in Washington, D.C.). But it also expressed the distance caused by her gender in the exclusively male community of entrepreneurs, engineers, and, more often than not, workers with whom she worked.

Hamilton was not blind to the specific vulnerability created by her femininity. While her institutional position, and the political support she enjoyed, should have protected her from any attempt to denigrate or circumvent her authority; there were, on the contrary, numerous episodes in which she was put in a difficult position by men who belittled the scientific nature of her discourse or considered her to be a troublemaker, accusing her personally of being the weak link in a disruptive institutional or political situation. As evidenced by internal correspondence of the Barrett Company, a New York-based manufacturer of benzol (a by-product of coal tar used as a fuel), Alice Hamilton's insistent recommendations to abandon the manufacture of this dangerous industrial compound were flagrantly ignored for years. Ten years after her first interventions within the company, its former owner still wrote: "I know that back in the old days some of our boys used to think that she was a plain nuisance".[19]

This contempt would become relentless in some cases. This occurred, for example, in the conflict that arose in 1932–1933 between Alice Hamilton and the owners of the Pyrene Phosgene Extinguisher Company, who were themselves supported by the Chemical Fire Extinguishers Association. The latter held the doctor responsible and wanted to sue her for having, in their opinion, caused their exclusion from the market for fire extinguishers for the Eighth Avenue subway line in New York, because she had denounced the dangerous effects of the manufacture and use of phosgene.[20] The incident lasted throughout the winter of 1932–1933, considerably undermining Hamilton, who was only the expert who had provided toxicological information to the Federal Bureau of Mines, which had decided not to ask the Pyrene

Company to equip the subway line. Confronting a federal institution was far more complex than trying to intimidate a woman doctor into denying her claims. Insightful, Hamilton wrote to her New York lawyer: "There is no use telling myself that I will not be intimidated by [the contractor]. I realize now that I have been, from the beginning". And further: "It is not simply a question of frightening one woman into submission but that, even with me quite demolished, the facts would remain as they are".[21] While the industrialist intended to obtain her surrender due to her supposed vulnerability, Hamilton remained confident that the scientific truth would prevail.

Can a Woman Stand in a Male World? Performing as a Professor at Harvard

Hamilton's most challenging experience as a woman scientist was her institutional integration into the Harvard Medical School, where she was appointed assistant professor in 1919. This new institutional position was clearly an exceptional platform for her to try to share her interest in industrial toxicology and medicine with the medical profession and to teach doctors-to-be. The appointment to the prestigious Harvard University of this petite, slightly severe woman in her fifties, who came from the prosperous Presbyterian society of Indiana and had passed through the active reformist circles of Chicago, who had just traveled thousands of miles across the Great Plains and the desert to go and crawl in the Arizona copper mines, constituted a belated but unquestionable legitimization of her scientific discourse, largely established thanks to the practice of field investigation. Her appointment was the academic consecration of her brilliant career as a medical and public health expert. Hamilton thus became the first woman to enter the exclusively male world of the prestigious medical school, a fact widely emphasized by the press. In fact, newspaper headlines attested to the unusual nature of her appointment, whether it was celebrated or whether doubts about the validity of this choice were already surfacing: "Very feminine is Dr. Alice Hamilton, the first woman to break down the sex barrier and join Harvard Faculty", a *Boston Globe* headline announced on this occasion, alongside a photographic portrait full of humbleness and restraint[22]: a portrait that showed Dr. Hamilton in her assigned place, reinforced by the title of the article, the latter insisting at least as much on the exceptionality of the career than on the confirmed femininity of one who crossed social and gender thresholds.

Despite this recognition, Hamilton always struggled to fully integrate into academia. Regarding daily academic life, she reported in her autobiography that despite her position, the Harvard Faculty Club remained closed to her, as was the annual graduation procession, spaces of sociability and publicity reserved for men alone, and she was not given tickets to football games. Furthermore, in spite of her national reputation and public office, she was never appointed full professor and received no promotions during the 16 years she held her position. She faced an institutional glass ceiling that is entirely consistent with what Margaret Rossiter notes from several examples during the same period in the United States: "an associate professorship was the highest rank to which most women faculty could aspire realistically, regardless of how strong their aging mentor had been or how highly acclaimed their own work was" (Rossiter 1982: 188). Divided between admiration and aversion, the medical profession had an interesting ambivalence toward Dr Hamilton. Shaken in its certainties by the new perspective proposed by Hamilton, part of the medical world did not conceal a certain contempt for the topic of her research, an attitude whose misogyny is not to be overlooked: the health of workers, even male workers, was indeed a "woman's" concern. As Robert Nye has argued (1997), the capacity for violence, deeply rooted in the eighteenth-century "field of honor", remained intact as an instrument of masculine authority well into the twentieth century: medicine was widely regarded as demanding as much bravery as military combat. This approach influenced social relations in a way that continued to seal women off from full participation in the public sphere, despite their increasing numbers in the medical profession. Some ethicists were openly opposed to women joining the profession, holding them to be lacking in energy, courage, judgment, personal authority, and firmness. Whether some conceded female physicians' place in acceptably "feminine" areas of medical practice such as public health, hygiene, gynecology, pediatrics, and obstetrics, there was a widely shared obsession with portraying the medical profession as uniquely dangerous and endowing practitioners with qualities of great courage, which women *obviously* lacked (Nye 1997). Throughout her life, Dr. Hamilton recalled the (non-)choices she made in her career, which she owed to prejudice against women doctors, such as the memory of one of the professors of bacteriology under whose authority she worked in Munich in 1896, who forbade her to work in certain fields because they required experiments on animals that were not suitable for her sex.

As a woman within the massively male-dominated medical profession, Hamilton was often stigmatized as belonging to the female care professions, with her colleagues denigrating her "all-female sentimentality toward the poor" that was supposed to reflect her interest in social action and industrial labor, in the morbidity and mortality of workers (Hamilton 1943: 115). This stigma showed through on many occasions, and can be seen again in a cartoon published in the press in 1957 (Figure 10.1) on the occasion of her nomination as "top medical woman of the year" (even though she was already 89 years old!): Hamilton, a prestigious doctor, academic and expert for international organizations, was depicted as a nurse who was somewhat bewildered and at a loss as to which remedies to administer to the poisoned patients laid out before her. Invariably, Hamilton countered this stigma with her status as a prominent scientist, and with seldom discussing her role as a woman in her private papers or public declarations.

The growing interest in Hamilton's writings went hand in hand with the medical world's ambiguity toward her: the innovative character of her approach, combined with the scientific caution of her always well-founded

SIGMA DELTA EPSILON NEWS April 1957 Page five

DR. ALICE HAMILTON TOP MEDICAL WOMAN OF THE YEAR

Dr. Alice Hamilton of Hadlyme, Conn., headed the list of the 10 medical women of the year as named by the American Medical Women's Association in December 1956. She has been well known both nationally and internationally for her work in industrial medicine, especially for her books "Industrial Poisons", "Industrial Toxicology" and "Exploring the Dangerous Trades". SDE is pleased to have Dr. Hamilton as an Honorary Member and extends congratulations.

Figure 10.1 "Dr. Alice Hamilton, Top Medical Woman of the Year", April 1957, *Sigma Delta Epsilon News*: 5. Schlesinger Library, Radcliffe Institute, Harvard University. Public Domain.

assertions, contributed to a considerable increase in the reputation of her work. As such, she occupied a fundamental place in the diffuse world of social reform, in the same way as her colleagues who denounced child labor or the legal discrimination against European and Mexican immigrants, in particular within associations such as the American Association for Labor Legislation or the consumers' leagues that grew more numerous throughout the country during the first half of the twentieth century.[23] It cannot be denied that in the long run, the opening up of the new field of industrial medicine, the methodological innovations, and, above all, the transformations brought about by Hamilton's work on the physician's view of the patient and the clinical picture, probably had their part in the evolution of professional practice and the medical profession itself. While her appointment to Harvard University in 1919 undoubtedly testified to the place she had managed to carve out for herself in the American medical landscape over the previous decade, the toxicology manual she published in 1934, and updated regularly thereafter, became the seminal textbook on the subject for the medical school's students for several decades (Hamilton 1934). The book was updated in 1945, then in 1949 in collaboration with Dr. Harriet Hardy, Hamilton's female *protégée*, and then republished in 1975, 1983, and 1998, under the title *Hamilton and Hardy's Industrial Toxicology*, which reflects Hamilton's fundamental impact on the discipline.

Conclusion

Alice Hamilton's entire career was unwillingly marked by the issue of gender. Born into a family where the only son married while the four daughters remained single their whole lives, Alice Hamilton experienced alternative forms of domestic life: 22 years within the community at Hull House, and after retirement, in what can be called a "Boston marriage"[24] with a female friend, soon joined by her sister Norah Hamilton, a painter. Although she participated in the first International Women's Peace Congress in The Hague in 1915, Alice Hamilton never defined herself as a feminist. While she sometimes privately took offense at the subordinate position in which academic institutions held her, she more often denied the interplay between gender and socio-professional position. When she visited car factories in Detroit or copper mines in Arizona, when she was the only woman acknowledged expert at international conferences (as in Munich in 1938), she acted as if being a woman was negligible, pretending most of the time not to see the contempt that industrialists, doctors, and trade unionists felt for her female status. In her autobiography, Hamilton recounted a few episodes in which she was confronted with the gender issue when her expertise was devalued because she was a woman in a man's world. Surprisingly, however, it was not in a professional setting, in a male workspace, or at Harvard Medical School that she noted feeling the widest gender gap, but, according to her own account, at a memorial ceremony for the execution of Sacco and Vanzetti, in New York City, two years after the trial of the two Italian anarchists. Speaking before a large audience that had come to commemorate the 1927 miscarriage of justice, Hamilton noted with ironic terror the manliness of the police officers present to oversee the demonstration: "I think I have never felt *so small and feminine* as I did when I squeezed through the crowd of huge, formidable creatures, assembled, so they said, to protect me against communist violence" (Hamilton 1943: 279).

Lastly, being a woman probably allowed Hamilton to hold a peripheral position, and thus to take a critical look at American society: by relentlessly shining an uncompromising spotlight on the disastrous sanitary conditions in a huge portion of the country's industrial workplaces, Alice Hamilton demonstrated that American factories had morbidity and mortality rates generally higher than their European counterparts, especially in Great Britain and Germany. In doing so, she made a major contribution to challenging the contemporary myth of American industrial omnipotence.

Witnessing the importance of women's roles in American social reform in the Progressive Era, the existence of a reformist locus such as Hull House reflects a sexual division of labor within the intellectual world which some female social reformers seized upon to legitimize their presence on the social front: "As long as the science of man was a science of wealth, rest, and self-interest, there was little incentive for women to take it up. The new social sciences have a human dimension and cannot do without the help of women", wrote the lawyer Florence Kelley as early as 1882.[25] This human dimension meant: interest in poverty, tireless work, and altruism. Thus, as the acknowledged repositories of the human qualities assigned to their sex (humanity, compassion, self-sacrifice, etc.), women who embraced social reform could rely on very largely female institutions that were the bearers of social innovation—community houses, associations, etc.—to bolster their still fragile legitimacy and to gain access to the male fields of academic knowledge and political power. Consequently, by using the medical investigation field as her venue for building knowledge and taking action, Alice Hamilton alternately played on two levels: on the one hand, she remained in her assigned role as a woman, focusing her work on giving a voice to the weak and the sick; on the other hand, by confronting the industrial world, which was eminently masculine and physically harsh, and by elaborating a scholarly discourse, she blurred the boundaries, transgressed codes and shifted gender boundaries.

Acknowledgement

The author gratefully acknowledges Christopher Mobley's assistance in finalizing the English version of this text.

Notes

1 Georges Mény in 1909, quoted in Chessel (2012: 154).
2 Hamilton made several surveys or speeches on the female workforce (Hamilton 1919, 1926) but they were definitely not at the core of her research interests.
3 Letter from Alice Hamilton to her cousin Agnes Hamilton, Minneapolis, July 1893 (Sicherman 1984: 61).
4 The life and work of Alice Hamilton is the subject of her autobiography (Hamilton 1943), as well as a biography to be published by the author of this article.

5 Industrial medicine gradually took shape after the First World War, focusing on occupational health issues. Papers emphasizing the chronology and growth of industrial medicine can be found in many volumes of *Journal of Occupational and Environmental Medicine*.
6 Entitled *Exploring the Dangerous Trades*, Alice Hamilton's autobiography was published in 1943 (Hamilton 1943); a small part of her private (mainly family) correspondence was published by Barbara Sicherman (1984), an annotated edition of about 100 letters covering the period 1888–1965. As for Madeleine P. Grant's book (1967), it is a hagiographic biography based on informal discussions between the author and Hamilton, who was then about to celebrate her one-hundredth birthday.
7 Records are mainly: Alice Hamilton Papers (hereinafter AHP), held at the Schlesinger Library at Radcliffe College, Harvard University; the Alice Hamilton Collection and the Hull House Collection at the University of Illinois at Chicago; the A. Hamilton Papers in Connecticut College, New London (Conn., US). More archives in Geneva (ILO and League of Nations Archives) are part of the main biographical book currently being written.
8 Elizabeth Blackwell was, in 1849, the first female medical graduate in the United States. When Alice Hamilton began her medical studies, there were already some 5,000 women doctors in the country: they made up about 4% of the American medical profession, a proportion that remained stable until the 1960s. See Regina Morantz-Sanchez (1985).
9 However, Johns-Hopkins University did not yet award a doctorate to women.
10 (Rossiter 1982: 1). This scheme fit clearly in what historian Linda Kerber has called "Republican Motherhood" (Kerber 1976).
11 It was after visiting Toynbee Hall in London's East End, the first of the British settlements, that Jane Addams (1860–1935), American radical, feminist and pacifist (and future Nobel Peace Prize winner in 1931), decided to establish this community house on the outskirts of Chicago. By 1913, there were more than 400 settlements in 32 states in the United States. Hull House had about 250 resident members (partial or full time) over the period 1889–1929: see "List of Hull House Residents, 1889–1929", University of Illinois at Chicago, Hull House Collection, folder 294. On the Hull House experience, see Jane Addams' autobiography (1910), Eleanor Stebner (1997) and Kathryn Kish Sklar's contribution: "Hull House Maps and Papers", in Martin Bulmer et al. (2011): 111–146.
12 The expression, now part of the common vocabulary of historians and sociologists, is from Christian Topalov (1999).
13 Letter from Alice Hamilton to Agnes Hamilton, Fort Wayne, June 1902 (Sicherman 1984: 143).
14 On the lawyer Florence Kelley (1859–1932), see Sklar (1995); on Julia Lathrop (1858–1932), see Lindenmeyer (1997)
15 Among exceptions to this picture were studies of female workers subjected to the toxic fumes of benzol in can-making or rubber goods factories.
16 Hamilton recounts this seminal experience in various notes, conferences and in her autobiography (Hamilton 1943: 208–222). She was accompanied by her friend and colleague at Hull House Clara Landsberg.
17 Alice Hamilton Papers (hereinafter AHP), Schlesinger Library, Harvard University, box 1, File 37, "Arizona Copper Mines, 1917–1919".
18 Report of Visit to Miami Copper Company Mine, Miami, 18 January 1919, 1 (emphasis added).
19 AHP, Box 1, File 40/1–4, "Benzene, Aniline, etc. Correspondence, 1913–1942".
20 AHP, Box 1, File 48, "Dispute with Chemical Fire Extinguishers Association, 1932–1933".

21 Letters from Alice Hamilton to Benjamin Cohen, November 5, 1932 and January 4, 1933, AHP, Box 1, File 48, "Dispute with Chemical Fire Extinguishers Association, 1932–1933".
22 *The Boston Globe*, April 6, 1919.
23 Hamilton was an early member and a president (1944–49) of the US National Consumers' League.
24 In late nineteenth and early twentieth centuries, "Boston marriages" were relatively formalized friendships or life partnerships between educated single women with careers and usually feminist shared values, frequently involved in social and cultural causes, who wanted to live self-sufficiently. Until the 1920s, this kind of arrangement was widely regarded as respectable in American society.
25 Quoted in Kathryn Sklar, "Hull House Maps and Papers..." in Bulmer et al. (2011): 111.

References

Addams, Jane. 1910. *Twenty Years at Hull House. With Autobiographical Notes.* New York: MacMillan.

Bulmer, Martin, Kevin Bales, and Kathryn Kish Sklar. 1991. *The Social Survey in Historical Perspective, 1880–1940.* Cambridge: Cambridge University Press.

Chessel, Marie. 2012. *Consommateurs engagés à la Belle Époque: La Ligue sociale d'acheteurs.* Paris: Presses de Sciences Po.

Fitzpatrick, Ellen. 1990. *Endless Crusade: Women Social Scientists and Progressive Reform.* New York: Oxford University Press.

Grant, Madeleine P. 1967. *Pioneer Doctor in Industrial Medicine.* New York: Abelard.

Hamilton, Alice. 1919. "Women in the Lead Industries". *Bulletin of the United States Bureau of Labor Statistics,* no. 253. Washington: Government Printing Office.

Hamilton, Alice. 1926. "Women Workers and Industrial Poisons". *Bulletin of the Women's Bureau,* no. 57. Washington: Government Printing Office.

Hamilton, Alice. 1934. *Industrial Toxicology.* New York: Harper.

Hamilton, Alice. 1943. *Exploring the Dangerous Trades. The Autobiography of Alice* Hamilton, M.D. Boston: Little, Brown and Company.

Haydu, Jeffrey. 2014. "Consumer Citizenship and Cross-Class Activism: The Case of the National Consumers' League, 1899–1918". *Sociological Forum* 29 (3): 628–49.

Kalaora, Bernard and Aantoine Savoye. 1989. *Les inventeurs oubliés. Le Play et ses continuateurs aux origines des sciences sociales.* Seyssel: Champ Vallon.

Kerber, Linda. 1976. "The Republican Mother: Women and the Enlightenment. An American Perspective". *American Quarterly.* 28 (2): 187–205.

Leiby, James. 1978. *A History of Social Welfare and Social Work in the United States.* New York: Columbia University Press.

Lindenmeyer, Kriste. 1997. *A "Right to Childhood": The US Children's Bureau and Child Welfare, 1912–1946.* Urbana: University of Illinois Press.

Morantz-Sanchez, Regina M. 1985. *Sympathy and Science.* New York: Oxford University Press.

Muncy, Robyn. 1991. *Creating a Female Dominion in American Reform, 1890–1935.* New York: Columbia University Press.

Nye, Robert. 1997. "Medicine and Science as Masculine 'Fields of Honor'". *Osiris.* 12: 60–79.

Ringenberg, Matthew, William Ringenberg, and Joseph Brain. 2019. *The Education of Alice Hamilton: From Fort Wayne to Harvard.* Bloomington: Indiana University Press.

Rodgers, Daniel T. 1998. *Atlantic Crossings: Social Politics in a Progressive Age*. Cambridge: Harvard University Press.

Rosenthal, Naomi, Meryl Fingrutd, Michele Ethier, Roberta Karant, and David McDonald. 1985. "Social Movements and Network Analysis: A Case Study of Nineteenth-Century Women's Reform in New York State". *American Journal of Sociology*. 90 (5): 1022–1054.

Rossiter, Margaret. 1982. *Women Scientists in America: Struggles and Strategies to 1940*. Baltimore: Johns-Hopkins University Press.

Sellers, Christopher C. 1997. *Hazards of the Job. From Industrial Disease to Environmental Health Science*. Chapel Hill: University of North Carolina Press.

Sicherman, Barbara. 1984. *Alice Hamilton. A Life in Letters*. Cambridge: Harvard University Press.

Skinner, Carolyn. 2014. *Women Physicians & Professional Ethos in Nineteenth-Century America*. Carbondale: Southern Illinois University Press.

Sklar, Kathryn K. 1995. *Florence Kelley and the Nation's Work: The Rise of Women's Political Culture, 1830–1900*. New Haven: Yale University Press.

Stebner, Eleanor. 1997. *Hull House: A Study of Spirituality, Vocation, and Friendship*. New York: State University of New York.

Tichi, Cecelia. 2011. "Alice Hamilton, M.D. The Dangerous Trades". 29–88. In *Civic Passions: Seven Who Launched Progressive America*. Chapel Hill: University of North Carolina Press.

Topalov, Christian. 1999. *Laboratoires du nouveau siècle. La nébuleuse réformatrice et ses réseaux en France (1880–1914)*. Paris: Éditions de l'EHESS.

Zach, Kim K. 2002. *Hidden From History: The Lives of Eight American Women Scientists*. Greensboro: Avisson Press Inc.

11 Feminism Behind Science in the United States

Women's Governance and Leadership in Reproductive Sciences

Angeline Durand Vallot

The history of birth control rights in the United States spans over 60 years, from the end of the nineteenth century until the advent of the first oral contraceptive.[1] In 1960, a new era arose for women's rights with a landmark event: the release of the first birth control pill that would become, for many women, one of the greatest and most significant scientific and medical breakthroughs of the twentieth century. Universally known as the Pill, it transformed women's lives by the emancipatory power it wields. This simple and effective method of contraception increased women's individual and reproductive freedom by allowing them to control their reproductive capacities and to prevent unwanted pregnancy.

The invention of the oral contraceptive inscribes itself in a long history of advocacy and activism for women's rights and emancipation. This chapter examines feminist activism in relation to reproductive rights within the context of scientific innovations. Indeed, the development of a new form of contraception was made possible by the collective dynamics involving several influential individuals. Among the driving forces behind its scientific elaboration, two women, Margaret Sanger (1879–1966) and Katharine McCormick (1875–1967), were able to set new funding priorities for science and brought considerable advancements in reproductive technologies by fostering innovative research. Both were driven by the need and desire for social change and by their commitment to the advancement of women's rights. While they envisioned science as a strategy and reproductive sciences as a crucial resource for women's empowerment, they led the way for advances in contraceptive research. In this chapter, we will focus on their fight to secure contraceptive access rather than on their connections to eugenics that cannot be ignored however. Although this divisive alliance helped to legitimate women's rights to contraception, it also overshadowed the ideology of the birth control movement underlying Sanger and McCormick's actions. Their correspondence, which attests to the strong ties of friendship they had forged, offers a significant glimpse into their lives and missionary work to provide and promote birth control. It also highlights their commitment to the advancement of science and the use of scientific knowledge to improve women's lives.

DOI: 10.4324/9781003562597-15

Often presented as a symbol of women's liberation and empowerment, the Pill did not mark the beginning of women's experience with contraception. Prior to 1960, birth control methods existed, and in particular the diaphragm which had been the most popular and effective contraceptive option throughout the twentieth century.[2]

Early Birth Control Activism

In the 1910s, Sanger started a campaign for birth control and spent the first half of the twentieth century lobbying for women's access to contraception. A pioneering member of the American birth control movement, she envisioned birth control as crucial to women's rights and gender equality (Sanger 1920). Born into a working-class family, she insisted on voluntary motherhood (Sanger 1938: 108) and the necessity for women to be in control of their own bodies in order to control their lives. Thus, the birth control movement emerged to provide women with information about the prevention of unwanted pregnancy and to improve access to contraceptive methods. Her lifelong struggle started in the early twentieth century, when she worked as a visiting nurse on the Lower East Side of New York City, confronted with uncontrolled childbearing and unsafe abortions due to the lack of access to contraception.[3] She sought to remedy this situation and took on the mission of providing women with birth control information and reliable methods of reproductive control.[4] Therefore, she launched a campaign for birth control in 1914 and illegally opened, with her sister Ethel Byrne and Fannie Mindell, the first birth control clinic in Brooklyn, New York, in 1916. The clinic offered information on birth control and ensured the distribution of diaphragms, but was raided by the police ten days after it opened and Sanger was arrested for the illegal distribution of contraceptive information. While progressively gaining support from the medical profession in order to legitimize birth control, she established the American Birth Control League in New York in 1921 and the Clinical Research Bureau in 1923 to conduct research on the efficacy of contraceptives.[5] She also formed the National Committee on Federal Legislation for Birth Control in 1929 to lobby Congress for legislation that would permit doctors to prescribe birth control. Sanger was extremely active in the 1920s and 1930s in expanding the reproductive rights movement but had to change her strategy to continue her mission. She tried to find a strategic ally in the eugenics movement, as its theories were very popular in the 1920s, supported by prominent scientists, and several feminist reformers, and widely accepted in the American academic community. Moreover, the American eugenics movement received extensive funding from various organizations and foundations.[6] Eugenics was understood as a way to improve the human race and to limit population growth. This ideology merged with the population control policies of the time and dissociated birth control from women's sexuality. Yet, for Sanger, eugenics measures consisted in reducing hereditary disabilities and diseases

(Sanger 1921, 1922: VII) to limit the degeneration of the race by the genetic transmission of mental and physical defects.[7] She justified birth control to preserve national interests, and to benefit American society and women. Under her leadership, the movement insisted on the prevention of degeneracy and the promotion of racial betterment to achieve social evolution and engaged in eugenic discourse to demand birth control in the communities they believed needed it most. Nevertheless, her ideas on eugenics have been subject to questionable interpretations and have led to persistent criticism about her true intention. Instead of challenging obscenity laws directly, Sanger sought allies among physicians and scientists to join her action and turned away from the radicalism that dominated the early birth control movement. Therefore, she formed an alliance with the medical profession by focusing on public health and population control. From the 1930s onwards, the role of physicians increased, leading to the medicalization of birth control as the ideology of population control was making its way through American society (Reed 1978: 143–148). In 1936 physicians were legally authorized to prescribe and distribute information about contraceptives, and the American Medical Association officially recognized birth control as an integral part of medical practice and education (Chesler 1992: 374).[8] In 1939, the American Birth Control League merged with the Clinical Research Bureau to become the Birth Control Federation of America (Chesler 1992: 391) which was renamed the Planned Parenthood Federation of America in 1942. This organization, led mostly by male physicians, grew rapidly in the 1940s, transforming the birth control movement into a more respectable organization, a necessary stage in the perspective of birth control legalization.[9] In the 1950s, Sanger insisted on the availability of safe and effective birth control that would allow for women's sexual liberation and autonomy and would represent a means to reduce overpopulation which was a serious social and political issue. For years, she had been searching for an effective, affordable, and reversible method to prevent conception that would be controlled by women and also regulated by the medical profession.[10] As she envisioned "a magic pill", Sanger turned to science to find an answer to women's lack of access to reliable contraceptive methods.[11] According to her, "The requirements of the perfect Pill must undergo careful and painstaking scrutiny from various scientists in biology, biochemistry, as well as physical and psychiatric experts (Sanger 1952:104–105)". Already in her book, *The Pivot of Civilization*, published in 1922, she appealed to science:

> My interest in Birth Control was awakened by experience. Research and investigation have followed. Our effort has been to raise our program from the plane of the emotional to the plane of the scientific. Any social progress, it is my belief, must purge itself of sentimentalism and pass through the crucible of science. We are willing to submit Birth Control to this test.
>
> (Sanger 1922: 26, IX)

The final objective of her feminist mission was the development of hormonal birth control. Nevertheless, in the early 1950s, contraception was still highly controversial and many pharmaceutical companies were reluctant to engage in hormonal contraceptive research and development. Although the discovery of female hormones at the beginning of the twentieth century had opened the way toward advances in research on hormonal contraception, the complexity of the development of an oral contraceptive, which relied on scientific knowledge, technology, legal regulation, financial resources, and scientists willing to take on innovative research was a major obstacle.

Premises of a Scientific Revolution

Margaret Sanger and her fellow activist, Katharine McCormick, joined forces in the 1950s to make the development of a hormonal contraceptive possible. While Sanger had been fighting for decades to legalize birth control and was still eager to drive her campaign forward, she crossed paths with McCormick, a major feminist and biologist. Born in 1875 into a wealthy American family, McCormick was one of the first women to graduate with a degree in biology from the prestigious Massachusetts Institute of Technology in 1904.

She married Stanley McCormick, son of the inventor and manufacturer of the mechanized reaper and multi-millionaire. McCormick was also a fervent women's rights activist who had a leading role in the suffrage movement (Fields 2003: 121). As an early feminist, she began volunteering with the Women's Suffrage Movement and spoke at the first rally for women's suffrage in Boston in 1909. She worked with Carrie Chapman Catt (Fields 2003: 170) and other leaders in the suffrage movement and was a prominent member of the National American Woman Suffrage Association. In 1920, after the ratification of the Nineteenth Amendment that gave the right to vote to women, McCormick served as Vice President of the League of Women Voters.[12] Convinced that the right to vote for women was essential, she also firmly believed that a woman's right to control her body was fundamental, and that science could liberate women from sexual oppression by giving them the possibility to limit their reproductive capacities. Thus, Katharine McCormick was very active in the reproductive rights movement, and she had the opportunity to meet Margaret Sanger for the first time in 1917 when she attended one of Sanger's conferences in Boston. The same year, McCormick served on the Committee of 100 of the National Birth Control League, an organization of wealthy benefactors promoting birth control, and she supported Sanger after the latter's arrest in 1916 for establishing the first birth control clinic in the United States. In 1920, with the victory of the women's suffrage movement, McCormick was eager to find a new cause to support. Consequently, she devoted herself to reproductive rights and began a long-term cooperative relationship with Sanger that would give both of them the power to catalyze change through sciences, to break down barriers, and to challenge institutions with the social evolution of the roles of women by affirming their commitment to giving women the ability to control their fertility, thus fully developing women's potential within society. During her summer trips abroad in the 1920s, McCormick unconventionally helped Sanger many times by smuggling diaphragms from Europe into the United States to supply birth control clinics. Sewn into her packed clothing, the diaphragms were secretly circulated for several years.[13] Not only did McCormick break the law in her defense of the birth control cause, but she also served on the National Council of the American Birth Control League,[14] founded by Sanger in 1921, and lent financial support to the Planned Parenthood Federation of America. McCormick was a very engaged woman: in the battle for the vote, in the founding of the League of Women Voters, in the birth control movement, in the Planned Parenthood Federation of America, and later, in the development of hormonal contraception. As a scientist, McCormick explored the field of endocrinology mainly because her husband had been diagnosed with schizophrenia (Marks 2001: 53). Her personal situation had a direct influence on her position on birth control as she did not have any children, just as Sanger had been confronted with the premature death of her mother due to too many pregnancies. In 1927, she hosted the World Population Conference on the family estate outside of Geneva (Sanger 1938: 383–384). Committed to science and the birth control cause, McCormick turned to the possibility of a scientific answer to family planning and the problem of overpopulation. When Stanley McCormick died in 1947, she inherited his considerable fortune and it proved to be a turning point in her life. Determined to advance the birth control cause, she wrote to Sanger in 1950 in order to identify the priorities for contraceptive research that would require financial support and to encourage experiments to develop new means of birth control.[15] Sanger replied that the priority was to find a simple, cheap, safe contraceptive and that contraceptive research needed to be expanded.[16]

It is the association between these two women's rights activists, resolute to give women more control over their own lives, which started the process that led to the development of one of the century's most revolutionary scientific and medical breakthroughs. They had created an intermittent correspondence at the end of the 1920s and continued as McCormick, now widowed, was willing to devote her energies to reproductive science and to finding a hormonal oral contraceptive. As early as 1948, McCormick declared: "I still feel, as I have felt from the first, that there is nothing more important than birth control"[17] before confirming her intention in 1950: "I have long wished to be of more constructive assistance to the Birth Control movement, especially along the lines of contraceptive research".[18]

In the winter of 1950, Sanger met a leading specialist in the field of female reproductive biology and discussed the possibility of developing an oral contraceptive that would solve the problem of overpopulation, but would also offer many opportunities for women to control their childbearing. At a time when a majority of states still had laws restricting the use of contraceptives, such a project was very audacious. She needed a physician willing to undertake research on reproduction when few were ready to take it on, as it remained transgressive. Dr. Gregory Pincus was the ideal scientist Sanger was expecting to concretize her vision of modern contraception. Upon graduating from Cornell University, he had started his career as an assistant professor at Harvard University and conducted research in mammalian sexual physiology. The first scientist to achieve in vitro fertilization of a mammalian egg in 1934, Pincus was a visionary in reproductive medical research.[19] In 1944, Pincus and a colleague, the endocrinologist Hudson Hoagland, founded the Worcester Foundation for Experimental Biology in Shrewsbury, Massachusetts, which soon gained an expertise in reproduction. They pursued research on the physiological effects of steroids, conducted experimental studies for pharmaceutical companies, and specialized in the field of reproduction and hormonal regulation of ovulation. In 1951, along with Min Chueh Chang, a colleague at the foundation, Pincus showed that the injection of progesterone into rabbits inhibited ovulation. Sanger financially participated in this study on hormonal contraception led by Pincus by offering a grant from the Planned Parenthood Federation of America. After having experimented the effectiveness of different dosages of hormones to keep rabbits from ovulating, Pincus had found the appropriate combination that also permitted them to simulate the menstrual cycle. This initial phase of the project ended by April 1951 and the challenge for Pincus was now to move forward with a clinical trial to prove the same effect in humans. In March 1952, Sanger wrote to McCormick, who typically had made regular financial contributions to the Planned Parenthood Federation, to inform her about Pincus' research on hormonal contraception at the Worcester Foundation.[20]

Conducting clinical trials required the supervision of a clinical physician. Therefore, Pincus had to find a specialist who would agree to take part in such a controversial project and would be able to enroll women in the trials. John Rock, a Harvard researcher who specialized in fertility, had been working on the mechanisms of ovulation and conception (Marsh and Ronner 2008). Although a fervent Roman Catholic, Rock was also a long-time supporter of birth control. In 1926, he became the director of the sterility clinic at the Free Hospital for Women in Boston while holding the position of Clinical Professor of Gynecology at Harvard. In 1931, he defied the Catholic Church and risked his reputation by petitioning the Massachusetts Legislature for the repeal of the Commonwealth's ban on contraception. This act marked one of his numerous conflicts with the Catholic Church's position over contraception. Rock firmly believed that women had to have access to birth control methods through clinics and physicians for medical reasons, and later became an advocate of population control which he assumed would reduce poverty (McLaughlin 1983). In 1936, Rock founded the first Rhythm Clinic in the United States to offer women advice on identifying the best time for conception with the rhythm method, and in the 1940s, he experimented in vitro fertilization of human ova. At the time Pincus approached Rock, the latter was conducting his own research on infertility by stimulating ovulation with hormonal treatment for sterile women. Rock's study consisted of injecting women who suffered from reproductive disorders with doses of progesterone and estrogen that would be gradually increased in order to reproduce the menstrual cycle and induce ovulation. The results of the experiments demonstrated that the hormones inhibited ovulation and that within several months after the end of the treatment, women were able to become pregnant.[21] For Rock, it was the proof that the hormonal treatment could be prescribed for sterility, but also for its contraceptive effect. In 1952, Rock and Pincus met at a scientific conference and started collaborative work on the activity of progesterone alone in order to verify its contraceptive effect on women. The next step was to extend research to human clinical trials for an oral contraceptive.

Two Women, One Goal

In June 1953, Margaret Sanger and Katharine McCormick met Gregory Pincus and Hudson Hoagland at the Worcester Foundation for Experimental Biology to discuss birth control research. This meeting convinced both women to commission the Worcester Foundation, as they were confident that this research program would succeed in providing a safe and effective contraceptive within a few years. McCormick provided the foundation with an initial grant of $10,000 (Chesler 1992: 432) and decided that other contributions would follow once the procedures and the agreement were established.[22] A week later, a contract was signed defining the terms for the project and the research schedule spanning over ten years. Spearheaded by McCormick, the Worcester Foundation began to expand its research program. Through the Planned Parenthood Federation of America and the Worcester Foundation for Experimental Biology, she provided a total of at least two million dollars in support of contraceptive research and testing.[23]

As a consequence of the Comstock Laws that imposed restrictions on contraception in Massachusetts, Pincus proposed a method of fertility treatment for sterile women in order to implement the test. Together, Rock and Pincus drove the research process and set up a series of clinical trials of progesterone: "the Pincus Progesterone Project". Beginning in 1953, Rock planned a three-month human trial of oral progesterone with a limited number of infertile women patients chosen from the Free Hospital in Boston, along with volunteers from the nursing staff at the Worcester State Hospital.[24] After having confirmed that they had a normal ovulation, women were administered high doses of progesterone for 20 days each within several menstrual cycles.[25] The objective was to keep the women from ovulating while permitting monthly menstruation. Study subjects also had to follow strict and constraining protocols about taking their daily temperature, vaginal smears during and after menstrual cycles, and enduring monthly endometrial biopsies. The results confirmed that progesterone compounds had inhibited ovulation for most women included in the study except 15% of them who did ovulate during progesterone treatment. Therefore, it established the contraceptive efficacy of progesterone on a temporary basis as its inhibiting action stopped after the end of the treatment.

The first clinical trials carried out by Pincus and Rock were based on a small clinical sample of 60 women over a few months.[26] Because some of these women suffered undesirable side effects,[27] and the procedure was very constraining, only half of the women enrolled in the study were able to complete the clinical trial. However, even if the initial clinical trial was a success, long-term studies needed to be performed.[28]

Pincus informed McCormick that he would need to test hundreds more women and this meant additional funds were required.[29] So be it, McCormick agreed to mobilize the financial resources needed.[30] Not only did she become the main financial support of Pincus' research project, but she also turned out to be its scientific manager. Indeed, she moved from California to Massachusetts to keep a close eye on the research and follow every stage of the project. She would visit the foundation frequently and ask to receive a report on the ongoing research activities every two weeks. Thoroughly involved in the fundamental aspects of these scientific investigations, she also became instrumental in sending regular updates to Sanger as the work progressed and informed her of any major advances:

> Dr. Pincus and Dr. Rock having completed progesterone tests on women for a one-year period have just now been reviewing their data. [...] They find that in these 30 cases progesterone has given them satisfactory contraceptive results.[31]

As a biologist, McCormick wanted to exert scientific leadership and became closely involved in this research. She repeatedly demanded further detailed information, reports on laboratory results, and clinical planning; addressing her questions and requests to Pincus. Under these conditions, McCormick would continue providing financial support to the program. In addition to regular progress reports, more formal meetings were organized with McCormick and Pincus to oversee the research and to review the budget allotted to these studies. Interestingly, McCormick truly immersed herself in contraceptive research and technology to the point of exemplifying another way for a woman to practice science and to contribute to scientific activity.[32] To a certain degree, the intersection of feminism and technosciences had a real impact on the organization, the direction of research, and the production of scientific knowledge. McCormick was often frustrated with the slow pace of progress, and she pushed the researchers to accelerate the project. Thus, she influenced and accompanied significant contributions to reproductive science by establishing a successful partnership with leading scientists in the field. Although she did not concretely carry out experimental investigations, McCormick set up a scientific and financial collaboration with these scientists, and directed the Pincus project, following every scientific step meticulously. Because she believed the issue of birth control could be solved by science, McCormick invested in this specific project which she believed would lead to scientific innovation and have a lasting impact on society. This effective woman's governance driven by feminist goals must be seen as a strategy to empower women.

Setting the Stage for a Scientific Breakthrough

Through hormonal research, John Rock and Gregory Pincus aimed at developing a method of inhibiting ovulation using a chemical compound with strong progestogenic activity and needed to evaluate the safety and efficacy of such an oral contraceptive (Slechta et al. 1954: 282; Pincus 1955; Rock et al. 1956; Chang et al. 1956; Marks 2001: 221–223). In 1955, Sanger announced that research for an oral hormonal contraceptive was under way and exposed the motivations for such investigations by insisting strategically on the problem of overpopulation and arguing for birth control as benefiting society:

> A growing awareness by government leaders and the public of the world's dangerous imbalance between overpopulation and natural resources was becoming apparent in 1955. ... Attention was focused on the need for developing a new, simplified and inexpensive contraceptive method–one which would be acceptable to people of diverse cultural backgrounds. ... M.C. Chang, research biologist at the Worcester Foundation for Experimental Biology under the direction of Gregory Pincus, predicted that such a pill might be available to the general public within five years.
>
> (Sanger 1956: 104–105)

The promising results of the first investigations set the stage for large-scale human trials to test the hormones on fertile women. However, birth control was still illegal under the Comstock Laws and contraceptive clinical trials could not be conducted in the United States. Therefore, the trials on hormonal contraception began in 1956 in Puerto Rico, a territory of the United States in the West Indies not subject to restrictive regulations (Gordon 2007: 287). This location had been selected as a site for the study of a new oral contraceptive due to several driving factors: overpopulation, poverty, and legal access to birth control.[33] Women who enrolled in the clinical trials were desperate to avoid pregnancy and eager to receive reversible and effective means of contraception. The motivation of the scientific investigators may appear suspicious and raise ethical questions, but professional standards of the time did not prohibit such practices as few regulations existed for the testing of drugs. Although it can be pointed out that these studies were experimental and that exposing women to high doses of hormonal treatment could have serious side effects, informed consent standards were minimal and no legal requirement existed. However, by modern standards, these clinical trials present a controversial aspect in the development of the Pill, particularly the trials in Puerto Rico and those which were carried out in asylums.

Although the first results were very encouraging and showed the significant contraceptive effect of the synthetic hormone, it was soon discovered that the progesterone was contaminated with small quantities of estrogen. Realizing that estrogen was necessary to reproduce the menstrual cycle, Pincus set the dosage at 10 milligrams of norethynodrel and 150 micrograms of estrogen.[34] The results of the next trials proved to be positive and demonstrated the success of the oral contraceptive formula. By 1957, Pincus and Rock were both confident that the birth control pill functioned safely and effectively.[35]

While Sanger was at that time traveling around the world campaigning for birth control, McCormick was scrupulously following the project and always agreed to provide the additional funding necessary to carry the research forward. McCormick, who was continually updating Sanger, about the research progress wrote, in January 1957: "I think we have gained the oral contraceptive we have been seeking and now it must be implemented as fully as possible. Nothing else matters to me now that we have got the oral contraceptive".[36] Both women exchanged many letters and their assiduous correspondence in the 1950s related their meetings, strategic actions, and the project's ongoing activities and findings. There is some evidence that they very much appreciated each other and shared their frustrations, optimism, and enthusiasm for scientific research on contraception.

The first oral contraceptive was submitted for approval to the Food and Drug Administration in 1957 as a therapeutic treatment for menstrual disorders and infertility, but not as a contraceptive, and required the prescription of a physician. Moreover, no mention of contraception was made to the Food and Drug Administration. As McCormick wrote to Sanger:

> Of course, this use of the oral contraceptive for menstrual disorders is leading inevitably to its use against pregnancy; and to me, this stepping stone of gradual approach to the pregnancy problem via the menstrual one is a very happy and fortunate course of procedure.[37]

Deeply disappointed by the Planned Parenthood Federation of America, which had chosen a different direction and declined its support for contraceptive research, McCormick complained to Sanger, in 1958: "Shall we have to wait another fifty years for the implementation of this vast and essential reform?"[38] But in October 1959, a new application to the Food and Drug Administration was submitted and this time, the contraceptive pill, named Enovid, was approved for contraceptive use (May 2010: 32, McLaren 1990: 240). Enovid was produced by G. D. Searle, the first company to manufacture an oral contraceptive pill.[39] Furthermore, it is worth mentioning that as birth control was still a controversial issue in 1960, the media did not take the risk of presenting the Pill as a symbol of reproductive freedom, but rather insisted on its medical benefits (Watkins 1998: 6).[40]

Feminism and Philanthropy

After nearly ten years of contraceptive research (Chesler 1992: 434), the Pill represented the culmination of a campaign that dated back almost half a century. The most prominent figure in the early American birth control movement, Sanger, and her active and greatest ally, McCormick, were the key driving forces in the process of governing reproductive sciences to give women control of their own bodies. Although the development of the oral contraceptive can be attributed to eminent scientists, philanthropists, birth control advocates, and social activists in response to the problem of population control, it also gave women control of their lives. The Pill remains Margaret Sanger and Katharine McCormick's legacy as feminists. While Sanger invented the concept of "birth control" and laid the foundations for its growing social acceptance, McCormick demonstrated feminist leadership and philanthropic engagement.

They both drove the research and the elaboration of this medical breakthrough, which was also the fulfillment of a dream Sanger and McCormick shared (Chesler 1992: 430). Through their determination and commitment, they strove to advance the birth control cause, breaking down barriers to give women access to a better future. Sanger truly appreciated having McCormick by her side who provided a great deal of moral and financial support from this fervent activist in their shared governance to move forward and secure scientific contraception. Furthermore, McCormick's personal financial donations were essential because she devoted most of her wealth to the funding of the entire research project on contraception as no government funds nor other resources were available for such an ambitious and controversial cause.[41] Her philanthropic actions to bring about social change and to gain access for women to reproductive rights were a crucial tool to scientific progress, offering women new opportunities to achieve equality.

Sanger's visionary spirit and McCormick's philanthropic endeavors have left a mark on women's history and contemporary society as millions of women's lives have been transformed by this discovery. As their work fundamentally redefined contraceptive practice, it was considered as an instru ment of women's equality and empowerment. Interestingly, in their mission, these two singular women with very different backgrounds and personalities were united by one common goal: women's autonomy over their bodies. They believed science was the solution to the social problem of birth control. Beyond their initiatives and successes, they created a feminist synergy and leadership that allowed them to raise ambitions for transformation with decisive implications for many aspects of women's lives in the twenty-first century. Even though the Pill still remains a very controversial invention nowadays, mainly due to potential health risks (Asbell, 1995: 301-312),

Sanger's and McCormick's original contributions to reproductive science as well as the unique role they played in shaping the future of women deserve to be recognized. They both dedicated their lives to women's emancipation until their death, respectively in 1966 and 1967, at the dawn of the legalization of contraception in American society.

Notes

1 The Food and Drug Administration (FDA) approved The Pill on 9 May 1960, granting American women reproductive freedom.
2 Birth control methods included intra-uterine devices, vaginal suppositories, abortifacients, condoms, and diaphragms.
3 Sanger's early life and career were marked by personal tragedies that altered the course of her life. See Chesler (1992: 63). As a nurse, she decided to engage in a struggle for the right to contraception: "I was resolved to seek out the root of evil, to do something to change the destiny of mothers whose miseries were vast as the sky" Sanger (1938: 92).
4 Sanger lived in the Progressive Era, a period generally defined as beginning in the 1890s and continuing into the 1920s, characterized by rapid social, demographic, and economic change in society. In 1873, Congress passed the Comstock Act. Led by Anthony Comstock, founder of the New York Society for the Suppression of Vice, this crusade for moral repression and social purity resulted in the passage of laws criminalizing birth control information and practices, associated with sexual promiscuity, immorality and obscenity. Moreover, many states adopted laws criminalizing abortion. Indeed, by 1880, every state had passed criminal laws prohibiting abortion that were to stay unchanged until the 1970s. See Kopp (1972:124).
5 Judge Crane's decision led to the founding of the American Birth Control League in 1921 and the Clinical Research Bureau in 1923. The Clinical Research Bureau dispensed contraceptives to women under licensed medical supervision. Its objective was also to study the effect of contraception on women's health. See Chesler (1992: 279).
6 Among these were the Rockefeller Foundation, Carnegie Institution, and the Harriman railroad fortune.
7 The aim was essentially to focus on restricting the reproduction of the feeble-minded.
8 A major legal victory for the birth control movement occurred with the decision in *U.S. v. One Package of Japanese Pessaries*. A federal appeals court ruled that the federal government could not interfere with physicians providing contraceptive information or contraceptive devices to their patients.
9 The name of the organization marked a strategic shift in the movement. It did not include the expression "birth control" in order to focus on overpopulation and to make birth control socially acceptable.
10 Sanger envisioned the ideal contraceptive as safe, affordable, easy to use, readily available, reliable, and in women's control. See Watkins (1998: 14).
11 Letter: Margaret Sanger to Clarence Gamble, 15 August 1939, The Selected Papers of Margaret Sanger, https://sanger.hosting.nyu.edu/publications/book/.
12 The League of Women Voters did not want to support birth control as a controversial issue.
13 McCormick smuggled diaphragms past customs officials each summer from 1922 to 1925 (May 2010: 22).
14 The National Council was composed of physicians, scientists, and prominent New York-area society women.

15 Letter: McCormick to Sanger, 19 October 1950, MS-SS.
16 Letter: Sanger to McCormick, 27 October 1950, MS-SS.
17 Letter: Katharine McCormick to Margaret Sanger, 15 November 1948, MS-SS.
18 Letter: McCormick to Sanger, 18 November 1950, MS-SS.
19 In 1936, he published "The Eggs of Mammals" that received wide acclaim in the international scientific community. Nevertheless, Pincus was denied tenure at Harvard University in 1937 certainly due to his highly controversial research but also antisemitism. The endocrinologist Hudson Hoagland, invited him to Clark University in Worcester, MA.
20 Letter: Sanger to McCormick, 10 March 1952, MS-SS.
21 The effect was called "The Rock Rebound." See Rock et al. (1956).
22 Letter: McCormick to Sanger, 4 June 1952, MS-SS. She also offered $50,000 to build a new housing for the lab animals.
23 The equivalent of more than 20,000,000 in 2021. She gave between $150,000 and $180,000 each year until her death to the Worcester Foundation for Experimental Biology. See Tone (2001: 214).
24 Twenty-seven women were included in the study.
25 They used norethynodrel as the only compound of the Pill.
26 Letter: McCormick to Sanger, 14 June 1954, MS-SS. They also ran a series of experiments on insane women at the Massachusetts State Hospital. Letter: McCormick to Sanger, 31 May 1955, MS-SS. As Pincus needed more women in the clinical trials, he asked a gynecologist, Henry Kirkendall to join the experiment and to enroll some of his patients in the study. See Eig (2014: 133–134). At that time, American researchers had no formal obligation to obtain informed consent to conduct clinical trials.
27 Such as nausea, breast tenderness, dizziness, absence of menstrual periods, and other symptoms
28 The results were published in Rock et al. (1956).
29 After a year of work, Gregory Pincus and John Rock had completed research on only about thirty women. Letter: McCormick to Sanger, 14 June 1954, MS-SS.
30 $245,000 in 1957 and $230,000 in 1958 for these tests.
31 Letter: McCormick to Sanger, 14 June 1954, MS-SS.
32 Gregory Pincus dedicated his book, *The Control of Fertility,* to McCormick: "This book is dedicated to Mrs. Stanley McCormick because of her steadfast faith in scientific inquiry and her unswerving encouragement of human dignity." Pincus (1965: dedication page).
33 Some trials were also carried out in Haiti and Mexico. Puerto Rico already had a well-established network of family planning clinics. There was also a poor community with high fertility and low educational standards, and mostly demanding for simple effective methods of contraception. See Marks (2001: 102). On the other hand, the Pincus' team was looking for women willing to participate in these experiments and who would be able to follow the instructions rigorously. See Marks (1998: 221). On site, there was also a medical school and two physicians to supervise the trials: Dr Edris Rice-Wray, a faculty member of the Puerto Rico Medical School and medical director of the Puerto Rico Family Planning Association, and Dr Celso Ramon-Garcia, a gynecologist. The synthetic hormone used in these large-scale trials was norethynodrel produced by Searle.
34 When the dose of mestranol (estrogen) in the norethynodrel product was decreased, women experienced breakthrough bleeding.
35 In 1958, two hundred women had been included in the testing program since its beginning. See Fields (2003: 283).
36 Letter: McCormick to Sanger, 3 January 1957, MS-SS.
37 Letter: McCormick to Sanger, 17 April 1957, MS-SS.
38 Letter: McCormick to Sanger, 28 January 1958, MS-SS.

39 G. D. Searle was initially a small American pharmaceutical company, founded by Gideon D. Searle in 1888 in Omaha, Nebraska. Searle was the only company to risk the controversy (Marks 2001: 7, 35).
40 Those who approved the Pill for contraception were hesitant to express their public support (Seaman 1969).
41 Including Planned Parenthood Federation of America.

References

Asbell, Bernard. 1995. *The Pill: A Biography of the Drug that Changed the World.* New York: Random House.

Chang, Min C., Elsayed S. Hafez, Anne Merrill, Gregory Pincus, and Meyer X. Zarrow. 1956. "Studies of the Biological Activity of Certain 19-nor Steroids in Female Animals". *Endocrinology.* 59 (6): 695–707.

Chesler, Ellen. 1992. *Woman of Valor: Margaret Sanger and the Birth Control Movement in America.* New York: Simon & Schuster.

Eig, Jonathan. 2014. *The Birth of the Pill: How Four Crusaders Reinvented Sex and Launched a Revolution.* London: Macmillan.

Fields, Armond. 2003. *Katharine Dexter McCormick: Pioneer for Women's Rights.* Westport, CT: Praeger Publishers.

Gordon, Linda. 2007. *The Moral Property of Women: A History of Birth Control Politics in America.* Chicago: University of Illinois Press.

Kopp, Marie E. 1972. *Birth Control in Practice: Analysis often Thousand Case Histories of the Birth Control Clinical Research Bureau.* (1934. McBride) New York: Arno Press.

Marks, Lara. 1998. "A 'Cage of Ovulating Females': The History of the Early Oral Contraceptive Pill Clinical Trials, 1950–1959." In *Molecularizing Biology and Medicine: New Practices and Alliances 1910s*–1970s, ed. Soraya de Chadarevian and Harmke Kamminga, 221–247. Amsterdam: Harwood Academic Publishers.

Marks, Lara. 2001. *Sexual Chemistry: A History of the Contraceptive Pill.* New Haven, CT: Yale University Press.

Marsh, Margaret, and Wanda Ronner. 2008. *The Fertility Doctor.: John Rock and the Reproductive Revolution.* Baltimore, MD: Johns Hopkins University Press.

May, Elaine Tyler. 2010. *America and the Pill: A History of Promise, Peril and Liberation.* New York: Basic Books.

McLaren, Angus. 1990. *A History of Contraception: From Antiquity to the Present Day.* Cambridge, MA: Basil Blackwell.

McLaughlin, Loretta. 1983. *The Pill, John Rock, and the Church: The Biography of a Revolution.* New York: Little Brown.

Pincus, Gregory. 1965. *The Control of Fertility.* New York: Academic Press.

Pincus, Gregory. 1955. "Some effects of progesterone and related compounds upon reproduction and early development in mammals", in International Planned Parenthood Federation, *The 5th International Conference on Planned Parenthood,* Tokyo. Report of the Proceedings, 175–184.

Reed, James. 1978. *From Private Vice to Public Virtue. The Birth Control Movement and American Society since 1830.* New York: Basic Books.

Rock, John, Gregory Pincus, and Celso-Ramon Garcia. 1956. "Effects of Certain 19 – Nor Steroids in the Normal Human Menstrual Cycle". *Science.* 124 (3227): 891–893.

Sanger, Margaret. 1920. *Woman and the New Race*. New York: Brentano's.
Sanger, Margaret. 1921. "The Eugenic Value of Birth Control Propaganda", *Birth Control Review*. 5 (10): 5.
Sanger, Margaret. 1922. *The Pivot of Civilization*. New York: Brentano's.
Sanger, Margaret. 1938. *The Autobiography of Margaret Sanger*. New York: W. W. Norton & Co. Maxwell.
Sanger, Margaret. 1952. "Birth Control Research on the March", October 1952, MS-SS.
Sanger, Margaret. 1956. "Birth Control, 1955." *Britannica Book of the Year* 1956: 104–105. Retrieved 5 May 2021 from https://archive.org/details/britannica-book-of-the-year_1956/page/104/mode/2up
Sanger, Margaret. Papers. SSC-MS-00108, Smith College Special Collections, Northampton, Massachusetts.
Seaman, Barbara. 1969. *The Doctors' Case Against the Pill*. New York: P.H. Wyden. Reprint 1995. Alameda, CA: Hunter House.
Slechta, Robert F., M. C. Chang, and G. Pincus. 1954. "Effects of Progesterone and Related Compounds on Mating and Pregnancy in the Rat." *Fertility and Sterility* 5(3): 282–293.
Tone, Andrea. 2001. *Devices and Desires: A History of Contraceptives in America*. New York: Hill and Wang.
Watkins, Elizabeth; 1998. *On The Pill: A Social History of Oral Contraceptives 1950–1970*. Baltimore, MD and London: The John Hopkins University Press.

12 Women with Transmitters

Female Engineers and the Gendering of Technology in the Soviet Union

Ekaterina Rybkina

Since the first decades of development of radio engineering during the nineteenth century, the experimental and accessible nature of wireless communication was every year becoming a more affordable hobby for technically oriented people, attracting more adherents, yet remaining a male-dominated practice (Haring 2012; Gessler 2014; Birdsall 2023). The absence of women in electrical engineering and communications development, linked to the limited opportunities they had, was seen as a natural state of affairs (Gessler 2014; Schafer and Thierry 2015; Ellis 2017).

At first sight, the history of radio engineering in the Russian Empire and in the Soviet Union is a classic example of a male corporate world that took shape long before the beginning of the early experiments with the transmission of wireless signals. Recent historiography associates radio's development with a military paradigm, and with women's limited access to engineering education in general (Gouzévitch and Gouzévitch 2000; Kojevnikov 2002; Valkova 2008). The long-standing state monopoly on communications (Zakharova 2020) through research and development, popularization, and education kept this field highly governed, in terms of access, by outsiders to the technosciences, mainly military and politicians. The narrow circle of specialists who engaged in wireless communication experiments consisted of a small number of laboratories run by men.

Even though wireless telegraphy ceased to be an exclusive purview of the military after the Bolshevik Revolution and the establishment of a new political regime, the participation of women in this field did not increase, since women's access to the discipline was not significantly affected by the new socio-political context. The Soviet government was quick to proclaim radio communication as its own achievement (Zakharova 2011, Lovell 2015) and willingly promoted the heroic image of radio's inventor among male pioneers of science and technology. With the Soviet state's official proclamation of Alexander Popov as the inventor of the radio in 1895, his name joined the ranks of the most notable male figures who headed radio research laboratories and institutes, who participated in the construction of large-scale socialist infrastructure, who participated in risky expeditions, and who thus brought glory to the state (Shatelen 1949).

DOI: 10.4324/9781003562597-16

Apart from its focus on inventors and their heroic deeds, the state tended to be silent about the names of those people who, on a daily basis, ensured the operation of crucial elements of technological infrastructure. In this respect, the Revolution and the change of political regime did not immediately change the existing order of things in radio engineering. Despite proclaimed Soviet values of gender equality and the emancipation of women's labor (Klots 2018), women were still rare in certain engineering professions and occupied positions of invisible technicians. Only in exceptional cases, they were noticeable actors who conquered their professional field (Kalemeneva and Lajus 2018).

While historians of radio have studied how the technology literally amplified women's voices—for there was an assumption that the female voice was naturally less powerful than the male and therefore needed broadcast amplification (Mitchell 2001; Skoog and Badenoch 2020)—this chapter suggests a metaphorical amplification of women's voices through gender performance in order to communicate their relationship with technology and access engineering professions. Thus, this chapter serves two purposes. First, the paper discusses a possibility for writing a history of technology from the perspective of marginalized subjects. It contributes to historicizing trajectories of seldom-seen women in radio in the Soviet Union; where, along with scientifically based social transformations and progressive emancipated claims, women occupied secondary roles in the engineering hierarchy.

Second, this chapter explores the opportunities that ego-materials might offer to the studies of technology, by shifting to the storytelling of individual experiences. Focusing on the personal reflections of participants can provide a different perspective on political and societal developments and shed light on unfamiliar aspects that might otherwise seem irrelevant (Griesse 2009). Confronting radio magazines discourses—written by men—that shaped the social representation of engineering professions among their audience to the intimate writings of female radio engineers stresses the gender imbalance and the normative frameworks in radio engineering. These public narratives governed the construction of gendered hierarchies, excluding women from leading positions.

This history focuses on the narrative of Natalia Freichko, a Soviet engineer who made her career in radio technology in the post-World War II years. A previously unconsulted private archive, which consists of carefully kept personal correspondence with Freichko's family, provides insight into the inner world of a Soviet female professional. The personal experience's account embodies the actions the state took to govern the communication technology, highlighting in the process gender discrimination in the professional sphere.

In Search of Female Figures in Radio Engineering

The male-dominated Soviet narrative of radio engineering, shaped through the press, did not exclude women altogether but rather testified to their actual insignificance in the field. Amateur radio practice promoted by the state in the early Soviet period through a network of radio clubs and through a radio popularization campaign made the hobby an alternative path into the profession. Periodical editors and popular science book authors about radio engineering—often professional radio engineers and representatives of government agencies—built a hierarchy within their community. As experts, they stressed the role of radio amateurs as active promoters of a new technology and valuable participants in a modernization process, but, at the same time, they perceived such amateurs as technicians in constant need of education. In this hierarchy of technical knowledge distribution, women were relegated to more marginal positions: they were excluded from the category of recipients of knowledge and instead were used as a narrative foil in a male-produced radio history, in which women were seen as backward foreigners incapable of understanding such a progressive technology as communications. Despite the state using popularization as a tool to minimize intellectual outflow from the professional community, amateur radio practices as a mass activity also became a window of opportunity and a possibility for upward and gender mobility.

Women portrayed and described in radio magazines and literature in the 1920–1930s could be grouped into several categories. One of the most widespread images was that of a grumpy and feisty wife who threatened her radio-enthusiast spouse with divorce because of his passion for the technical hobby, which she could not possibly understand (see Figure 12.1).[1] Groups of amateur radio operators' wives were playfully called "radio widows". Women were particularly lonely at night, when signal quality was higher, and there was less shortwave interference to lessen their husband's enjoyment of their hobby. Such descriptions reinforced the notion that only men could be involved in radio practices.

Fictional caricatures of women, created by men, sometimes wove themselves bizarrely into real narratives about women's careers in radio engineering. Another category of female stereotypes referred to the canonical image of the indignant neighbor, concerned about the suspicious activities of a male neighbor (Galperin 1925: 76). This *female neighbor* was usually portrayed as an elderly, illiterate, Orthodox woman. She could also be an older relative of a promising young male amateur radio operator. She did not know much about radio and was suspicious of radio amateurs, condemning their strange activities. Such images were used by radio magazines to criticize people's ignorance about radio. To a large extent, this repertoire of images attributed certain attitudes toward technology to certain age categories of people conducting a dialogue about differences. Youth was identified with progress, whereas older people were characterized as backward. This perfectly illustrates the phenomenon of "technophobia"—a process related to people's reactions to new technology (Bauer 1995), which also ascribes gendered positions toward technology (Wajcman 2004; Kelan 2007).

Groups of like-minded enthusiasts emerged sporadically in the 1920s in the wake of increased interest toward science and technology that eventually

Figure 12.1 A cover of the radio magazine *Radioliubitel'* no. 3 (1930). With *Radio*'s authorization.

coalesced into larger paramilitary associations. For example, Osoaviakhim, an organization controlled by the Department of the Military, brought groups of people together in the 1930s based on interests in chemistry, military training, aviation, and radio. The Soviet government gradually and consistently enlisted young people's technological interests and transformed them into military buffs. Soviet newsreels made by Dziga Vertov captured footage in which young women belonged to radio circles and were creating their first radio receivers. Along with their classmates, schoolgirls in pigtails were attaching parts to their new devices. Those cadres for the first time showed women with radio devices in their hands—the earliest female radio subjects of the 1920s.[2]

In the 1920–30s, there were several state attempts "to bring women to radio" in the Soviet Union.[3] Radio magazines published information about various initiatives to attract women to radio practice. One such article specifically focused on women's absence from radio clubs. In 1925, *Radioliubitel'* launched a series of publications titled "Letters to the radio propagandist". Their author, a representative of the Moscow Council of the Trade Union, noted that working women were rarely present among the members of radio amateur clubs because of their need to spend afterwork time at home to do the cleaning and the laundry, cook food, and provide childcare.[4] "All this, taken together, eliminates female workers from the country's social and political life"[5] the author highlighted. Comparing passive listening to radio broadcasts with active design and construction of radios, the author stressed that, although radio listening could "awaken a woman from social hibernation", the aim of the popularization of radio was to turn her into "a designer familiar with radio engineering and knowing how to take apart, reassemble, and rebuild a radio receiver without any help".[6] Thus, wireless technology itself gave agency to some individuals since it relied on their abilities to access radio waves with hand-made devices. Despite proclaimed objectives to bring women to radio in the Soviet Union, women remained underrepresented. Only about one percent of registered short-wave radio operators were women prior to World War II (Chliiants 2008).

Rare positive images of women in radio engineering were either that of a younger sister or of a female friend who, despite difficulties, was ready to help her more enlightened, and more technologically advanced male comrades. Sisters could follow in the footsteps of their brothers and join radio clubs. The image of the radio amateur's sister repeatedly appeared in the press. This was the case of Maria Giliarova from Leningrad, sister of Pavel Giliarov, who had graduated from the Polytechnic Institute, and who had become the eighth registered radio amateur in the city.[7] Brother and sister became delegates to the first All-Union Conference of Shortwave Workers of the USSR in 1928.[8] The radio enthusiast's sister also frames the image of Valentina Podzorskaia, who at the age of 38 became a short-wave radio amateur. In 1927, Podzorskaia, like many other people, bought a radio receiver to listen to radio broadcast programs. One day, when her receiver broke down, she both literally and metaphorically opened the black box of her receiver and became a devoted member of the local radio club.[9]

Most of the time radio newspapers and magazines ignored images of women. One exception was *Radiolubitel'* which wrote about women who practiced short wave in various cities of the USSR. The first mention of women in radio magazines coincided with the beginning of a government campaign to militarize the civilian population.[10] In 1928, the first radio transmitter was registered in the name of a woman in Tomsk under the call-sign 38RA, to V.V. Shumilova.[11] Despite the state, through radio clubs and magazines, encouraging the emergence of shortwave operators, it was primarily interested in creating collectives of any gender. The militarization of the population began through the system of clubs of the Osoaviakhim. For example, one such club in the city of Vladimir under the leadership of a male shortwave radio amateur brought together 12 people in 1929, all of whom were women who had undergone professional radio training.[12]

The late 1920s saw the emergence of a new myth centered on military professions, in which the one of radio operator, an occupation that would be of crucial importance in case of war, figured prominently. As a result, the image of radio operator acquired heroic proportions. The literature from this period provides numerous examples of the romanticization of amateur radio life; examples which also allow us to track gender roles. Ernest Krenkel, who earned fame for his participation in Arctic expeditions, became the undisputed star of amateur radio. His example served as a model for the newly formed image of the heroic radio operator as a figure of paramount importance in polar explorations. Selflessness, endurance, and heroism were portrayed as characteristic features of male Arctic heroes. In contrast to heroic male radio engineers, women did not leave published memoirs (Krenkel 1973; Khurges 2012). Although a woman's career success could be considered as a breakthrough because it deviated from the traditional narrative about the heroic fathers of radio, only exceptional cases made it to the pages of radio magazines. Among the heroes of the Arctic radio operators was a woman, Liudmila Shreder (see Figure 12.2).

The little we know of Shreder's life is fragmentary. Her biography has not yet been written. Unlike Krenkel, she did not publish her memoirs. She was born in Ekaterinoslav (Dnipro in today's Ukraine) in 1911, to parents who were the head of a railway station and a telegraph operator. Shreder had a craving for technical knowledge and became a telegraph operator herself. Later, she decided to go to Leningrad, where she took courses in radio technical training. There she met her future husband, who was an aerologist (a technician who developed methods of measuring atmospheric properties) and she moved with him all the way to Uelen, a settlement on the Arctic Circle at the far easternmost tip of Siberia. There she worked as a radio operator, and her frequent broadcasts drew the attention of the press.[13] Information about Liudmila Shreder is only available through the published recollections of her male colleagues, who described her as "a small, cheerful young woman", and from her few short comments for the press made after her participation in the rescue of the crew of the expedition ship *Chelyuskin* in 1934.[14] Shreder was not the only female radio operator working in the harsh conditions of the Arctic, but according to a radio operator with whom she had studied in Leningrad, "Shreder is no hero, she is just a disciplined short-wave radio

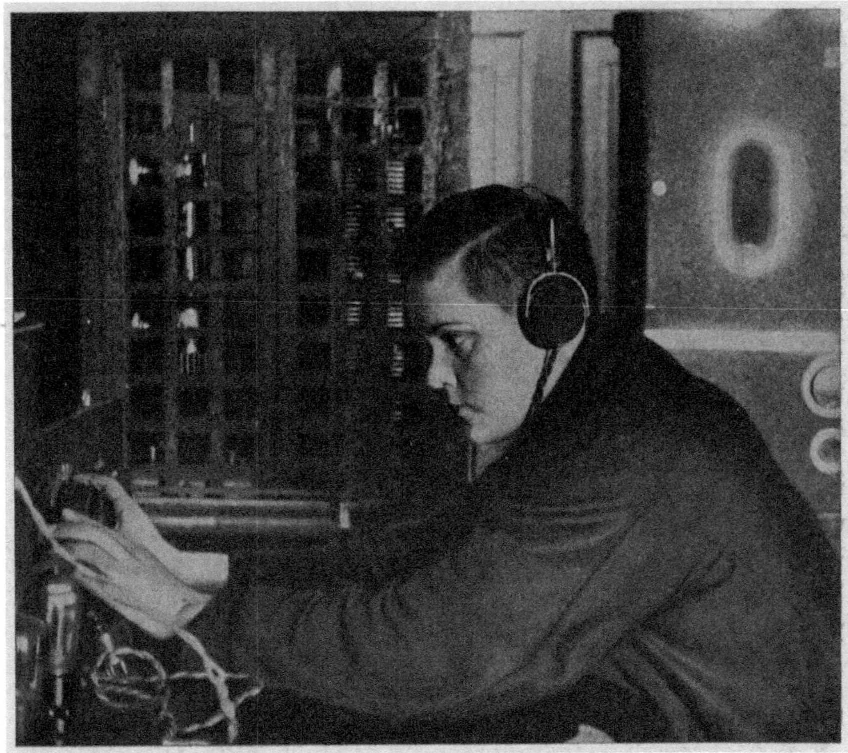

Figure 12.2 Liudmila Shreder, a radio operator from Uelen.

Source: Pokhod "Cheliuskina". Vol. 1 (Moscow: "*Pravda*", 1934): 375. Public Domain.

operator, unwilling to talk about herself and not differentiating herself from the family of Soviet polar radio peers".[15]

In his report to the newspaper *Izvestia*, Ernest Krenkel also appreciated Shreder as a skillful operator who did "truly striking work" that ensured the transmission of radio messages from the Shmidt expedition camp in the Arctic, set up to rescue the crew of the *Chelyuskin*, through Uelen to Moscow. In another of his books, Krenkel described Shreder by using the diminutive version of her name, "She, poor thing, had to bear on her shoulders the whole burden of emergency transmissions the burden of work, which would be enough for half a dozen men, fell on Liudochka's girlish shoulders".[16] Thus, the late 1930s witnessed the emergence of professional female radio operators—women that worked in communication departments and who used the chance to get training in radio telegraphy as a means of climbing the career ladder.[17] Published sources do not allow us to know what those women thought about themselves and how they planned their lives. Sources of more intimate nature would let the voices of these women be heard.

For the Sake of a Career

"I will not get married because I do not want to spoil my autobiography", Natalia Freichko, a student at the Leningrad Electrical Engineering Institute of Communications, unequivocally stated in a postscript to a letter to her parents in December 1948.[18] Thus she made a choice of not getting married for the sake of her future career. Her clear statement, made in response to her parents' concerns about her marital status, was less a reaction to a long-running topic of conversation than it was an emphasis on the interdependence of her personal life and career prospects. Half a year later, she would finish her studies at the Institute of Communications in Leningrad, and she would be making plans for her future life. At some point, she realized that all of the students from the previous cohort that the department had promoted for admission to graduate school were male. Thus changing her marital status before graduation, in her opinion, could affect her dream of going to graduate school to pursue cutting-edge research in television engineering, and the prospect of working in the most vibrant branch of electrical engineering at that time. With that in mind, she adjusted her personal life goals to her career perspectives.

This preoccupation with autobiography can be grasped in a Soviet context of writing as a career tool and as a type of public self-identification (Hellbeck 2001: 342). Thus, individuals accepted a mechanism of social governance and adapted to its practices of formalized self-presentation. In Natalia Freichko's case, this could also be a commitment to career aspirations in accordance with the existing academic system. Historians tend to differentiate the modes of self-performance of Soviet subjects during the Stalin era from those after it. The post-Stalin era shift is often argued as subjectivities allowing people to consciously make decisions no longer based on a strict code of rules but on a relatively wider set of available practices (Pinskii 2018: 26–27). Freichko's example blurs this boundary in several respects. She expressed herself as an independent subject and, simultaneously, she followed socially acceptable practices without having any superimposed written rules that would limit her choices.

The collection of ego documents to which Freichko's letter belonged was discovered by the author in a private collection of radio memorabilia in the provincial town of Pereslavl'-Zalessky in Russia.[19] The owner of the collection inherited these materials from the relatives of Natalia Freichko, along with other items from her amateur radio life. Written materials consist of some official documents, memorial awards, and a collection of QSL cards. The latter are arranged neatly in a photo album since these special postcards, used by radio amateurs to confirm communication with other enthusiasts, correspond to the size of a standard photograph.

Unlike her makeshift radio devices, which the collector displays, Freichko's personal correspondence and other documents from her career in

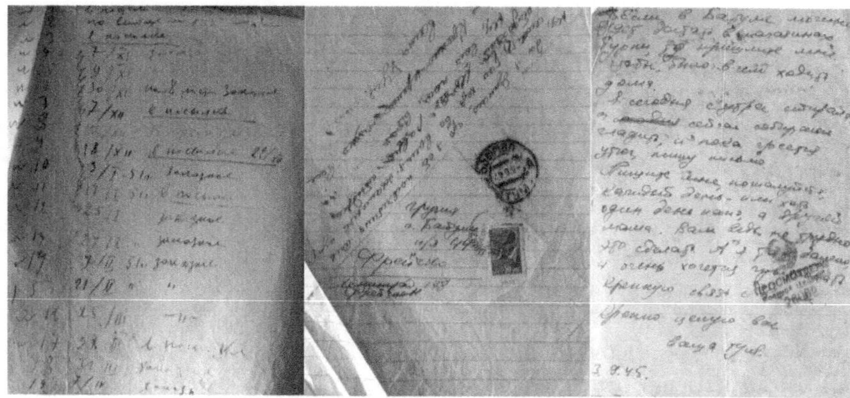

Figure 12.3 From left to right: a fragment of a hand-written list of sent letters indicating the dates; A letter from Natalia Freichko to her parents sent without an envelope and wrapped in triangle as "a front letter", dated June 3, 1947; An excerpt from a letter with the seal "viewed by military censorship". Museum of Radio, Pereslavl-Zalessky, Russia, https://музей-радио.рф. Private collection. With the owner's permission. © Ekaterina Rybkina

radio engineering have remained hidden away in a cardboard storage box in the collector's improvised archival room. The box was full of personal correspondence covering roughly the decade from August 1945 to the end of 1955. During her studies at the Institute, which lasted years, Natalia Freichko and her parents sent letters to each other regularly and continued to do so in subsequent years. About one hundred total letters, received and sent, uncover many facets in the life of one family, with their problems and doubts, expectations and joys (see Figure 12.3). They sent each other letters to share family news, to make plans, to discuss everyday issues, and to support each other.

Evidence of Freichko's amateur radio practices and personal correspondence can serve as a kind of keyhole through which a historian can glimpse a portion of a person's life. Taken together with other sources, ego documents can focus on previously unasked questions. The research task here is to look differently at the historical reality surrounding this family, and through their unique evidence, grasp their experience in order to juxtapose a particular case with the existing narrative and challenge some of the historiographic cliches about the place of women in engineering and Soviet society.

Natalia Freichko was born in 1924 in Saratov, to a family of musicians who worked in the Political and Educational Committee of the People's Commissariat of Education.[20] This family first received some public attention in 1941, when the major national radio magazine *Radiofront* published a page-long article about "The Freichko Family", introducing a mother, her son, and her daughter who practiced shortwave radio communication (see Figure 12.4). The article showed how they became interested in the hobby and their local fame in Batumi among other radio enthusiasts. According

Figure 12.4 The Freichkos.

Source: "Sem'ia Freichko", *Radiofront*, no. 11 (1941): 32. With *Radio*'s authorization.

to the article, Natalia's older brother, Boris, was the first in the family to become interested in radio and to enroll in radio operators' courses, which were organized by the local branch of the Soviet paramilitary organization Osoaviakhim. Boris was keen to experiment with radio transmitter and receiver design, and "received a lot of QSL cards", a sign that he had an individual callsign and was an active radio operator who frequently transmitted messages on the air. Freichko's mother, also called Natalia, enrolled in shortwave radio operator courses with her son, after which she joined a local group of amateur short-wave radio operators at the Batumi collective radio station[21]. Unlike Boris, Natalia and her mother did not have callsigns that would allow them to transmit messages on the air. Instead, they were certified as "radio observers", people who listened to other transmissions and learned how to decode them. However, the Batumi collective's callsigns enabled them to work as radio operators at the radio station (Chliiants 2019: 51).

In 1947, the Freichkos were once again promoted as an exemplary radio family in the same magazine (renamed *Radio*). The new version of their family story, told by the shortwave radio popularizer Vladimir Burliand, added that Natalia Freichko became interested in short wave radio under the influence of her brother when she was less than 15 years old.[22] In contrast to the earlier article, Burliand portrayed Natalia's mother as the one who followed in the footsteps of her children.[23] Burliand contrasted the "young radio operator" with her mother, "an older woman, the same age as radio". Burliand's words reveal the conviction that middle-aged people, especially women, were not able to become fond of novel technologies such as radio on their own. At the same time, Burliand mentioned that Natalia Freichko's mother had been a telegraph operator for a few years when she was younger. Burliand noted that Natalia Freichko dreamt of becoming a radio operator in polar aviation. Like her brother, she studied at a radio club and by the beginning of World War II, she had become a qualified radio operator and parachutist.[24] However, because of her young age, she was not accepted to an aviation club, so Natalia decided to enroll in a short-wave radio section, "located in the same club and therefore closer to planes".[25]

When WWII began, Freichko was 16 years old, but she was willing to go to the front, as is shown in military commissariat records.[26] Freichko was drafted into military service as a radio operator at the Military Intelligence Department in the North Caucasus and stayed there until she finished her service with a Medal for Military Merit in November 1943, and the rank of Lieutenant-Engineer. Any other information about this period is unknown since her correspondence after the war was silent about her military experience.[27]

In the aftermath of the war, Freichko decided to become a radio engineer and went to study in Leningrad. Upon arriving there in August 1945, she wrote to her parents with excitement. Natalia Freichko's letters were sent rolled up military-style with an obligatory censor's seal. Despite the lack of privacy, she described in detail her institute life and living conditions. In general, everyday life and financial expenses were the main themes of her letters. In the second month of her studies, she began commenting on her courses, writing about the main subjects she studied, ballroom dancing courses, and lectures on behavior in society required for all male and female students. From these letters, it is evident how dire her financial situation was as a student in Leningrad during the first post-war years. Despite the fact that she had a part-time job in a radio repair workshop[28] and received increased financial aid as a front-line soldier; Freichko was malnourished, had one pair of stockings, and hoped for a pair of new shoes.[29]

In a letter from 1946, Freichko regretted that due to exams she could not devote time to short wave radio, while she also admitted that her short wave experience had helped her during her studies.[30] At the radio operators' competition in Leningrad, she received fourth prize for constructing a radio transmitter. The prize was an annual subscription to the *Radiofront* magazine, and her fourth-place finish allowed her to participate in an All-Union radio competition. In a short interview for the journal, Freichko acknowledged that she hoped to receive her own callsign, but this did not happen for several years.[31]

The magazine *Radio* frequently updated readers about Natalia Freichko's career.[32] She may have been flattered by this attention, but she nevertheless was shy of her appearance in the photo that accompanied the article, writing to the parents that it "disgraced us in front of the whole Soviet Union".[33] In fact, the small black and white passport-like images hardly gave any idea they were pictures of a woman to the magazine's readers (see Figure 12.5). Notably, later publications were illustrated by better photographs. One photo was

Figure 12.5 Photos of Natalia Freichko (left) and her mother (right).

Source: Vladimir Burliand, "Semiia korotkovolnovikov", *Radio*, no. 4 (1947): 39. With *Radio*'s authorization.

taken in 1950 and was probably from her personal archive, while the second one was taken by a professional photographer during an interview in 1976 (see Figure 12.6).

Freichko was preoccupied with her appearance. Among other things, in letters to her mother she discussed dressmaking in the context of choosing a career strategy.

> Now I have seen that knowledge, of course, is useful to an engineer. The main thing is, first, a diploma, and second, clothing; and clothing is more important since a diploma plays a role only in finding a job. This is a very short-term factor, while clothes are permanent.[34]

This seemingly insignificant conversation in the midst of her career planning can serve as an argument on the performativity of gender (Opitz and Van Tiggelen 2025: 9). The issue of dressing, especially when it comes to the appearance of female specialists, which rarely comes to historians' attention, mattered in private lives of female Soviet engineers.

That her bonds with her family were strong is evident in Freichko's invitations to them to come and live with her.[35] She shared her desire to go to

Figure 12.6 Natalia Freichko in the laboratory of Sverdlovsk Radio Club in 1950 (left) and in 1976 (right).

Source: N. Grigor'eva, "Odna zhizn", *Radio* 5 (1976): 15. With *Radio*'s authorization.

graduate school because it was considered to be "the only opportunity to escape this place [Sverdlovsk]… I have two roads I can take: one is to agree to everything and stay here forever and another is to go to graduate with all of its difficulties".[36] Despite her efforts to leave Sverdlovsk, the plan did not materialize.[37] Her "only consolation" was "to receive a call-sign", which became a real outlet for her.[38]

Apparently, Freichko endured her failure to enter graduate school and, apart from her main job, she began to teach algebra, geometry, and physics at a local evening school in 1951.[39] The school administration appreciated her diligence and rewarded her with perfume on International Women's Day.[40] The same year, she became secretary of the local party organization and no longer had enough time for shortwave radio.[41]

In a letter to her parents, Freichko also discussed the status of female engineers when she started working with a team of radio technicians at a radio station near Sverdlovsk. Evidently, the hierarchy of positions between engineers that made decisions and technicians that were tasked with their implementation led to tensions at the working place. In the first year of her work at the station, Freichko had conflicts with her mostly male colleagues, which she reported to her parents:

> We turned the entire center upside down here. It all began when administration attacked us for being engineers, while they are technicians,

> and they applied the most unfair methods. At one point it appeared they would have the upper hand and break us, but as long as we had nothing to lose, we appealed to the party's district committee and even higher, and commission after commission followed. In the end we turned out to be right in every aspect, and now they went quiet, but we caused them so many troubles that we now must be careful. Everyone is waiting for us to stumble. [42]

In the same letter, she complained that "because of all these quarrels I am no longer working on short wave equipment".[43]

Being a female engineer, Freichko often found herself in a disadvantaged position. She attributed numerous conflicts within the team to the fact that she was a woman.[44] After yet another letter full of Natalia's complaints about her colleagues, her father wrote back, "If your relationship with the guys is going bad only because they consider you a suck-up, I think this is just envy, and there is no need to do them a favor by spoiling the relationship with the administration".[45] This response shows that Natalia's parents provided her with emotional support, and were certain that she needed this support because she was a woman in a male collective. During 20 years of her work, the Soviet Union actively built a network of radio and television stations across the country and abroad. The readers of *Radio* met with Freichko again in 1976, when she was 52, as part of a series of articles entitled "One Life", dedicated to the fiftieth anniversary of the paramilitary defense society Osoaviakhim, which had changed its name to DOSAAF in 1951.[46] The idea of the series was to show different professional paths of those who started their radio practice in Osoaviakhim radio clubs in the 1930s. The correspondent accompanied Freichko (by then a senior engineer-tuner at Radiostroi, a special trust within the Ministry of Communications of the USSR tasked with the design and construction of radio infrastructure) on an urgent site visit to once again "breathe life into the metal hearts of radio devices". In the series, Freichko herself admitted that amateur radio practice had been a step into the profession. By following along on the trip, readers learned that Freichko had become a respected specialist, but they could hardly discern what work she actually did as an adjusting engineer. The head of her department, Isaak Dumer, emphasized Freichko's level of experience, noting that since her job required hard technical and physical work, women were rare among hardware tuners, "but she demanded not to be discounted as a woman and never put her personal interests above public ones". Dumer also noted that the collective regularly chose her as head of their party organization. From the series of articles, readers learned that Freichko provided technical assistance at radio centers that were under construction in the USSR and in sister socialist and developing countries as the head engineer on those projects. Thus Freichko's status as a skilled specialist and a trustworthy citizen allowed her to go abroad. Some evidence confirming her trips abroad can be found among the documents in the private collection. Freichko, despite all of the challenges early in her career, did not give up her job, and 30 years later spoke happily of working with technology and the romance of a nomadic life.

Conclusion

Upon any close examination, amateur radio activity in the Soviet Union was not a gender-neutral pastime. Remaining largely within a paramilitary system and, thus, governed by the state, amateur radio was actively promoted as an accessible hobby for large groups of people. Radio magazines, the editorial staff of which consisted exclusively of men, shaped the perception of the engineering profession among their readership, setting up the normative frameworks for gender imbalance. Thus, images of amateur women radio operators in the 1930s did not correspond with images of male representatives. Female radio enthusiasts were presented either as rare exceptions and curiosities or as objects of satire. The very mention of women's names in the press served as an element of governance and the construction of hierarchies, by stressing the curious and exceptional nature of such cases.

Freichko's path may be seen as a success story of a radio enthusiast who became a professional technician, even though she strove for a different career path. It also shows that gender largely determined the boundaries of career opportunities. Despite her efforts to adjust, she was not able to pursue her career in research and to access decision-making positions. Meanwhile, her successful professional activity and the fact that she was a woman made her visible. Women radio engineers became notable figures for their achievements on an equal basis with men, but the latter ones continued to be the main witnesses and storyteller of the achievements. The difficulties experienced by Freichko as a woman were not easy to grasp by reading official narratives only. Various types of sources, including personal correspondence and private collections, allow the human dimension of the history of technology to be more pronounced. Freichko's letters allow us to have a glimpse at how a Soviet female engineer perceived her career prospects, or what obstacles she faced in a predominantly male environment.

Notes

1 1925, "Domashnie sovety", *Radioliubitel'* 3: 6.
2 Dziga Vertov, *Kinopravda (Radiopravda)* 23, Russian State Documentary Film and Photo Archive (RGAKFD), no. 314.
3 "Sbornik materialov po rabote ODR i radiokruzhkov 1929", TsMS, f. Radio, op. 1, d. 1505, l. 8.
4 D. Kositsyn. 1925, "Pis'ma radiopropagandistu. Agitatsiia za radio sredi rabotnits", *Radiolubitel'* 17–18: 358.
5 Kositsyn, "Pis'ma radiopropagandistu".
6 Kositsyn, "Pis'ma radiopropagandistu".
7 1928, *Radioliubitel'* 5; V. A., 1937, "Odna iz pervykh", *Radiofront* 5: 7.
8 1928, *Radioliubitel'* 5.

9 V. Podzorskaia. 1936, "Moi put' v efir", *Radioliubitel'* 5: 9.
10 1929, "Sbornik materialov po rabote ODR i radiokruzhkov". TsMS, f. Radio, op. 1, d. 1505. l. 11.
11 1928, "Spisok zaregistrirovannykh korotkovolnovykh radioperedatchikov v SSSR" *RA-QSO-RK 1*. Supplement to *Radio Vsem* 1.
12 V. Akminskii. 1929. "Vladimirskie istoricheskie khroniki 1921–1939". Available from: http://rw3va.qrz.ru/history/his.html [Accessed 21 July 2022].
13 Pokhod, "Cheliuskina". Vol. 1 (Moscow: "Pravda", 1934), 375.
14 I. Chiviliov. 1934. "Krasnoznamennaia radistka—Liudmila Shreder" *Radiofront* 14.
15 I. Chiviliov. 1934. "Krasnoznamennaia radistka ".
16 Ernst Krenkel. 1935. "Razgovory s mirom", in *Pokhod "Cheliuskina"*. Vol. 2, 370–379. Moscow: Izdatel'stvo redaktsii "Pravdy".
17 1936. *Radiofront* 5.
18 N. Freichko to her parents, 13 December 1948.
19 The website of the Pereslavl'-Zalessky museum of lamp radio is accessible at: https://музей-радио.рф
20 https://centrasia.org/person.php?st=1434959804.
21 The daughter Natalia Freichko will now on be designated by Feichko.
22 Vladimir Burliand. 1947, "Sem'ia korotkovolnovikov" *Radio* 4: 39.
23 Burliand, "Sem'ia korotkovolnovikov".
24 Burliand, "Sem'ia korotkovolnovikov".
25 N. Grigor'eva. 1976, "Odna zhizn'" *Radio* 5: 15.
26 N. Freichko was drafted to the army on June 23, 1941, see https://pamyat-naroda.ru/heroes/memorial-chelovek_vpp31399790/ [Accessed 21 July 2022].
27 N. Freichko to her parents, 5 October 1945.
28 N. Freichko to her parents, 2 September 1945.
29 N. Freichko to her parents, 1 September 1945.
30 N. Freichko to her parents, 15 December 1946.
31 N. Freichko to her parents, 11 September 1950.
32 Burliand, "Sem'ia korotkovolnovikov", 39.
33 N. Freichko to her parents, 26 May 1947.
34 N. Freichko to her parents, 13 December 1948.
35 N. Freichko to her parents, 29 June 1949, and 4 October 1949.
36 N. Freichko to her parents, 17 December 1949.
37 N. Freichko to her parents, 12 February 1950.
38 N. Freichko to her parents, 26 March 1950.
39 N. Freichko to her parents, 7 December 1950, 24 December 1950, and 23 March 1951.
40 N. Freichko to her parents, 23 March 1951.
41 N. Freichko to her parents, 9 June 1951.
42 N. Freichko to her parents, 12 February 1950.
43 N. Freichko to her parents, 12 February 1950.
44 N. Freichko to her parents, 17 December 1949.
45 N. Freichko's father to N. Freichko, 24 September 1955.
46 N. Grigor'eva. 1976. "Odna zhizn'" *Radio* 5.

References

Bauer, Martin W. 1995. "Technophobia: A Misleading Conception of Resistance to New Technology". In *Resistance to New Technology: Nuclear Power, Information Technology and Biotechnology*. ed. Martin W. Bauer. Cambridge University Press.

Birdsall, Carolyn. 2023. *Radiophilia*. Bloomsbury Academic.

Chliiants, Georgii. 2008. *Listaia Starye Call Book i ne tol'ko*. Lviv: Spolom.
Chliiants, Georgii. 2019. "Zhenschiny-korotkovolnoviki (1928–1941)". *Radio* 3: 49–51.
Ellis, Heather. 2017. *Masculinity and Science in Britain, 1831–1918*. Palgrave Macmillan UK.
Galperin, Mikhail. 1925. *Tiotkina antenna*. Moscow: ODR.
Gessler, Anne. 2014. "'Purifying the Upper Atmosphere': Women's Work in Early Radio, 1905–1913". *American Studies in Scandinavia* 1: 87–102.
Gouzévitch, Dmitri and Irina Gouzévitch. 2000. "A Woman's Challenge: The Petersburg Polytechnic Institute for Women, 1905–1918". In *Crossing Boundaries, Building Bridges*, ed. Annie Canel, Ruth Oldenziel, and Karin Zachmann. London: Routledge.
Griesse, Malte. 2009. "Enjeux historiques des journaux et de la correspondance dans la réécriture de l'histoire de la révolution sous Stalin", *Cahiers du monde russe* 50 (1): 93–123.
Haring, Kristen. 2012. *Ham Radio's Technical Culture*. Cambridge, MA: The MIT Press.
Hellbeck, Jochen. 2001. "Working, Struggling, Becoming: Stalin-Era Autobiographical Texts". *Russian Review* 60 (3): 340–359.
Kalemeneva, Ekaterina and Julia Lajus. 2018. "Soviet Female Experts in the Polar Regions." In *The Palgrave Handbook of Women and Gender in Twentieth-Century Russia and the Soviet Union*, ed. Melanie Ilic. London: Palgrave Macmillan.
Kelan, Elisabeth K. 2007. "Tools and Toys: Communicating Gendered Positions Towards Technology, Information." *Community and Society* 10 (3): 358–383.
Khurges, Lev. 2012. *Moskva–Ispaniia–Kolyma: iz zhizni radista i zeka*. Moscow: Vremia.
Klots, Alissa. 2018. "The Kitchen Maid as Revolutionary Symbol: Paid Domestic Labour and the Emancipation of Soviet Women, 1917–1941." In *The Palgrave Handbook of Women and Gender in Twentieth-Century Russia and the Soviet Union*, ed. Melanie Ilic. London: Palgrave Macmillan.
Kojevnikov, Alexei. 2002. " The Great War, the Russian Civil War, and the Invention of Big Science". *Science in Context* 15 (2): 239–275.
Krenkel, Ernst. 1973. *RAEM—moi pozyvnye*. Moscow: Sovetskaia Rossiia.
Lovell, Stephen. 2015. *Russia in the Microphone Age. A history of Soviet Radio, 1919–1970*. Oxford: Oxford University Press.
Mitchell, Caroline, ed. 2001. *Women and Radio: Airing Differences*. London: Routledge.
Opitz, Donald L., and Brigitte Van Tiggelen. 2025. "Gender and Governance in the Historiography of Science and Technology".
Pinskii, Anatolii. ed. 2018. *Posle Stalina. Pozdnesovetskaia subjectivnost' 1953–1985*. St. Petersburg: Izdatel'stvo Evropeiskogo universiteta v Sankt-Peterburge.
Schafer, Valerie, and Benjamin G. Thierry, eds. 2015. *Connecting Women. Women, Gender and ICT in Europe in the Nineteenth and Twentieth Century*. Cham: Springer.
Shatelen, Mikhail. 1949. *Russkie Elektrotekhniki Vtoroi Poloviny XIX Veka*. Leningrad, Moscow: Gosenergoizdat.
Skoog, Kristin, and Alexander Badenoch. 2020. "Women and Radio: Sounding out New Paths in Women's History". *Women's History Review* 29 (2): 177–182.

Valkova, Olga. 2008. "The Conquest of Science: Women and Science in Russia, 1860–1940". *Osiris* 23 (1): 136–165.

Zakharova, Larissa. 2011. "Le socialisme sans poste, télégraphe et machine est un mot vide de sens: Les Bolcheviks en quête d'outils de communication (1917–1923)". *Revue historique* 660 (4): 853–73.

Zakharova, Larissa. 2020. *De Moscou aux terres les plus lointaines. Communications, politique et société en URSS*. Paris: EHESS.

13 How to Save a Soviet Nature Reserve

The Strategies of Dr. Vera Varsanofieva

Olga Valkova

Bolshevik government awarded full legal equality to women in 1918, which included equal rights to professional careers. As a result, already in the 1920s, there were a number of women in the USSR heading scientific expeditions, laboratories, and different departments in the universities, sometimes even whole institutes, although usually they were newly organized departments, laboratories, and institutes (Valkova 2008). In the soviet historiography, these female scientists have long been absent. Only a few biographies of outstanding female scientists have periodically been published, and topics such as career-building, discrimination in the workplace, motherhood, and its complicated relationship with work were never studied. Interest in such topics began to arise in the 1990s, among sociologists first, whose articles were mainly based upon available statistical sources (Mirskaya and Martynova 1993). To this day, little research has been published in which the history of Soviet female scientists is discussed by historians working with primary sources (Valkova 2008; Dolgova and Streltsova 2019; Pushkareva 2011; Stepnov 2019), and even less on their leadership roles in directing the technosciences. So, we don't know how high-ranked female scientists managed their department, how gender influenced the decision-making process, and how they implemented their decisions in a male-dominated scientific world.

These issues can be addressed through the life and career of Vera Varsanofieva (1889–1976). This outstanding geologist was, in 1935, the first woman to be awarded a doctorate in geology and mineralogy in the USSR. She belonged to the first generation of Soviet female scientists. Having received a higher education, she was awarded a state degree in geology before the 1917 Revolution, on the wave of which she began her career. Varsanofieva earned a reputation as a field geologist while simultaneously pursuing an academic career as a teacher and administrator. Very early in this career, she organized and led the Department of Geology at the Second Moscow State University. She had frequent contacts with the Ministry of Education and was involved in the process of preparing curricula in geology for schools and universities. In addition to these responsibilities, she was a member of several scientific societies, was elected vice-president of the Moscow Society of Naturalists—one of the country's oldest and most prestigious naturalist societies—and

DOI: 10.4324/9781003562597-17

edited one of its scientific magazines. She also dedicated a lot of her time and efforts to protecting nature in the wild. Like other female scientists of her station, she eventually worked as a public servant, becoming a people's deputy. Elected and re-elected several times, Varsanofieva became a member of the Moscow City Council, and the Soviet bureaucracy used her professional success as an ideological example.

All this gave her the knowledge and position that she needed to successfully navigate the bureaucratic system on behalf of Soviet nature reserves. These were nearly all destroyed by a government decision of 1951 to liquidate 49 of the existing 128 state nature reserves. Varsanofieva worked around the clock and was selflessly devoted to science—qualities that could just as easily be found in a male scientist. She also had a clear moral code that she strictly adhered to—again, a trait not exclusive to women. But it seems to me that, as a woman, she met more resistance than a man in all of her professional activities, forcing her to learn how to act in order to achieve her goals more easily. Three qualities—the fearlessness of a woman used to fighting for her rights from early youth, a devotion to ideals instilled since childhood, and an ability to manipulate the system acquired by necessity—made her an ideal candidate for the role of leader.

Varsanofieva's actions went unmentioned in her biographies by colleagues and historians of science ("Vera Aleksandrovna Varsanofieva" 1951; "Varsanofieva" 1960; "Varsanofieva" 1971; Kalashnikov 1990), and her efforts to restore the system of Soviet nature reserves were never evoked in the relevant historical literature (Brain 2012; Coumel 2013). The only exception is a book by Douglas Weiner (1999), which frequently mentions her name, but which doesn't refer to her specific role. My chapter will precisely study how Varsanofieva saved Soviet reserves by examining a small, but very important episode of the clash between Soviet bureaucracy and environmental activists, an episode during which she played a key role and one that demonstrates how a high-ranking female scientist used her position and her "capital" to defend a cause. The strategy she employed reveals the way in which, to paraphrase Bourdieu, she incorporated an academic, political, and social habitus. The analysis relies upon primary historical sources found in the collection of Varsanofieva's private papers held by the Russian State Archive of Economics; her letters to colleagues held by the Archive of the Russian Academy of Sciences and its Saint Petersburg Branch; various documents of scientific societies held by the State Archive of the Russian Federation, and the Archive of the Moscow Society of Naturalists; various documents of state institutions concerning matters of environmental protection held by the Russian State Archive of Economics; and papers dedicated to Varsanofieva by her contemporaries.

Education and Scientific Career

Varsanofieva was born in 1889 in Moscow into a military family. In 1907, she graduated with a gold medal (the highest possible honor) from a women's gymnasium in Ryazan, one of Russia's oldest towns, roughly 180km from Moscow, where her father was stationed. From 1907 to 1915 she was a student at the faculty of physics and mathematics of the Department of Natural Sciences of the Higher Women's Courses in Moscow (*Moskovskie Vysshie Zhenskie Kursy*), known to contemporaries and historians alike as the "Women's University". At the time, Russian women were still not allowed to attend actual universities. There were some exceptions, but Varsanofieva was not one of them. She graduated in 1915 and in 1916 successfully passed her state exams before the Moscow University Commission, gaining a state diploma from the University.[1] She then became a professional geologist, despite this not being considered a suitable profession for women at the time. Her mother was Belgian, and Varsanofieva inherited from her not only her delicate beauty but also her native language. Bilingual from childhood, later in life all her private correspondence and her diary were written in French. Her mother was her first teacher and instilled in her a real love of science.

During the early years of her career, her time was divided between teaching at the university during the winter months and fieldwork during the summer. Female lecturers in university classrooms were still a new and somewhat strange occurrence in Russia—but the idea of a young woman joining a geological expedition through wild and remote regions, accompanied by typically rough men and without the protection of her father or husband, was absolutely unheard of. But none of this concerned her. As early as in 1918 she had been teaching a dynamic geology course at her alma mater, the Higher Women's Courses in Moscow, which would be renamed Second Moscow State University in 1918, and her first geological expedition dated back even earlier, to 1911. In fact, from then onwards, she had explored the northern Urals almost every year, actively publishing the results of her research. Her choice of that region was no coincidence, as her father served there and was well-known and respected by the locals. He spoke very highly of the place. One of her professors and lifelong friend, the famous Soviet geologist Alexander Chernov (1877–1963), also loved the Urals and regularly took his female students and future colleagues with him. In 1925, Varsanofieva began to carry out a geological survey of the 124th sheet of the Geological Map of the USSR, which included territory in the northern Urals, at the head of her own expedition (Varsanofieva 1928: 733–768). She specifically researched the stratigraphy of palaeozoic deposits and tectonics in the Upper Pechora basin as well as the geomorphology and history of the Pechora Region and adjacent areas during the Quaternary period. The papers that she published on these topics were considered remarkable.[2] Also in 1925, she was made a professor and given responsibility for the geological department at the Second Moscow State University—a position that she would hold for several decades. In 1935, following the restoration of scientific doctorates in the USSR in 1934, Varsanofieva became the first woman to be awarded a doctorate in geology and mineralogy (Kalashnikov 1990: 13–19). In the 1930s she began to be quite an important figure in the geological community, and a variety of public positions became open to her.

Public Service

The idea of using high-ranking female scientists as a living advertisement for the Soviet state's achievements emerged in the early 1930s. Pelageya Kochina (1899–1999), for example—another outstanding female scientist, physicist, and future member of the USSR Academy of Sciences (from 1958)—was from 1931 to 1934 elected as a People's Deputy of Leningrad City Council. Later, from 1947 to 1955, she was a member of the Moscow City Council of Deputies and in 1951–1959 a deputy of the Supreme Soviet of the RSFSR. Other similar examples can be found. The idea behind this was originally formulated in the preface of a book about Soviet female scientists published in 1933 (Gelstein 1933: 8), and in Spring 1935 it was introduced by Vasily Kramer (1876–1935), Lenin's physician, in an official letter to a member of the Soviet government—Nadezhda Krupskaya (1869–1939), who was not only Vladimir Lenin's widow but also Deputy People's Commissar of Education.[3] Krupskaya liked the idea very much, as her reply makes clear,[4] and high-ranking female scientists began to be recruited for political purposes. Varsanofieva was one of them. Although she never tried to use politics to boost her career—indeed she was not even a member of the Communist Party—later in life she was invited to join various official i.e. government-supported public organizations, and was subsequently elected to be a people's deputy with the full support of the authorities.

During World War II (from 1942 onwards), Varsanofieva was an active member of the Anti-Fascist Committee of Soviet Women (*Antifashistskii Komitet Sovetskikh Zhenshchin*). In 1947 and in 1959 she was elected as a deputy of the Frunzensky District Council of Moscow, becoming a deputy of Moscow City Council in 1953. Several days a week, she dutifully held visiting hours and spoke with people. This took a lot of time from her research and other duties, but worst of all she knew that it was all in vain. As she wrote to her old friend, the fellow geologist and academician Vladimir Obruchev (1863–1956) on 11 March 1954, "My parliamentary work... brings its own emotions and worries, since I cannot be indifferent to the difficult situation of so many voters, mainly in relation to the absolutely tragic issue of [the lack of] housing".[5] Varsanofieva tried to do as much as she could for the people that she represented, but it was not nearly enough. She did not have the tools that she needed for this task, and the situation troubled her greatly.

By all Soviet standards, Varsanofieva was a very successful person. She held a high position in her chosen profession, and her salary allowed her some everyday luxuries that were unavailable to many of her contemporaries—e.g. better housing, food, and clothes; the use of a car when needed; the means to travel; the ability to support her relatives; to have domestic help and not to burden herself with housework, etc.[6] And her position as an elected deputy gave her even more privileges. Varsanofieva appreciated all this and understood that it was a form of payment for her mostly symbolic public role and service, yet although she was deeply dissatisfied with this role, she played it well and could not give it up. More importantly, however, by fulfilling her various public and professional administrative duties, she learned well how to communicate with various officials, how to manipulate them and achieve the results that she needed them to produce. And Varsanofieva used this knowledge for the benefit of her environmental causes when the need arose.

Protecting Nature

The environment and in particular the protection of nature reserves were issues that were of great concern to Varsanofieva. Her love (she explicitly used the word) for wild nature was one of the most important things in her life. She wrote about it in her private correspondence and in her papers and spoke about it during her presentations (Brovina and Valkova 2021).[7] Years when she had no opportunity to join an expedition to some wild place were miserable. In the most difficult times of her life, wild, untamed nature was a source of solace, comfort, and joy.[8] Varsanofieva spent a lot of time and effort fighting to protect nature. Throughout her life, she published papers on the topic, gave speeches, wrote expert opinions, and organized scientific conferences and seminars (Valkova 2019). In 1954, she was among the handful of scientists who began the process that would lead to the restoration of the Soviet Union's nature reserves and to the rescue from destruction of her favorite one, the Pechoro-Ilychsky State Nature Reserve (*Pechoro-Ilychskii Gosudarstvennyi Zapovednik*).

In 1951, when the Council of Ministers of the USSR published Resolution No. 3192 (29 August) "On nature reserves", Varsanofieva was already 62 years old. This resolution stated that "a network of nature reserves has unreasonably expanded in a number of areas". At the time, the Soviet Union had 128 nature reserves, and in 1949–1950 the state spent 54 million rubles on their upkeep. The resolution also stated that most of the work on state nature reserves was "unsatisfactory and not fulfilling the tasks that had been assigned. Many nature reserves have no scientific or cultural value or are redundant". It abolished 49 state nature reserves "as unnecessary and of no scientific significance".[9] Several reserves survived, but their size was drastically reduced. In addition to this tragedy was the fact that the economic exploitation of former nature reserve land was barbaric and unprofitable. More resources were being destroyed than were being generated. Among the nature reserves whose territory was reduced to almost nothing was Varsanofieva's beloved Pechoro-Ilychsky State Nature Reserve in the northern Urals.

The reform was unexpected. Nobody was consulted—neither the people working in the reserves nor even prominent members of the scientific community. For Varsanofieva, this event was a personal tragedy. On the 6th of November 1951, she wrote to fellow geologist Vladimir Obruchev: "The destruction of almost all the nature reserves in our country struck me like a thunderbolt. I cannot recover from this blow, which is hard for all naturalists who love nature".[10] Varsanofieva could not simply stand idly by, she had to do something, anything, to remedy the situation. While she did not have direct access to the people who had made this decision, still she had to be heard. After some thought, she drew up a plan for possible action.

In 1952, two government bodies were appointed to oversee all the remaining Soviet nature reserves: the Commission on Nature Reserves of the USSR Academy of Sciences (*Komissiya po Zapovednikam Akademii Nauk SSSR*) and the Department of Nature Reserves and Hunting Management of the Ministry of Agriculture (*Upravlenie po Zapovednikam i Okhotnich'emu Khozyaistvu Ministerstva Sel'skogo Khozyaistva*). Eleven reserves belonged to the Academy of Sciences, and 28 were given to the Ministry of Agriculture.[11] But the staff of the Ministry of Agriculture were not to be trusted with environmental protection, as it was they who had orchestrated the devastating reform of the nature reserves, and they were mostly interested in the exploitation of natural resources.

How to Rescue the Nature Reserves: The First Ideas

Varsanofieva considered it of utmost importance to rescue the nature reserves from the Ministry of Agriculture and to place them into the care of someone interested in their protection. The Academy of Sciences was a good candidate for this job, particularly as it was already responsible for 11 similar reserves. As Varsanofieva wrote a little later, in 1954: "Undoubtedly, the Academy of Sciences of the USSR, the scientific community, the All-Union Geographical Society and all our country's natural science societies should be in charge of environmental protection".[12] So the idea was to somehow have the nature reserves transferred away from the Ministry of Agriculture and given to the Academy of Sciences. To obtain the Academy's support was a logical first step. Weiner mentions a meeting "in the early summer of 1952" between Vera Varsanofieva, Vladimir Yanshin (1911–1999)—a younger colleague of hers who became an academician (in 1958) and later the vice-president of the USSR Academy of Sciences (1982–1988)—and Alexander Nesmeyanov (1899–1980), who was at the time the president of the USSR Academy of Sciences (Weiner 1999: 205). Varsanofieva and Yanshin had come to ask the Academy to help with the Soviet nature reserves, but Nesmeyanov refused to assist them in any way (Weiner 1999: 205). But the Academy's unwillingness to help did not stop Varsanofieva and her like-minded associates. They just needed to change Nesmeyanov's opinion on the matter. This took some time and effort.

On 30 April 1954, a group of Soviet scientists wrote a letter to Georgy Malenkov, Chairman of the Council of Ministers of the USSR, about the tragic and unacceptable situation of the USSR's nature reserves. A copy of the letter was also sent to Nikita Khrushchev, the head of the Communist Party. Of the letter's 16 signatories, Varsanofieva was the third in the list and the only female scientist.[13] This letter is quite a famous event in the history of Soviet nature reserves. Soviet scientists and public figures were asking the government to remove the nature reserves from the Ministry of Agriculture, whose main goal was to exploit their natural resources and to transfer them to the Ministry of Culture or some new institution reporting directly to the Council of Ministers of the USSR.[14] It is obvious that at that point Varsanofieva and her colleagues had not obtained Nesmeyanov's support, but when the letter was sent in late April 1954, Varsanofieva had already set in motion some other preparations.

The Natural Societies

Even with the signatures of 16 more or less influential people, Varsanofieva knew that a single letter would not be enough to make a difference. Something more would be needed to persuade the government—public support—so Varsanofieva decided to launch one of the "public initiatives" (i.e. public discussions) that were so popular in the USSR. These were often used to demonstrate public support for various governmental decisions, and the government could not completely ignore them. The problem was that organizing anything without official permission was nearly impossible and could even be dangerous. It was obvious that the government did not want any real public discussion of the question of the nature reserves, and that it would therefore withhold permission for a public initiative if asked. But Varsanofieva thought that the risk was worth it, and in March 1954, before the letter to Malenkov was sent, she began her preparations. She decided to organize a surprise meeting of three prominent scientific societies in Moscow to discuss the question of the nature reserves. The main aim of this meeting was to draft and adopt a resolution similar to the first letter and to send this to Malenkov as well. This time, however, there needed to be 150 signatories. Varsanofieva never asked for official permission for this meeting, but she did take steps to give it an air of legality.

Early on in her career, when scientific societies in Russia were still an important element of the wider scientific landscape, Varsanofieva became a very active member of several of them—notably the oldest one, the Moscow Society of Naturalists (*MOIP—Moskovskoe Obshchestvo Ispytatelei Prirody*), and one of the most influential, the Russian Geographical Society (*Russkoe Geograficheskoe Obshchestvo*). Membership in these societies was an excellent way for an up-and-coming scientist to gain access to informal scientific networks. When the All-Russian Society for the Protection of Nature (*Vserossiiskoe Obshchestvo Okhrany Prirody*) was founded in 1924, Varsanofieva, who was at the time a lecturer, became a member as she was a staunch defender of nature. These learned societies were among the first democratic institutions in the Russian Empire, and for that reason, the imperial government was traditionally suspicious of them and watched them closely. As it turned out, the Soviet government was even more suspicious, and as soon as the early 1920s it had established strict controls over all such societies. In the late 1920s, most of them were liquidated or were transformed into bodies that only imitated public associations and were devoid of resources or influence in fact they became a variation of state organizations—something tame, harmless, predictable, well controlled. Yet, Varsanofieva continued to work diligently within the societies that she was a member of, and over time and thanks to her professional career she achieved high positions in all of them. From 1946, she led the department tasked with studying the Earth's crust at the All-Russian Society for the Protection of Nature, and was a member of its board. She was also a member of the board of the Moscow branch of the Russian Geographical Society. Most importantly, however, in 1942 she became a vice-president of the Moscow Society of Naturalists (MOIP)—a highly prestigious but usually rather formal role. MOIP had a president: at the time, Nikolai Zelinsky (1861–1953), a famous chemist and academician; and 3 vice-presidents, all elected by the society's members. Although Varsanofieva was the third one, from 1942 onwards it was she who ran the Society on a day-to-day basis. Zelinsky was already too old to do much, and Varsanofieva was formally given the right to sign important papers, including financial documents, on his behalf. She prepared the society's reports on its activities, solved any problems when they arose and organized new elections of the society's officials.[15] Overall, holding a position within a scientific society was not seen as very important. It could be time-consuming, and although perfectly appropriate for a scientist of Varsanofieva's caliber, it was a role devoid of practical influence beyond the society itself—or so it was presumed. In her fight for the Soviet nature reserves, Varsanofieva decided to use the scientific societies within which she had built up her influence over time. Nobody could have predicted that Soviet scientific societies in the 1950s could be used as a political and economic weapon, but Varsanofieva did it.

Preparations

On 31 June 1953, MOIP president Zelinsky died. He was not replaced for some months, and on 10 February 1954, the term of office of the society's council and vice-presidents expired.[16] Varsanofieva was, for the time being, the only person left with the authority to act on the society's behalf, and the MOIP became the main organizer of the impromptu meeting of Soviet natural societies that she had planned.

Simultaneously, the All-Russian Society for the Protection of Nature had been in disarray since September 1953, when the government had decided to reorganize it.[17] By 1954 its only ruling body was an organizing committee,[18] but unlike most of the committee's members, Varsanofieva put a lot of time and effort into it. Its chairman, the biologist Gurgen Avetisyan (1905–1985), shared her ideas about the future of the nature reserves, and in the circumstances did not need anyone's authorization to join the meeting on behalf of his society. The third participant was from the Moscow branch of the Russian Geographical Society. It seems that its official chairman, the polar explorer Ivan Papanin (1894–1986), was not involved. All the documents were signed by the geographer David Armand (1905–1976), who would be elected vice-president in 1958. Both Varsanofieva's co-conspirators were much younger than her—they quite literally belonged to the next generation.

On 8 March 1954, the Presidium of the Council of the Moscow Society of Naturalists, MOIP's ruling body, founded a Standing Commission on environmental protection issues that would operate under the Presidium of the Council itself. The idea officially came from a group of the society's members, but Varsanofieva was the chairperson of the meeting, took an active part in discussions, and "strongly supported the need to *urgently* organize a commission on environmental protection issues".[19] The members of the new Commission were given just a couple of weeks to prepare an action plan.[20] One month later, when the letter to Malenkov had already been sent, the Presidium of the Council held a further meeting (once again under the chairmanship of Varsanofieva) on 5 April during which the new Commission recommended that a meeting of scientific societies be organized. It was accordingly decided to convene a meeting of three societies at the end of the month.[21]

With the legitimacy of the future meeting established, Varsanofieva sent invitations to two different groups of participants. The first one was made up of officials from the Commission on Nature Reserves of the USSR Academy of Sciences and from the Department of Nature Reserves and Hunting Management of the Ministry of Agriculture. To make the meeting an official event, representatives of both these state bodies had to be present. The USSR Academy of Sciences took the invitation seriously; they decided to take part and even prepared a report.[22] The Department of Nature Reserves and Hunting Management of the Ministry of Agriculture sent its director, Alexander Malinovsky (1900–1981), and its chief scientist, Alexey Korol'kov (1905–1985).[23] The second group of invitees were people who cared about environmental protection so much that they were not afraid to speak their minds and perhaps even risk their careers or freedom—or more. There were subsequently complaints that not all those who were interested were invited,[24] but Varsanofieva dismissed the matter by saying that the invitations had been sent. The official foundation of the future meeting had been established.

The Meeting

The meeting was held on 12 and 13 May at the Moscow House of Scientists (*Moskovskii Dom Uchenykh*), a traditional place for scientists to convene for different purposes. Different documents give conflicting numbers of attendees. The meeting's minutes from 12 May mention 84 people present, but the meeting's final report counts 148.[25] Varsanofieva chaired the meeting.[26] She gave an opening speech about the importance of the wild nature reserves and also spoke about the meeting's aims.[27] She argued that the meeting would help the Department of Nature Reserves and Hunting Management of the Ministry of Agriculture to formulate "ways of further developing" environmental protection in the Soviet Union. References to the Communist party and government resolutions were expertly included in the appropriate places. Everything was very polite and quite ordinary.

The representatives of the Academy of Sciences (Nikolay Kabanov, 1905–1992) and the Ministry (Korol'kov) then presented perfectly appropriate papers on the management of the remaining nature reserves.[28] No surprises there. But the 33 questions that followed seemed less appropriate. Of these, fully 32 were addressed to Korol'kov and Malinovsky from the Ministry—the latter was considered to have been one of the main authors of the unpopular reform, and indeed one of its main beneficiaries.[29] Weiner describes this meeting in some detail in his book (Weiner 1999: 217–239) and evaluated it very accurately:

> Moscow was witnessing the first public protest of the scientific intelligentsia, a meeting that affirmed a social identity of scientists-as-citizens sharply at odds with the Kremlin's definition of 'citizens' and opposed the dictatorship in science imposed by Lysenko and the Party bosses.[30]

He mentions Varsanofieva's opening speech and the fact that she was the meeting's chairperson, but he seems to have neither known nor cared that she was its main organizer. He does remark that she was cautious and wary and would "cut off" speakers she found too reckless,[31] that she understood "her role as a voice of reason".[32] But Weiner was wrong on two points. Firstly, the meeting was not held as a form of protest or revolutionary action: it was organized to give the government a nudge in the right direction, to create a document that could be accepted by the government, and not to alienate officials. And secondly, in those days Varsanofieva was quite literally risking her life and the lives of her supporters. Barely a year had passed since Stalin's death. The XXth Congress of the Communist Party and its astonishing revelations were still a few years away. Varsanofieva would not have known whether she would be going home or to prison after the meeting. Her age, gender, and social position would not have stopped prosecutors if it came to the worst. Barely a few years ago, in 1949, Lina Stern (1875–1968)—a famous biochemist and physiologist, the first woman to be elected to membership of the USSR Academy of Sciences, the director of her own scientific institute—was arrested, imprisoned and sent to Kazakhstan for less than Varsanofieva was doing. Stern was released after Stalin's death, but the risk was real. It could be argued that the famous "thaw" made all the difference (Coumel 2013), but in May 1954 it was still too early for its effects to be felt. Varsanofieva had to be careful. Overall, however, Weiner is perfectly right: this was the first public protest in Moscow of the scientific intelligentsia for decades, and it was organized by a woman.

Results

On 13 May, the meeting's participants realized that tempers were rising a little too much and that more time and fewer people were needed to draft a proper resolution. The suggestion was made by Varsanofieva[33] and a Commission was elected for this purpose.[34] This final resolution—the meeting's main goal—was too important to be left to the increasingly angry participants. The Commission met twice, chaired by Varsanofieva. The second of these was held on the 2 June. Besides Varsanofieva and her allies, also present were the representatives of the Academy (Kabanov) and the Ministry (Korol'kov and Malinovsky).[35] They somehow agreed to the resolution, which stated that "the existing network of reserves and the total area of preserved land are inadequate for the needs and requirements of the USSR and need to be expanded".[36] The resolution also recommended the creation of a new government agency for managing the nature reserves that would report directly to the Council of Ministers of the USSR.[37] It required the ruling bodies of the three societies to "*urgently* appeal to the government" with a request to examine the question of the nature reserves and environmental protection in general in the USSR with the help of the scientific community.[38] How Varsanofieva managed to convince all those present is unknown. Once the resolution had been signed by all parties, it became an official document.

On 15 June, the ruling bodies of the three societies got together to prepare their appeal to the government. Fifteen people were present, including Varsanofieva reaching the main result: the drafting of a new letter to Malenkov. The discussion was long and somewhat heated, but in the end a new, concise letter on behalf of the three societies was prepared.[39] Three main points were outlined. First, the existing network of nature reserves was inadequate for a country as large and as geographically diverse as the USSR. Second, all organizations engaged in environmental protection or in the rational usage of natural resources should belong to a separate institution placed under the Council of Ministers of the USSR. Third, environmental advocacy was insufficient and in need of intensification.[40] Varsanofieva was the first to sign this letter on behalf of MOIP, followed by Avetisyan from the All-Russian Society for the Protection of Nature and Armand on behalf of Moscow branch of the Russian Geographical Society.[41] The letter was sent immediately.

Overall, the meetings had several outcomes. On 11 May 1954, Malenkov received a letter that "a group of scientists" had sent to him on 30 April. He immediately ordered that the letter be forwarded to Nesmeyanov at the Academy of Sciences and to Ivan Benediktov (1902–1983), the USSR Minister of Agriculture, instructing them "to look into this question and to report suggestions".[42] The second letter was also forwarded to Nesmeyanov later on.[43] On 17 May, Kabanov wrote a report to Nesmeyanov about recent developments concerning the nature reserves in which he admitted that the matters mentioned in the first letter were of the utmost importance and deserved attention.[44] He also argued that state nature reserves should be taken away from the Ministry of Agriculture and transferred to the Ministry of Culture or to some new institution under the Council of Ministers of the USSR.[45] This was exactly the same idea that the meeting's participants had agreed upon.[46] He also reported on the meeting on 12–13 May itself. Kabanov's report was favorably received, and he was ordered by Nesmeyanov to draft a reply to Malenkov based on his report.[47] On 11 June, Nesmeyanov signed a letter to Malenkov in which he agreed with the "scientists" and recommended the creation of a Committee for Environmental Protection and Nature Reserves under the Council of Ministers. This committee would take over all the nature reserves, both those of the Ministry of Agriculture as well as those of the Academy of Sciences.[48] Agreeing with "public opinion",[49] he also recommended that the "network of nature reserves" should be reorganized in the interests of environmental protection.[50] With that, Nesmeyanov became a valuable ally. The first goal of what had been a long journey had been achieved.

After these events, Varsanofieva wrote a letter to Obruchev on 14 June 1954 in which she gave him a brief outline of what had happened:

> This whole year and especially the last few months have been very difficult... In May, another big public matter was completed: a two-day meeting of the activists of three scientific societies... dedicated to the state of the USSR's nature reserves... The debate on the reports took two evenings. *I had to preside* and it was quite a responsibility. This is an important matter, which I will tell you about in greater detail when I see you.[51]

With the support of the Academy of Sciences of the USSR, the tide had turned. Resolution No. 1004 (9 August 1955) of the Council of Ministers of the RSFSR established a Main Directorate of Hunting and Nature Reserves under the direct authority of the Council of Ministers.[52] On 14 January 1959, the Pechoro-Ilychsky State Nature Reserve recovered some of the lands that it had lost, increasing its area to 721,322 hectares.[53] This was still less than before the reform of 1951, but was all the same very satisfactory. Varsanofieva's treasured reserve still exists and is indeed listed as a World Heritage Site (*Virgin Komi Forest*) by UNESCO.[54]

Conclusion

Varsanofieva was an eminent and leading scientist who rose to various scientific high positions. Everything she did was made possible partly by the gender equality laws adopted by the Bolsheviks in 1918, and in equal part to her own hard work. But her success was hard won. Precisely because of these 1918 laws, gender equality was no longer a political issue, and the scientific world remained male-dominated. So Varsanofieva frequently found herself in positions where her opinion did not matter, and she had to learn how to navigate with the scientific and the political systems in order to overcome the dismissive attitude of officials and to achieve her goals—not only personal and professional goals but also ones of national importance.

Notes

1 Central State Archive of Moscow (TSGAM) 363/4/4719: 1–129.
2 Full list of publications by Vera Varsanov'eva in an Index (Kushtysev 1977).

3 State Archive of the Russian Federation (GARF) 4737/1/722: 7.
4 State Archive of the Russian Federation (GARF) 4737/1/722: 5–6.
5 Archive of the Russian Academy of Sciences (ARAN). 642/4/316: 136.
6 Archive of the Russian Academy of Sciences (ARAN). 642/4/316: 107.
7 St. Petersburg Branch of the Archive of the Russian Academy of Sciences (ARANSP) 893/1/316: 27.
8 Archive of the Russian Academy of Sciences (ARAN) 642/4/316: 15, 17, 21, 23, etc.
9 http://oopt.aari.ru/sites/default/files/documents/Sovet-Ministrov-SSSR/N3192_29-08-1951.pdf.
10 ARAN 642/4/316: 96.
11 Russian State Archive of Economics (RGAE) 544/1/14: 1.
12 RGAE 3/1/144:77.
13 RGAE 3/1/144: 5–10.
14 RGAE 3/1/144: 10.
15 Archive of the Moscow Society of Naturalists (AMOIP) 1954. no. 1996v: 3–7.
16 AMOIP 1954. no. 1977: 2.
17 GARF 404/1/210:1–2.
18 GARF 404/1/11:52–53.
19 AMOIP 1954. no. 1992: 26.
20 AMOIP 1954. no. 1992: 27.
21 AMOIP 1954. no. 1992: 20.
22 RGAE 544/1/12: 4, 5.
23 AMOIP 1954. no. 1977–1979: 94.
24 RGAE 3/1/24: 70.
25 AMOIP 1954. no. 1977–1979: 94 and AMOIP 1954. no. 1996e: 37.
26 GARF 404/1/216: 2.
27 GARF 404/1/216: 2–5.
28 GARF 404/1/216: 6–55.
29 AMOIP 1954. no. 1996v: 46.
30 AMOIP 1954. no. 1996v: 217.
31 AMOIP 1954. no. 1996v: 233.
32 AMOIP 1954. no. 1996v: 237.
33 RGAE 3/1/24: 69–70.
34 AMOIP 1954. no. 1977: 96.
35 AMOIP 1954. no. 1977: 96.
36 AMOIP 1954. no. 1977: 1.
37 AMOIP 1954. no. 1977: 1.
38 AMOIP 1954. no. 1977: 1.
39 AMOIP 1954. no. 1996в: 36–42.
40 AMOIP 1954. no. 1996в: 51–52.
41 AMOIP 1954. no. 1996в: 55.
42 RGAE 544/1/13: 10.
43 RGAE 544/1/13: 19–20.
44 RGAE 544/1/13: 5.
45 RGAE 544/1/13: 5.
46 AMOIP 1954. no. 1996b: 48.
47 AMOIP 1954. no. 1996b: 1.
48 AMOIP 1954. no. 1996b: 9.
49 AMOIP 1954. no. 1996b: 48.
50 RGAE 544/1/13: 9.
51 ARAN 642/4/316: 138.
52 http://www.libussr.ru/doc_ussr/ussr_5021.htm.
53 https://www.pechora-reserve.ru/territoriya.
54 https://www.pechora-reserve.ru.

References

Brain, Stephen. 2012. "Novyj vzglyad na unichtozhenie zapovednikov v SSSR v 1950-e gg." *Istoriko-biologicheskie issledovaniya.* 4 (1): 57–72.

Brovina, Alexandra Alexandrovna, and Valkova, Olga Alexandrovna, (ed.). 2021. *Prirodoohrannaya deyatel'nost' Very Aleksandrovny Varsanof'evoj.* Syktyvkar: FIC Komi NC UrO RAN.

Coumel, Laurent. 2013. "A Failed Environmental Turn? Khrushchev's Thaw and Nature Protection in Soviet Russia." *The Soviet and Post-Soviet Review.* 40: 167–189.

Gelstein, Amalia. 1933. *Zhenshchiny v sovetskoj nauke.* Leningrad: Lenoblizdat.

Dolgova, Evgeniya Andreevna, and Streltsova, Ekaterina Alexandrovna. 2019. ""Welcome to the club": the status of women in Soviet science of the 1920s." *Sociologicheskie issledovaniya.* 2: 97–107.

Kalashnikov, Nikolai Vlasovich, ed. 1990. *Vera Aleksandrovna Varsanov'eva.* Syktyvkar: Tsentr UrO AN SSSR.

Kushtysev, Evgeny Alexandrovich, ed. 1977. *Vera Aleksandrovna Varsanof'eva 1890–1976. Ukazatel' literatury.* Syktyvkar: Komi knizhnoe izdatel'stvo.

Mirskaya, Elena Zinovievna, and Martynova, Elena Alexandrovna. 1993. "Women in Science." *Herald of the Russian Academy of Sciences.* 63 (8): 575–580.

Pushkareva, Natalia L'vovna. 2011. "Women in Soviet Science." *Voprosy istorii.* 11: 92–102.

Stepnov, Aleksey O. 2019. "The First Female Scientists in Soviet Physics – Professors of Tomsk University. V. M. Kudryavtseva, N. A. Prilezhaeva, M. A. Bolshanina: gender aspects of professional-adaptive practices." *Vestnik Tomskogo gosudarstvennogo universiteta. Istoriya.* 57: 193–192.

Valkova, Olga. 2008. "The Conquest of Science: Women and Science in Russia (1860–1940)." *Osiris.* 23: 136–165.

Valkova, Olga. 2019. "Dr. V. A. Varsanofieva Writings on the Protection of the Environment." *IOP Conference Series. Earth and Environmental Science*, Bristol: 350 (1, Nov 2019). DOI:10.1088/1755-1315/350/1/012018

Varsanofieva, Vera Alexandrovna. 1928. "Predvaritel'nyj otchet ob issledovaniyakh v severo-vostochnoj chasti 124 lista letom 1925 g." *Izvestiya Geologicheskogo komiteta.* 7: 733–768.

Weiner, Douglas. 1999. *A Little Corner of Freedom: Russian Nature Protection from Stalin to Gorbachev.* Berkeley: University of California Press.

"Vera Alexandrovna Varsanofieva". 1951. *Trudy Moskovskogo obshchestva ispytatelej prirody. Otdel geologicheskij.* 1: 5–18.

"Vera Alexandrovna Varsanofieva". 1960. *Byulleten' Moskovskogo obshchestva ispytatelej prirody. Otdelenie geologii.* 35: 3–13.

"Vera Alexandrovna Varsanofieva". 1971. *Trudy Instituta geologii Komi filiala Akademii nauk SSSR.* 14: 5–14.

Archives

Archive Moscow Society of Naturalists (Arkhiv Moskovskogo Obshchestva Ispytatelej Prirody AMOIP).

Archive of the Russian Academy of Sciences (ARAN).

Central State Archive of Moscow (TSGAM).

Russian State Archive of Economics (RGAE).

St. Petersburg Branch of the Archive of the Russian Academy of Sciences (ARANSP).

State Archive of the Russian Federation (GARF).

14 Forging a New Archaeological Discipline in the Kitchen

The Volunteer Career of Arlette Leroi-Gourhan

Gwendoline Torterat

Arlette Leroi-Gourhan was the wife of André Leroi-Gourhan, perhaps the most famous French archeologist, considered nowadays as one of the founders of the modern prehistoric ethnology. But she was also an archeologist who established pollen analysis within archaeology (palaeopalynology) as a specific scientific field. She created her own space within a field that men had largely abandoned—a space which later became a fully fledged field of scientific research within prehistoric archaeology. She worked as a volunteer with her assistant, Michel Girard, from the lab in her kitchen or the lab in the Musée de l'homme's basement. Without any institutional position, she was able to gain a scientific legitimacy and to create a new community of archeologists who inherited from her pioneer research.

From her early days working alongside her husband, whom she assisted, to her full recognition as a researcher, Arlette Leroi-Gourhan had to overcome a number of obstacles. First, she had to free herself from the model of the female assistant for museums, and later for her husband when he was in Japan. Second, she broke away from the collaborative tasks she had long performed for him, on his excavation sites, at their home, and in his laboratory. During the first twenty years of her scientific activity, she built up a research project that fitted in perfectly with her husband's broader project. Through him, Arlette Leroi-Gourhan benefited from material and human resources she would probably have been deprived of outside marriage.

It is precisely the aim of this chapter to understand how Arlette Leroi-Gourhan successfully introduced epistemological and methodological innovations, reshaping part of French archaeology. Leroi-Gourhan's case is not unique and adds up to previous examples of women scientists who were recognized by their peers, but who also chose to maintain their independence by remaining volunteers or without institutional affiliation (Rossiter 1982). But it gave the possibility to raise the question of the gendered government of technosciences in adopting an approach which tends to shift the focus away from institutions, to consider how individuals manage their own field of action and that of others. It will thus show that it was by renouncing a career in male scientific government that Arlette Leroi-Gourhan was able to exercise her leadership in archaeology.

DOI: 10.4324/9781003562597-18

To this end, this article will refer to two historiographies. First, the historiography of scientific couples which have overturned the traditional gender stereotype of the male genius and his female assistant that prevailed until then (Pycior, Slack, and Abir-Am 1996; Lykknes, Opitz, and Van Tiggelen 2012). It also revisited the classic dichotomy between the masculine sphere of the public domain and the feminine sphere of the private one (Rogers 2019). Having revealed just how intertwined these two worlds are, studies of women in science have explored the material spaces in which science was done as dynamic, ever-changing locations with blurred boundaries (Von Oertzen, Rentetzi and Watkins 2013; Opitz, Bergwick and Van Tiggelen 2016). The path of Arlette Leroi-Gourhan raises issues relating to couples in science, the assistant and genius team, and the gender division of the technoscientific spaces between private and public locations. The three protagonists were divided into two professional pairs: the genius husband André Leroi-Gourhan and his assistant wife Arlette Leroi-Gourhan (1913–2005), and the genius wife Arlette Leroi-Gourhan and her male assistant Michel Girard, both binomes working from home or/and the Musée de l'homme in Paris.

Second, this contribution mobilizes the historiography on archeology that shed light on the role of women in its scientific development (Diaz-Andreu and Stig Sorensen 1998; Coudart 1998; Smith 2000; Adams 2010; Coudart 2015; Diaz-Andreu 2021). Prehistorians such as Madeleine Colani (1866–1943) or Annette Laming-Emperaire (1917–1977), both pioneers in their disciplines, have remained as famous as their husbands, whom they followed in their field abroad. However, our research differs in that it focuses on Leroi-Gourhan's agency to reshape a part of archaeology from the periphery of institutions. It is based upon new and original sources: a collection of recently digitized photographic archives and a new collection of oral archives that we are building thanks to a series of filmed interviews. The information in this article was gathered during a first period of research in 2016 (the "2ARC" and "Labex LPP" projects coordinated by N. Goutas, L. Mevel, and P. Bodu) and during a second period in 2019 as part of the "Anthrop'Arc" project (DIM, IDF, Matériaux Anciens et Patrimoniaux, under the supervision of N. Goutas and B. Buob). The interviews that this paper is based upon were carried out during these two phases of research.

Espousing Another Man's Work

Arlette Royer, born into the industrial *bourgeoisie*, traveled far and wide during her youth—particularly in Europe and North Africa—and grew up in the intellectual and artistic circles of 1930s Paris. She studied under Marcel Mauss at the École Pratique des Hautes Études (EPHE) and, like many of her fellow students, contributed as a volunteer to the renovation of the Musée du Trocadéro where, in 1934, she met André Leroi-Gourhan.[1] During this

Figure 14.1 Arlette Leroi-Gourhan in Nojiri (Nagano Prefecture), 1937. MSH Mondes Archives. ©André Leroi-Gourhan.

period in Europe, some scientific institutions started to open their doors to women scientists. The two leading radioactivity research centers—in particular the Institut für Radiumforschung (Radium Research Institute) in Vienna and Marie Curie's radium laboratory in Paris—attracted female graduates as early as the 1920s, giving them the opportunity to join laboratories as researchers rather than assistants (Rayner-Canham 1997; Harvey 2012; Massiot and Pigeard-Micault 2016). In France, it wasn't mostly until the years following the creation of the CNRS in 1939 that the backstage areas of research institutions became more feminized (Sonnet 2004; Waquet 2022). However, women were mainly employed as secretaries, typists, or, more rarely, labora tory secretaries (Gardey 2008). Some joined their husbands' laboratories, either as assistants or collaborators or more rarely as research associates, with a position that enabled them to continue their own research (Gosztonyi Ainley 1986). Under the guise of professional opportunity, mar riage in fact reproduced, and consequently amplified, gender inequalities similar to those already in place in the private and public spheres (Figure 14.1).

This is exactly how Leroi-Gourhan's "career" started. The couple celebrated their wedding in 1936 before leaving for Japan on a two-year ethnographic mission, during which she taught at the Institut Franco-Japonais in Kyoto. "Arlette Leroi-Gourhan devoted herself entirely to her husband's work: as a photographer [developing and printing photographs] and as a secretary, constantly accompanying him on his research trips outside Kyoto" (Emery-Barbier, Leroyer, and Soulier 2006: 227). Together they gathered the materials that André Leroi-Gourhan subsequently used to prepare his doctoral thesis (Leroi-Gourhan 1946). But rising international tensions with Japan forced Arlette Leroi-Gourhan to return to France early in 1938. Over the next six months, the Musée de l'Homme tasked her to work with the Musée Guimet to receive the collections of objects gathered in Japan, where her husband had remained.[2] The two were reunited in Paris in 1939. Over the next 15 years, "Arlette Leroi-Gourhan devoted herself to the schooling of her four children and to her husband's career. As she had done in Japan, she dealt with the administrative side of his teaching work at the University of Lyons and the Musée de l'Homme" (Emery-Barbier et al. 2006: 228). Acting daily as her husband's secretary was particularly time-consuming, one of her daughters remembers: "It sometimes went on until two o'clock in the morning. She also did all the housework and dealt with everything else".[3]

From the 1950s onwards, Arlette Leroi-Gourhan stepped aside from her role of scientific assistant while continuing to deal with the household on a daily basis. In 1949, she was responsible for the management and documentation of the archaeological site of Arcy-sur-Cure (in Burgundy) that her husband led until 1963 (Figure 14.2).[4] In 1952, the couple bought a second home in Vermenton, only a few kilometers away from the site and linked to Paris by rail. This made it much easier for Arlette Leroi-Gourhan to reach Arcy, giving her the time to look after their two younger sons while remaining close to her husband's archaeological site and to their two elder daughters, who spent their days with the rest of the team. "My mother would come during the day. She did not drive, which made things a bit complicated. In fact, she never drove in her life, because in those days women were not supposed to drive", her daughter remembers.[5] Every year, the couple would leave all the equipment needed at the site in their second home in Vermenton instead of taking it back to their main residence in Paris. Buying a second home thus enabled the couple to reconcile family life with archaeology—but the gender-based division of domestic chores, for which Arlette Leroi-Gourhan was responsible, was dissymmetrical.

Distinguishing and Surrounding Herself

This proximity to the archaeological site enabled Arlette Leroi-Gourhan to devote some of her time to her own research. She chose to focus her efforts on archaeopalynology (also known as palaeopalynology), i.e. the study of fossilized pollens found at prehistoric sites. At the time, this field was considered relatively minor and was relatively unexplored. In 1954, she carried out her first control investigations in the kitchen of her family home in Avallon. At the time, she needed only three things to study the composition of her sediment: a microscope, a small twin-tube centrifuge, and three kinds of acids and coloring agents.[6] That year, over a period of a few days during the

Figure 14.2 Arlette Leroi-Gourhan at the archaeological site of Arcy-sur-Cure in Burgundy, 1950. MSH Mondes Archives. ©André Leroi-Gourhan.

Arcy-sur-Cure summer dig, the site's stratigraphy was verified based on the pollinic profile of each level (Figure 14.3).

In order to improve her knowledge and skills, she briefly studied under the palynologist Madeleine van Campo (1920–2017), who at the time was one of the few scientists to be interested in pollinic morphotypes.[7] Aline Emery-Barbier (born 1945), who in 1968 began to work closely with Arlette Leroi-Gourhan, explained that the training involved classifying, recognizing,

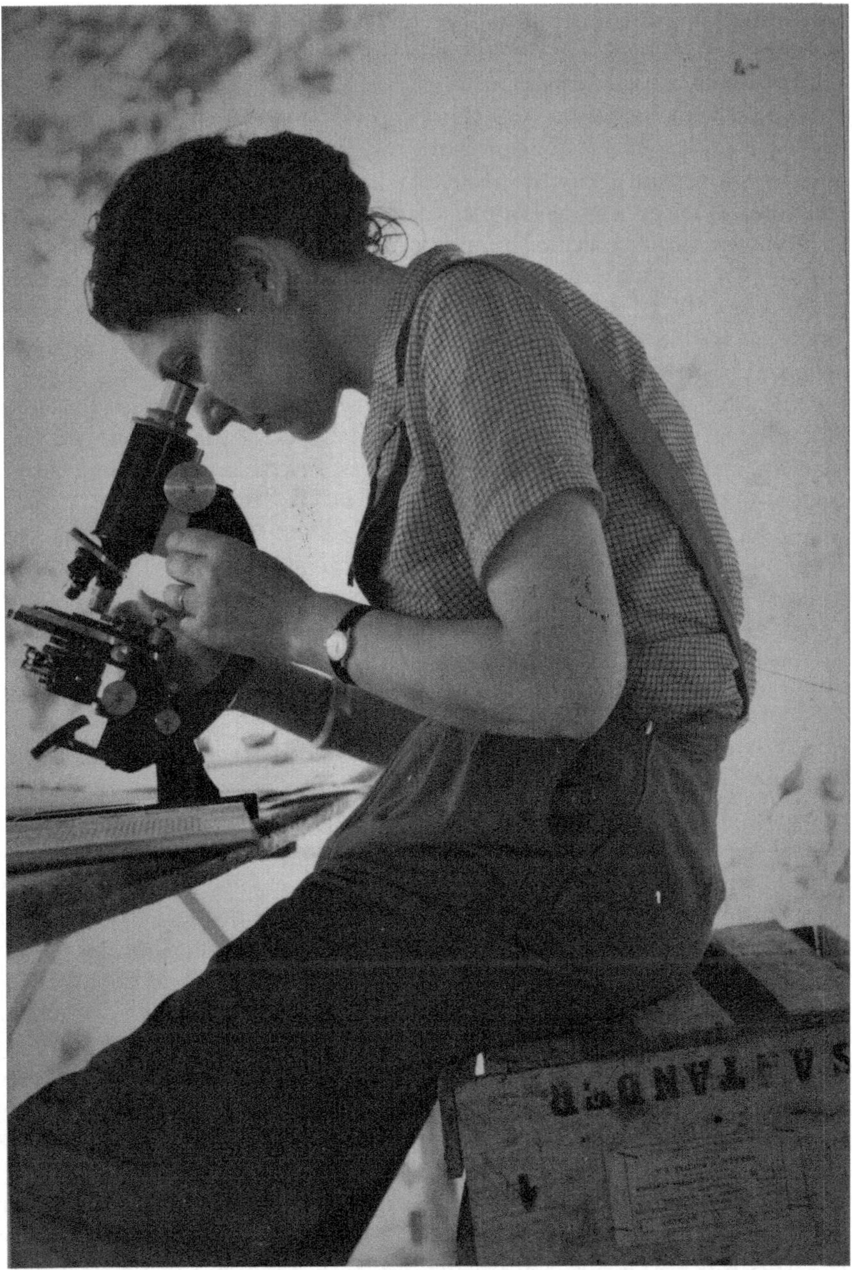

Figure 14.3 Arlette Leroi-Gourhan at the El Pendo archaeological site near Santander, 1955. MSH Mondes Archives. ©Luis Diego Couscoy.

and memorizing different pollens. Arlette Leroi-Gourhan also learned various laboratory techniques for the extraction of grains of pollen, such as extracting one or two flower stamens in order to subject them to acetolysis.[8] This collaboration with Madeleine Van Campo resulted in Arlette Leroi-Gourhan's first two publications, but their work together stopped there (Van Campo and Leroi-Gourhan 1956a, 1956b).

Initially, Arlette Leroi-Gourhan carried out her experiments alone, but she soon managed to secure the appointment of a technical assistant. Michel Girard arrived at the Arcy-sur-Cure site as a volunteer in 1955. Only around fifteen years old, he was one of the youngest members of the team. During his studies at the École Supérieure de Biologie-Biochimie in Paris, he was soon attracted by Arlette Leroi-Gourhan's work at the site to build up a series of pollinic reference slides. She suggested that he work with her as her assistant. His role was crucial, for he developed the new technique of acetolysis that made it possible to build up a set of references. Arlette Leroi-Gourhan mainly dealt with the preparation and analysis of fossil pollens:

> She needed fresh pollens, but she didn't really have the time. This became the work of her technical assistant. [...] At the beginning, her method wasn't powerful enough [and] her sediments contained too much clay. In order to obtain fifty grains of pollen, she needed to read ten slides. With the same sediment, I had 3,000 pollen grains per slide.[9]

In 1959, just before he left in order to begin his military service, the CNRS officially employed Michel Girard as a technical assistant. He was hired thanks to André Leroi-Gourhan. The members of the technical support division of the CNRS were subdivided into four categories: engineers and managers, technical assistants and forepersons, draftspersons, and administrative roles. Technical assistants enjoyed no official status until 1952, thirteen years after the creation of the CNRS.[10] Michel Girard was replaced for three years by a person recruited temporarily by the CNRS. Upon his return in 1961, he was given back his job. "The boss got me into the CNRS", explains Michel Girard, who over the next few years climbed the rungs of his professional ladder thanks to André Leroi-Gourhan. In parallel but without these advantages, Arlette Leroi-Gourhan was working as a volunteer to develop the quasi-unexplored field of palaeopalynology. She became Michel Girard's mentor, who never ceased to work at her side (Figure 14.4).

This period saw the growth of close personal and professional ties between the researcher and the technical assistant who had been appointed to help her. This was also a role that became rooted in the status of both persons, researcher and technical assistant, as Michel Girard explained later: "At the CNRS, the technical assistants are there to serve their superiors... You are a researcher's technical assistant. It's almost personal. You are being given a technical expert to help you with your research".[11] Researchers enjoyed the long-term technical support of their assistants. This was certainly the case

Figure 14.4 Michel Girard collecting pollens at the Arcy-sur-Cure site in Burgundy, 1962. MSH Mondes Archives. ©André Leroi-Gourhan.

with Michel Girard and Arlette Leroi-Gourhan. The working relationship between the volunteer researcher and the officially employed technical assistant was the material expression of a remarkable phenomenon: an inversion of the gender-based roles that were characteristic of the period. The originality of this asymmetrical gender configuration—the female volunteer researcher and the male institutional assistant—sheds light on how Leroi-Gourhan, as a woman scientist, could, even from the periphery of institutions, guide, direct, and structure a new field of knowledge.

We know very little about the career opportunities that Arlette Leroi-Gourhan may have enjoyed. She was apparently offered a job at the CNRS by Jean Dresch (1905–1994), who at the time was the Director of the University of Paris' Institute of Geography (between 1960 and 1970). She could have accepted it as long as she agreed to join his laboratory. Michel Girard remembers witnessing her refusal: she was saying that she did not want to deprive a student of this opportunity and that her husband was earning enough money.[12] On the other hand, the historian Philippe Soulier claims that André Leroi-Gourhan "was explicitly against his wife applying to join the CNRS. He thought it would be inappropriate for public money to be used to pay the salaries of both a husband and a wife. And it was for this very same reason that he disapproved of the situation enjoyed by his colleagues

François Bordes and Denise de Sonneville-Bordes" (Arlette Leroi-Gourhan in Soulier 2018: 398). This statement raises two questions about the internalization of gender-based norms as the origin of this refusal and about the often limited room for maneuver within which women must attempt to fulfill their professional ambitions. However, we need to remain vigilant about this phenomenon of self-censorship, and study the more discreet social mechanisms and their context that help internalize gender norms (Gardey 2000; Marry 2004).

Working at Home: A Gender-based Organization of Space

In some cases, the domestic site represented a space where women could direct their projects or experiments alone, and exercise their agency in the scientific field. This was the case for the German chemist Agnes Pockels (1862–1935) established the science of surfaces. By experimenting with soapy water in her own kitchen, she created the first apparatus capable of measuring surface properties (Van Tiggelen et al. 2019). The historiography has shown that the private home could become a scientific arena enabling women, be they married or single, to renegotiate gender-based norms (Kohlstedt and Opitz 2002; Lykknes et al. 2012; Opitz et al. 2016; Lykknes et al. 2019).

As for Arlette Leroi-Gourhan, she only occasionally used the house in Avallon to carry out her analyses during the digging seasons in Arcy-sur-Cure. That said, this house did offer her a space of her own for her research that was compatible with her obligations as a wife and mother. The family's flat in Paris was rearranged and reorganized so as to permit her to have her own laboratory alongside her husband's:

> To the right, Mom's rooms, and on the left, Dad's. I'm simplifying it, because my parents' bedroom was towards the children's part. But from time to time, when it was very late, Dad would sleep in his study. There was the corridor, Dad's study, the laboratory on the other side.[13]

The family's home in Paris was actually made up of two apartments linked by a corridor. The domestic spaces (kitchen, bedrooms, drawing room, etc.) that the couple's youngest daughter Martine Leroi-Gourhan recalls as "Mom's rooms" were separate from those reserved for work. More precisely, the left-hand (the "male side") consisted of two rooms separated by bookcases containing thousands of books. The first of these rooms contained André Leroi-Gourhan's very large desk, which also held books. He would sometimes exhibit objects from his private ethnographic collection here. His laboratory—smaller and containing both Leroi-Gourhan's workbench and another space dedicated to photography—was on the other side of the room, and, separated by cabinets containing his ethnographic collections, his wife's palynological laboratory. This home laboratory contained a desk for the microscope and shelving for the collection of pollen references and books.[14]

This "room of one's own" was of course what Virginia Woolf underlined as essential to a woman if she were to work (Woolf 1945). But Arlette Leroi-Gourhan's laboratory was in reality no more than a part of a multi-functional space that she shared with her husband and her assistant. Moreover, she was forced to perform some tasks in the kitchen as she had no tap in their shared study. "The centrifuge was in the kitchen, as was the hydrochloric acid! And Michel was there every day when I came home from school. I would say that he was practically part of the family", adds Martine Leroi-Gourhan.[15] She even remembers that he would regularly fix her ice skates! Michel Girard spent his working hours—six to eight hours per day—at the Leroi-Gourhans', more often than not in the room that served as Arlette Leroi-Gourhan's laboratory. This space was also shared with her technical assistant. This home laboratory was essentially unsuitable for the manipulation of chemicals and did not afford the necessary tranquility that research requires. Its installation within the family home, in these conditions, therefore maintained Arlette Leroi-Gourhan in a certain state of subordination. Contrary to that which some historians had attempted to prove in the early 1990s, modern family identity does not uniformly pass through a strict separation between private and public realms (Prost and Vincent 1991). A material analysis of the scientific work of women reveals that the borders of the domestic space are not clearly defined. Spaces used by women are given, as a minimum, a dual function, blending family and work. This is certainly the case of this laboratory, which is neither an individual space (she shares it with her husband and her assistant) nor one solely devoted to palaeopalynological work. And this is also the case of the kitchen, whose tap serves the entire family's needs but is also needed for laboratory chemicals. She experienced these flat-gendered temporal and material limitations of her agency due to her domestic obligations.

In these circumstances, being a married woman with children demanded important levels of commitment and adaptation in relation to the domestic organization of the family home. In 1962, Arlette Leroi-Gourhan's home palynological laboratory was reinstalled in the basement of the Passy wing of the Musée de l'Homme. Her working conditions evolved, giving her a space for her research. But this recognition of sorts did not extend to her being offered an official job, and she continued to work as a volunteer researcher in this new space. Jacques Millot (1897–1980), Director of the Musée de l'Homme from 1960 to 1967, gave André Leroi-Gourhan the use of an old electrical workshop. Two connecting rooms were available. Roger Humbert—technical assistant at the CNRS since 1946 and draftsperson for the museum's researchers—was seated next to the part of the room dedicated to microscopy. The second room, the laboratory, became a space for the preparation of chemical solutions in which Arlette Leroi-Gourhan worked, assisted by Michel Girard (Figure 14.5). This period saw no change in their professional status.

Hitherto superimposed within the couple's home, their domestic and work spaces were finally decorrelated following this move. The latter event

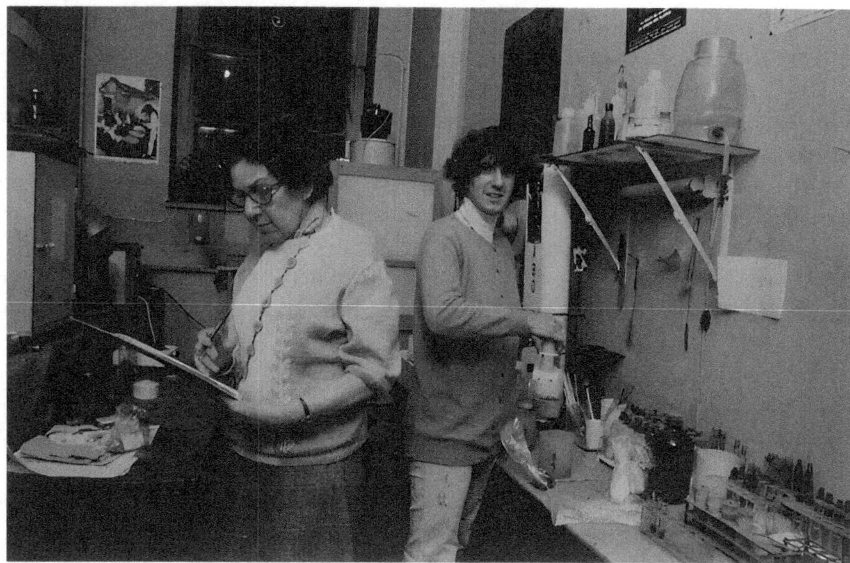

Figure 14.5 On the left, Arlette Leroi-Gourhan, and on the right, Monique Olive (a trainee), Palynology Laboratory in the basement of the Musée de l'Homme, Paris, 1980. MSH Mondes Archives. ©Trong Hieu Dinh.

probably had an impact on the family's domestic management, but the available archives are silent on this matter. Understanding the way in which relationships between men and women manage a domestic form of science can therefore also show how scientific research is carried out on a daily basis and reveal the dynamics of hierarchical powers whether formalized by the institution or implicitly acted upon by the individual's agency.

Toward a Recognition of Palynology

When Arlette Leroi-Gourhan began her practice of archaeopalynology, the field was relatively unexplored within archaeology (Emery-Barbier et al. 2006). Archaeopalynology is the study of fossil pollens collected from prehistoric sources. The first stage involves finding, identifying, and counting grains of pollen in a contemporary or fossil context. First experimented with in Scandinavian bogs and focusing upon periods more recent than prehistory, palynological methods were greatly improved in the 1940s. But in an archaeological context, much remained to be done, notably in the karst environments of caves and shelters (Dubois 1932; Lémée 1949, 1952; Kraehenbühel 1952). "Through her dealings with prehistorians, Arlette Leroi-Gourhan became convinced that the environmental approach was vital to the archaeological process, and she sought to learn from advances in palynology in Northern Europe".[16]

The caves of Arcy-sur-Cure, the archaeological site that her husband was exploring, became for her an area of exploration where she developed palynological methods applied to its quaternary levels. This site was a karst environment with traces of ancient human activity. But for contemporary specialists, this kind of context unfortunately offered no significant spectrum from an environmental point of view. The pollens had not been carried in by the wind as in open-air sites and had instead been deposited by the movements of various living things. It was mainly for this reason that Arlette Leroi-Gourhan's fellow palynologists did not contribute to her development of a new approach.

In 1955, Arlette Leroi-Gourhan for the first time presented the results of her work to members of the French Prehistoric Society. A year later, she called for a methodological synergy between the analysis of pollinic and zoological data, and radiocarbon dating. Her goal was to obtain common and significant climate curves (Leroi-Gourhan 1956a). She rapidly extended her first methodological developments to the wider questions being debated by prehistorians. Basing herself upon her husband's palaeoethnological approach, she proposed her own via a precursory palaeoethnobotanical focus: "the reconstruction of the botanical landscape make's Man's material existence directly perceptible" (Leroi-Gourhan 1956b).

With almost 180 publications between 1956 and 2003, including a book (1989), she never ceased during her career to demonstrate the feasibility and utility of her field. Among her publications, 35 were co-authored with various colleagues, including 7 with her technical assistant Michel Girard between 1971 and 1979, and only one with her husband, in 1965. In comparison, André Leroi-Gourhan's bibliography is much larger, with around a dozen books and 430 articles (Soulier 2018). Their joint publication of 1965 is an interesting case when seeking to understand the number of concrete difficulties linked to the process of recognition of Arlette Leroi-Gourhan's research. In it, she had decided to summarize her first results. The main publication on the Arcy-sur-Cure site having been delayed, she offered to submit the results of her palynological analyses to the prestigious journal *Gallia*, to which she proposed to add her methodology and a critical history of her research based upon work she had carried out at other archaeological sites (Figure 14.6).

Fearing that the palaeopalynological study of "his" archaeological site would be published without his signature, André Leroi-Gourhan decided to not only co-sign the article but also add "his" contribution. The 64-page article that resulted is thus divided into two parts, each with a bibliography. The first of these (pages 1–35), "Climates of the Recent Quaternary", presents Arlette Leroi-Gourhan's research and bears the following note: "ERRATUM At the end of page 32, restore the signature: Arlette Leroi-Gourhan". The second, "Industries of the Upper Palaeolithic", begins thus:

> Following this botanical chronology by Arlette Leroi-Gourhan, and until the detailed study of Arcy's various Upper Palaeolithic layers is published, it seems useful at this point to provide a summary of the main characteristics of the horizons opened up by the palynological analysis.
>
> (Arlette and André Leroi-Gourhan 1964: 36)

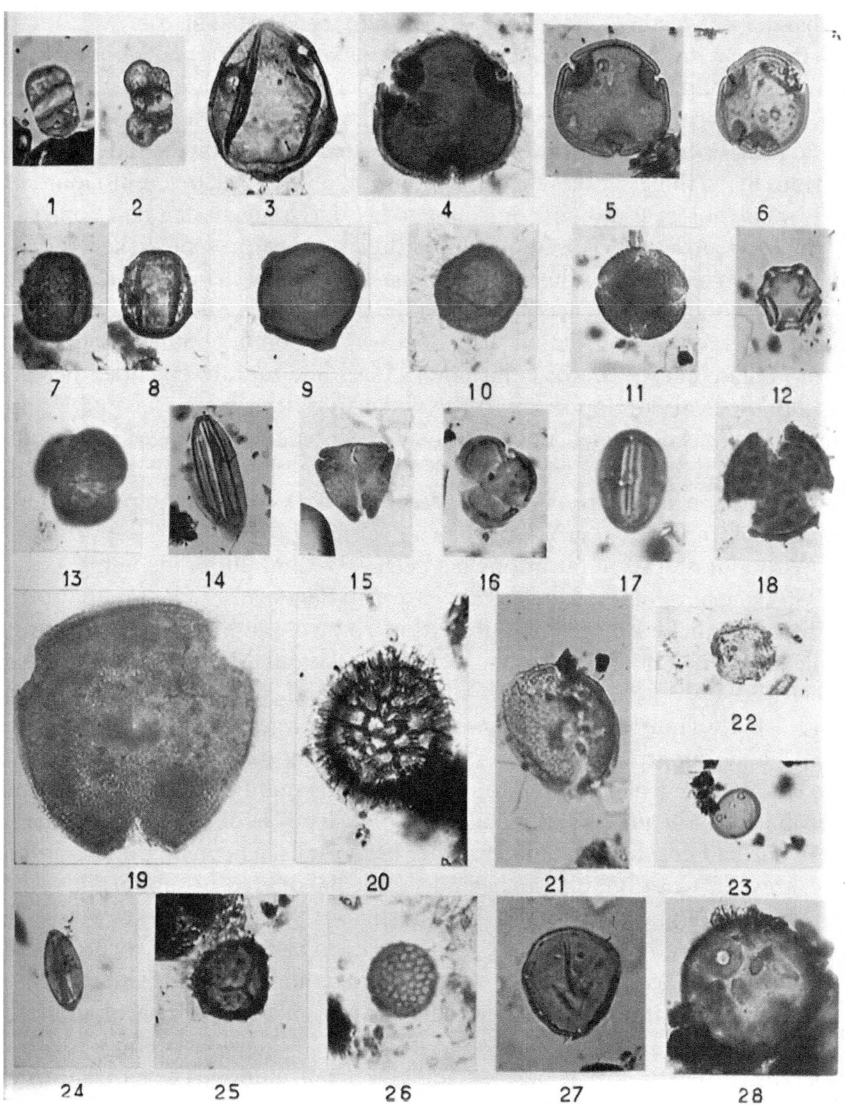

Figure 14.6 Examples of pollens found at Arcy-sur-Cure in Burgundy. In Leroi-Gourhan and Leroi-Gourhan 1964: 25.

This second half includes around 15 illustrations of the lithic and osseous materials collected from layers containing pollens and ends with André Leroi-Gourhan's individual signature on page 64. The issue of attribution is particularly important for scientists, especially when starting out in their career. It enables them to advertise the results of their research beyond the walls of their laboratory, and this was particularly important for Arlette Leroi-Gourhan, whose audience was made up of experts in the environmental sciences, especially palynologists working on contemporary periods. She had to distinguish herself from her husband, yet it is hard not to see evidence of a certain form of scientific subordination to him in this solitary co-authored article.

From the late 1950s onwards, Arlette Leroi-Gourhan opens up new fields of international study in North Africa and the Middle East (Figure 14.7). Her work in the Levant enables her to develop new areas of research, e.g. cereal proto-cultures in the 1970s. Over 50 years of research, she took part in numerous conferences and colloquia organized by influential scientific societies founded in the 1960s, such as the French Association for the Study of the Quaternary (Association Française pour l'Étude du Quaternaire), the International Union for Quaternary Research and the Association of Francophone Palynologists. Arlette Leroi-Gourhan also trained an entire generation of internationally recognized archaeopalynologists in her laboratory at the Musée de l'Homme.

One must also underline the contribution her later research would make. Her goal was not only to experiment with and improve various techniques, nor only to work directly with her husband, but also to develop a new field

Figure 14.7 Arlette Leroi-Gourhan at Mureybet in Syria, 1973. MSH Mondes Archives. ©Olivier Aurenche.

of scientific study. She lived for another 20 years after his death and never ceased to develop her research and to formulate ever bolder hypotheses. In an article published in 1997, for example, she proposed to combine climate curves obtained from Arctic cores with terrestrial palynological data over a period stretching from 60,000 to 15,000 BCE (Leroi-Gourhan 1997). She also described precise variations in the climate curves by comparing them to those obtained from pollens that she studied as well as radiocarbon datings. She freed palaeopalynology from the archaeological confines to which some devotees of the environmental approach sought to restrict it.

Arlette Leroi-Gourhan garnered only a few awards. She was elected Honorary President of the French Prehistoric Society in 1988, having initially joined it as Vice-Treasurer in 1967 and been its President in 1971. During a ceremony held at the Musée de l'Homme in 1990, i.e. fully 50 years after her career had begun, the famous French palaeontologist Yves Coppens presented her with the National Order of Merit (with the rank of Knight).

The Institutionalization of Palynology

André Leroi-Gourhan's work to broaden the spectrum of the kinds of artifacts that could be studied encouraged the multiplication of works on palaeoenvironments (Figure 14.8). This field was already of interest to other

Figure 14.8 Arlette and André Leroi-Gourhan at the El Pendo site near Santander, 1957. MSH Mondes Archives. ©Nicole Maurichaud.

prehistorians who also recognized the importance of analyzing sediments, fossil pollens, and other palaeontological traces in order to read certain climatic correlations (Audouze et Fiches 1993: 19; de Lumley 1976). Nevertheless, for André Leroi-Gourhan this desire first manifested itself during a period of reflection on the professional and institutional future of the wider field. His goal was to "restore the work of French prehistory to a level of scientific excellence worthy of the country" (Soulier 2018: 211). In order to achieve this, and in addition to his growing influence, he tied his efforts to those of the post-war drive to reorganize and restructure the CNRS and its representative and financial bodies. His influence grew further during the 1960s when he was promoted to Professeur des Universités de 1ère Classe (University Professor, 1st Class) and became a member of the CNRS's National Committee. He restructured and led the Centre de Recherches Préhistoriques et Protohistoriques that he had created a decade earlier, now reattached to the general chair of ethnology at the Sorbonne.

Two women helped André Leroi-Gourhan in this latter capacity: Annette Laming-Emperaire and Arlette Leroi-Gourhan. The first one's career evolved up the rungs of a CNRS that had already existed for 15 years. She began by being awarded a scholarship by the CNRS's Ethnological Commission (1946–1948) and was later made a Research Assistant in order to prepare her doctoral thesis (1948–1961) before becoming Maître-assistante [17] at the Sorbonne's chair of prehistoric ethnology and archaeology (1961–1966). Laming-Emperaire thus already held tenure before beginning to assist André Leroi-Gourhan. The second woman ran, as a volunteer, the palynological laboratory in its new incarnation in the basement of the Musée de l'Homme. Her name was featured in the organizational chart of this center's management, and later on, she would take part in the teaching of trainee palaeobotanists. At the time, the laboratory began to receive a growing number of requests for analyses from around the world. Arlette Leroi-Gourhan surrounded herself with part-time collaborators in order to live up to the growing fame of her work. This "sometimes irritated André Leroi-Gourhan, but he had no other choice than to accept the growing fame of her work" (Soulier 2018: 399).

In 1966, the CNRS requested from universities that they restructure their small research units by transforming those that employ researchers and receive funding into Équipes de Recherche Associée (Associated Research Teams). The Centre de Recherches Préhistoriques et Protohistoriques was implicated by this request, and André Leroi-Gourhan responded by redraw ing its organizational chart, with Arlette Leroi-Gourhan responsible for laboratory pollen analysis and Michel Girard as a CNRS technical assistant. Associated Research Team No. 52 was thus joined to the palynological laboratory, largely thanks to its indisputable national and international reputation and to the efforts of several important partners such as the Faculty of Sciences of the University of Paris.

In 1970, the laboratory became part of the Collège de France a year after André Leroi-Gourhan's election to its Chair of Prehistory. Yet, despite the growing importance of its research, the laboratory's location remained unchanged for as long as Arlette Leroi-Gourhan ran it. Over the years, dozens of students worked at the laboratory in order to complete internships or training, study Michel Girard's reference slides, work on other slides or discuss current analyses. Josette Renault-Miskovsky (1938–2018) and Aline Emery-Barbier were certainly the most dedicated of the palynologists that Arlette Leroi-Gourhan and Michel Girard trained. Indeed, when in 2001 the laboratory was transferred to the Maison des Sciences de l'Homme in Nanterre, Arlette Leroi-Gourhan handed over its direction to Aline Emery-Barbier.

In the midst of a far-reaching institutional reorganization of prehistoric research, André Leroi-Gourhan's influence had a number of repercussions on palaeopalynology. It enabled this laboratory to maintain a structured institutional framework, from the Centre de Recherches Préhistoriques et Protohistoriques that was established at the Sorbonne in 1961 to the palynological laboratory integrated within Associated Research Team No. 52 that was turned over to the Collège de France. The prehistorian succeeded in making good use of his various institutional ties in order to support the development of his wider ambitions for prehistory. On the other hand, however, working conditions in the field of palaeopalynology hardly ever evolved. Although the number of trainees and requests for analysis never ceased to grow, neither the laboratory itself nor Arlette Leroi-Gourhan's volunteer status ever changed. Only her job description evolved, from site supervisor to director of the palynological laboratory.

Conclusion

Arlette Leroi-Gourhan's research made major contributions to prehistoric archaeology. She was never formally employed, nor was she ever paid (with the exception of her costs when abroad)—presumably because she wanted it to be so, although we cannot be sure of her reasons. The fact that she remained a volunteer throughout her entire scientific career is certainly proof of the 1950s French gender norms, as much as her agency. She devoted her career to building a new archaeological discipline and to training an entire generation of future scientists. She always defended the importance of her research, as Aline Emery-Barbier underlines:

> I think that the house, the children, the cooking, was never enough for her. She also wanted to exist in another way. Palynology gave her the opportunity to do so... She often confided in me, telling me about the difficulties she faced in her attempts to establish herself as a scientist, notably in relation to her husband. Leroi-Gourhan didn't want to do her any favors, for it would have been very embarrassing. She fought for herself. She had to show that she could exist without simply being so-and-so's wife. It was Arlette who existed.[18]

By remaining on the periphery of scientific institutions, Arlette Leroy-Gourhan protected herself against the gendered mechanisms of exclusion within the scientific community. As a result, she was able to remain at the heart of collective scientific production and establish a new field of research for which she is still recognized today.

Notes

1 I am grateful to Aline Emery-Barbier, Michel Girard, Alfonso Ramirez-Galicia and Philipe Soulier for their support and advice. This museum later became the Musée de l'Homme in 1937 under the impulse of Paul Rivet. In addition to the recent and very complete biography of André Leroi-Gourhan by P. Soulier, see Gaucher (1987) and Soulier (2003).
2 These objects correspond to an important series of handicrafts of the daily life of this period and a large collection of Ainu objects. See Arl. Leroi-Gourhan and A. Leroi-Gourhan (1989).
3 Martine Solarek Leroi-Gourhan, interview with the author, Avallon, France, 19 July 2019.
4 Until then, Hélène Balfet was involved in the supervision of the Furtins site (1946–1949).
5 Martine Solarek interview, 19 July 2019.
6 Until 1955, Arlette Leroi-Gourhan used a NACHET monocular microscope—a particularly tiring system for continuous work—before obtaining a STIASSNIE binocular microscope. Michel Girard, interview with the author, Venoy, France, 15 September 2020.
7 M. van Campo's (*née* Duplan) career began in 1943 at the *Institut national de la recherche agronomique*. She joined the CNRS in 1944 as a research assistant and rose through the ranks to become a research director in 1964.
8 "Acetolysis is a rather complicated chemical manipulation that uses nitric acid, acetic acid and acetic anhydride in very particular proportions. It empties the pollen grain of its cytoplasm in order to keep only the envelope, i.e. the exine membrane. When the pollen is still full of cytoplasm, it is difficult to distinguish the external membrane. It is therefore necessary to make this preparation". Aline Emery-Barbier, interview with the author, Nanterre, France, 14 February 2020.
9 Michel Girard interview, 15 September 2020.
10 Statuses continued to evolve. The technical assistants became ITAs in 1970 (*Ingénieurs, techniciens et administratifs*) and ITs in 1999 (*Ingénieurs et techniciens*).
11 Michel Girard interview, 15 September 2020.
12 Michel Girard interview, 15 September 2020.
13 Martine Solarek interview, 19 July 2019.
14 This description of the place is possible thanks to the collaboration of Michel Girard.
15 Martine Solarek interview, 19 July 2019.
16 Aline Emery-Barbier interview, 14 February 2020.
17 A position equivalent to that of *maître* or *maîtresse de conférence* in the current French educational system.
18 Aline Emery-Barbier interview, 14 February 2020.

References

Adams, Amanda. 2010. *Ladies of the Field. Early Women Archaeologists and their Search for Adventure*. Vancouver: Greystone Books.

Audouze, Françoise, and Jean-Luc Fiches. 1993. "L'archéologie française et les paléo-environnements". *Annales* 48(1): 17–41.

Coudart, Anick. 1998. "Archaeology of French Women and French Women in Archaeology". In *Excavating Women: A History of Women in European Archaeology*, ed. Margarita Díaz-Andreu and Marie Louise Stig Sørensen, 59–83. London: Routledge.

Coudart, Anick. 2015. "Longtemps durant… le Genre ne fut pas un genre français sinon qu'il était du genre masculin… E pur si muove". *Les Nouvelles de l'Archéologie* 140: 9–15.

de Lumley, Henri. 1976. *La Préhistoire française, 1. Les civilisations paléolithiques et mésolithiques*. 2 vol. Paris: CNRS.

Díaz-Andreu, Margarita. 2021. "Las mujeres y la arqueología en Europa: de la aristocracia a las clases medias", in *Las mujeres y las artes. Mecenas, artistas, emprendedoras, coleccionistas.* ed. Beatriz Blasco Esquivias, Jonatan Jair López Muñoz, and Sergio Ramiro Ramírez, 783–803. Madrid: Abada Editores.

Díaz-Andreu, Margarita, and Marie-Louise Stig Sørensen. 1998. *Excavating Women: A History of Women in European Archaeology*. London: Routledge.

Dubois, Georges. 1932. "L'analyse pollinique des tourbes et son application à l'étude du quaternaire et de la Préhistoire". *L'Anthropologie* 42: 269–289.

Emery-Barbier, Aline, Chantal Leroyer, and Philippe Soulier. 2006. "Arlette Leroi-Gourhan (1913–2005): l'initiatrice de la palynologie appliquée à l'archéologie préhistorique". *ArcheoSciences*. 30. https://doi.org/10.4000/archeosciences.411

Gardey, Delphine. 2008. *Écrire, calculer, classer: comment une révolution de papier a transformé les sociétés contemporaines, 1800–1940*. Paris: Éditions La Découverte.

Gardey, Delphine, and, Ilana Löwy. 2000. *L'invention du naturel: les sciences et la fabrication du féminin et du masculin*. Paris: Éditions des archives contemporaines.

Gaucher, Gilles. 1987. "André Leroi-Gourhan, 1911–1986". *Bulletin de la Société Préhistorique Française (BSPF)* 84(10–12): 302–315.

Gosztonyi Ainley, Marianne. 1986. "D'assistantes anonymes à chercheures scientifiques: une rétrospective sur la place des femmes en science". *Cahiers de recherche sociologique* 4(1): 55–71.

Harvey, Joy. 2012. "The Mystery of the Nobel Laureate and His Vanishing Wife". In *For Better or For Worse? Collaborative Couples in the Sciences*, ed. Annette Lykknes, Donald L. Opitz and Brigitte Van Tiggelen, 57–77. Basel: Birkhauser Springer.

Kohlstedt, Sally G., and Donald L. Opitz. 2002. "Re-Imag(in)Ing Women in Science: Projecting Identity and Negotiating Gender in Science". In *The Changing Image of the Sciences*, ed. Ida H. Stamhuis, Teun Koetsier, Cornelis. De Pater, and Albert. Van Helden, 105–139. Amsterdam: Kluwer Academic.

Leroi-Gourhan, Arlette. 1997. "Chauds et froids de 60 000 à 15 000 BP" [Hot and cold from 60000 to 15000 BP]. *BSPF* 94(2): 151–160.

Leroi-Gourhan, Arlette. 1966. "L'analyse pollinique des coprolithes". *BSPF* 63(5): clxiii–clxiv.

Leroi-Gourhan, Arlette. 1957. "Notes sur l'analyse pollinique des sédiments quaternaires des grottes". 15e *CPF*. Poitiers-Angoulême. *SPF*: 671–675.

Leroi-Gourhan, Arlette. 1956a. "Analyse pollinique et Carbone 14". *BSPF* 53(5–6): 291–301.

Leroi-Gourhan, Arlette. 1956b. "Notes sur l'analyse pollinique des sédiments quaternaires des grottes". XXIIe Congrès préhistorique de France. Poitiers. *Société préhistorique française*: 671–675.

Leroi-Gourhan, Arlette, and André Leroi-Gourhan. 1964. "Chronologie des grottes d'Arcy-sur-Cure (Yonne)". *Gallia Préhistoire* 7: 1–64.

Leroi-Gourhan, Arlette, and André Leroi-Gourhan. 1989[1938]. *Un voyage chez les Aïnous: Hokkaïdo*. Paris: Albin Michel.

Leroi-Gourhan, André. 1952. "Discours de président sortant". *BSPF* 49(1): 5–8.

Lykknes, Annette, and Brigitte Van Tiggelen. 2019. *Women in their Element: Selected Women's Contributions to the Periodic Systems*. Singapore: World Scientific.

Lykknes, Annette, Donald L. Opitz, and Brigitte Van Tiggelen. 2012. *For Better or for Worse ?: Collaborative Couples in the Sciences*. Basel: Springer.

Marry, Catherine. 2004. *Les femmes ingénieurs. Une révolution respectueuse*. Belin: Paris.

Massiot, Anaïs, and Natalie Pigeard-Micault. 2016. *Les coulisses des laboratoires d'autrefois: vies et métiers à l'Institut du radium et à la Fondation Curie*. Paris: Éditions Glyphe.

Opitz, Donald, Staffan Bergwik, and Brigitte Van Tiggelen. 2016. *Domesticity in the Making of Modern Science*. Basingstoke: Palgrave Macmillan.

Prost, Antoine, and Gérard Vincent. 1991. *A History of Private Life. 5. Riddles of Identity in Modern Times*. Cambridge, MA: Belknap Press, Harvard.

Pycior, Helena, Nancy Slack, and Pnina Abir-Am. 1996. *Creative Couples in the Sciences*. New Brunswick NJ: Rutgers University Press.

Rayner-Canham, Marelene, and Rayner-Canham, Geoffrey. 1997. *A Devotion to their Science: Pioneer Women of Radioactivity*, Philadelphia and McGill-Queen's University Press, Montréal.

Rogers, Rebecca. 2019. "Le sexe de l'espace: réflexions sur l'histoire des femmes aux XVIIe et XIXe siècles dans quelques travaux américains, anglais et français", in *Les espaces de l'historien*, eds. Jean-Claude Waquet and Odile Goerg, 181–202. Strasbourg. Presses Universitaires de Strasbourg.

Rogers, Rebecca. 1982. *Women Scientists in America: Struggles and Strategies to 1940*. Baltimore, MD: Johns Hopkins University Press.

Smith, Pamela Jane. 2000. "Dorothy Garrod as the First Woman Professor at Cambridge University", *Antiquity* 74(283): 131–6.

Sonnet, Martine. 2004. "Combien de femmes au CNRS depuis 1939?", in *Les femmes dans l'histoire du CNRS*, ed. Mission pour la place des femmes au CNRS, Paris: Comité pour l'histoire du CNRS, 46–58.

Soulier, Philippe. 2003. "André Leroi-Gourhan, 25 août 1911–19 février 1986". *La Revue pour l'histoire du CNRS* 8: 54–68.

Soulier, Philippe. 2018. *André Leroi-Gourhan (1911–1986): une vie*. Paris: CNRS éditions.

Van Campo, Madeleine, and Arlette Leroi-Gourhan. 1956a. "Note préliminaire à l'étude des pollens fossiles de différents niveaux des grottes d'Arcy-sur-Cure". *Bulletin du Muséum National d'Histoire Naturelle* 28(3): 326–330.

Van Campo, Madeleine, and Arlette Leroi-Gourhan. 1956b. "Un paysage forestier rissien dans l'Yonne". *Bulletin de la Société Botanique de France* 103(5–6): 285–286.

Van Tiggelen, Brigitte, Annette Lykknes, Luis Moreno-Martinez. 2019. "The Periodic System, a History of Shaping and Sharing". *Substantia* 3(2, Suppl. 4): 9–12.

Von Oertzen, Christine von, Maria Rentetzi, and Elizabeth S. Watkins. 2013. "Finding Science in Surprising Places: Gender and the Geography of Scientific Knowledge Introduction to "Beyond the Academy: Histories of Gender and Knowledge"". *Centaurus* 55(2): 73–80.

Waquet, Françoise. 2022. *Dans les coulisses de la science: techniciens, petites mains et autres travailleurs invisibles*. Paris: CNRS éditions.

Woolf, Virginia. 1945. *A Room of One's Own*. Richmond: Hogarth Press.

Afterword

A History to be Continued

Grégory Dufaud and Isabelle Lémonon-Waxin

Revised by Robert Fyke

This book has offered an innovative historical overview of the entanglement of gender, government, and the technosciences: an overview that considered the technosciences as governed both from above and from within. Such an approach has highlighted the dispersion of power among multiple locations and actors that enabled men to dominate women in science. The contributions in this book have shown the variety and multiplicity in this dispersal pattern of male domination, as well as the capacity for action by scientific women who have endeavored to win more prestigious positions, to carve out spaces of their own, or to take control of their own bodies. Essays have highlighted how asymmetrical gender relationships were built, maintained, and challenged within the technosciences as a masculine domain. The goal here has been to examine from a gendered perspective how technoscientific institutions organize, regulate, valorize, and invisibilize their activities, and to look into how actors have affected the evolution of these institutions.

One of the main interests of comparing diverse countries and political regimes has been to underline a generally shared situation. In both the Soviet Union and Western democracies, the government of the technosciences remained in the hands of men, while women everywhere have encountered strong obstacles and resistance to their social and professional ascent, even after women have become increasingly active in the technosciences. If this point unfortunately notices that most of the contributions of this book concern the Global North, it also focuses attention on the profound difficulties encountered in some locations when developing these research topics. For example, in Russia it took a long time for Women and Gender Studies to become a legitimate academic field; and the hardening of the Putin dictatorship and the invasion of Ukraine will make such research more difficult. The progress of historical knowledge is not a one-way street: it can stop and go backward as well. Furthermore, the comparison of various disciplines, whether in well-established fields such as healthcare or in new ones such as nuclear power, demonstrates that the increased presence of women does not guarantee greater recognition or easier access to leading positions unless women take over marginalized domains neglected by their male colleagues. In this sense, identifying and analyzing the spatial, social, and disciplinary

DOI: 10.4324/9781003562597-19

peripheries of technoscientific production is crucial to implementing egalitarian and inclusive policies.

Women scientists have faced different kinds of obstacles. Scientists, researchers, and technicians have been seen as abstract, scientific persona, who made discoveries through individual and collective effort. This persona was characterized by a research ethic that determined a level of trust and probity that peers owed each other. Yet this abstraction implicitly referred to male scientists who generally regarded women as little more than clever assistants. This generalization was inherited from centuries past, when women had no access to education or to official positions, and often had to content themselves with assisting male scientists. This female social figure, dedicated to carrying out tasks and assuming functions considered less noble, kept sticking itself to women scientists because of its resonance with gendered prejudices that attributed care-giving to women. In this construct, women were performing gender as expected, making it difficult for them to climb the scientific hierarchy. Not only did women scientists have to overcome social and cultural barriers, but they also had to fight back against a government of the technosciences dominated by men. Indeed, the claim of reducing gender inequality and prejudice in the technosciences should have implied that states, government agencies, international unions, etc., had more fully taken responsibility as prescribers and regulators of the integration of gender issues into their own institutions. But until recently, the priority has been to build internationally competitive scientific systems where male domination was barely questioned. For example, in the Soviet Union, the low number of women in the highest positions of technoscientific government could not officially be an issue, since the Bolsheviks had claimed to have solved the problem of gender inequality. This theoretically gender-neutral stance enabled the tools, policies, and cultures of technoscientific government to remain largely masculine in practice. Globally, if men accepted that women occupied a place in science and technology, they managed to exclude these competitors from decision-making positions that they kept for themselves. Despite the obstacles, there were women whose individual or collective actions succeeded in gaining access to more prestigious positions, or in reshaping the government of technosciences. Such women scientists can be referred to as "change agents" who advocate, coordinate, and implement change within organizations (Dahmen-Adkins and Peterson, 2021).

Several chapters in this book identify change agents' strategies, performance, and communication that enabled greater gender inclusivity in technoscientific executive positions. Performance and communication are indeed tools of government and represent forms of commitment that challenge existing social ties. Here, ethical issues about achieving epistemic justice also highlight the political economy of science: by limiting access to leading positions in the technosciences, gender prejudice limits production, creativity, and innovation. Meanwhile, it is also necessary to explore the objects and spaces produced and used by the government of the technosciences: an exploration through the prism of gender in the history of the technosciences that some researchers have already started (Rentetzi 2023; Van Tiggelen, Rentetzi and Lykknes 2024). Spaces and things not only shape gender hierarchies, they also suggest efficacy for certain types of actions. As Maria Rentetzi underlines, objects "become sites where normative discourses are incorporated and, in conjunction with humans, they act to devalue manual tasks, impose gender divisions of labor, and construct normative masculinity and femininity. Yet, things are not only the inanimate objects that people act upon—and that in turn act upon people" (2023: 1–2). Furthermore, the geographical and social periphery of technoscientific institutions are spaces of interest for studying the entanglement of gender, government, and technosciences. The government of Citizen Science or Community Science, understood as the scientific work undertaken by the general public in collaboration with professional scientists and institutions, should be included in future research around its articulation of gender. With this in mind, future studies can adopt an intersectional approach toward ethnicity, sexual identity, disability, and class structures. These categories should play a much larger part in future discussions about the government of the technosciences. In this sense, this volume lays the foundations for a future history we hope others will continue to investigate and to develop.

References

Dahmen-Adkins, Jennifer, and Helen Peterson. 2021. "Micro Change Agents for Gender Equality: Transforming European Research Performing Organizations." *Frontiers in Sociology* 6: Article 741886. doi: 10.3389/fsoc.2021.741886.

Rentetzi, Maria, (éd.). 2023. *The Gender of Things: How Epistemic and Technological Objects Become Gendered.* London: Routledge.

Van Tiggelen, Brigitte, Maria Rentetzi, and Annette Lykknes. 2024. *The Gender of Things and Spaces in the Sciences.* Symposium 11th ESHS Conference, Barcelona.

Index

Note: **Bold** page numbers refer to tables; *italic* page numbers refer to figures and page numbers followed by "n" denote endnotes.

For Product Safety Concerns and Information please contact our EU representative GPSR@taylorandfrancis.com Taylor & Francis Verlag GmbH, Kaufingerstraße 24, 80331 München, Germany

Batch number: 10406157

Printed by Printforce, the Netherlands